Miseducation

Miseducation

A History of Ignorance-Making in America and Abroad

EDITED BY A. J. ANGULO

Johns Hopkins University Press
BALTIMORE

Printed in the United States of America on acid-free paper
9 8 7 6 5 4 3 2 1

Johns Hopkins University Press
2715 North Charles Street
Baltimore, Maryland 21218-4363
www.press.jhu.edu

Library of Congress Cataloging-in-Publication Data

Miseducation : a history of ignorance-making in America and abroad / edited by A.J. Angulo.
pages cm
Includes bibliographical references and index.
ISBN 978-1-4214-1931-2 (hardcover : alk. paper)
ISBN 978-1-4214-1932-9 (pbk. : alk. paper)
ISBN 978-1-4214-1933-6 (electronic)
ISBN 1-4214-1931-9 (hardcover : alk. paper)
ISBN 1-4214-1932-7 (pbk. : alk. paper)
ISBN 1-4214-1933-5 (electronic)
1. Ignorance (Theory of knowledge)—Cross-cultural studies.
2. Comparative education. I. Angulo, A. J., editor of compilation.
BD221.M57 2016
001—dc23 2015026979

A catalog record for this book is available from the British Library.

Special discounts are available for bulk purchases of this book. For more information, please contact Special Sales at 410-516-6936 or specialsales@press.jhu.edu.

Johns Hopkins University Press uses environmentally friendly book materials, including recycled text paper that is composed of at least 30 percent post-consumer waste, whenever possible.

To the memory of
Thomas George Dyer

Contents

Ignorance — 1
A. J. Angulo

PART I: Legalizing Ignorance

1 Slavery — 13
Kim Tolley

2 Sex — 34
Jennifer Burek Pierce and Matt Pierce

3 Sexuality — 52
Karen Graves

4 Evolution — 73
Adam R. Shapiro

5 Environment — 96
Kevin C. Elliott

PART II: Mythologizing Ignorance

6 Class — 123
Daniel Perlstein

7 Identity — 140
Eileen H. Tamura

8 **Religion** 161
Adam Laats

9 **History** 184
Donald Warren

PART III: Nationalizing and Globalizing Ignorance

10 **US** 217
Lisa Jarvinen

11 **Germany** 244
Lisa Pine

12 **USSR** 268
E. Thomas Ewing

13 **Israel** 295
Soli Vered and Daniel Bar-Tal

14 **China** 319
Dongping Han and Stephen Samuel Smith

Reflections 339
A. J. Angulo

Acknowledgments 351
Contributors 355
Index 363

Miseducation

Ignorance

A. J. ANGULO

Ignorance. It's something that pervades our lives, but we rarely give it a second thought. We hear it on the radio on the way to work, find it on social media throughout the day, and see it on the television in the evening. Like water, it finds its way into the cracks of our daily existence and it's a critical part of the human experience. We're born ignorant. We're taught and raised to believe ignorant things. Our lives take us in directions that require us to become less ignorant about some aspects of life and increasingly ignorant about others. These are inevitabilities, rather than value-laden observations, and they are ones that almost all of us spend little to no time thinking through.

Some spend their lives peddling ignorance without necessarily knowing it. Marketing and advertising agencies direct billions of dollars every year toward creating in children and adults expectations that fundamentally ignore or distort reality. Major movie studios do the same to offer us opportunities to suspend disbelief and escape the realities of everyday life. Informed bankers sell financial products to less informed consumers through what bankers like to describe as "asymmetric" information relationships. Ignorance comes to us from politicians on the campaign trail, where we can learn that "legitimate rape" rarely leads to pregnancy, and from elected officials when debating science, public health, and threats from abroad. It emanates from places of worship, too, where wide-ranging and contradictory values are treated as fact and where such treatments exist as protected forms of speech. News outlets sell us the hope of knowledge and along the way—through editorialists, pundits, and spin doctors—distort what, when, and how we know what we know. Our world is awash in ignorance, and some make a fine living from it.[1]

Governments typically have mixed views when it comes to ignorance. At times they promote it; at others, they combat it. They rely on ignorance to protect state secrets and classified information. Plans, procedures, and products necessary for creating weapons of mass destruction are fiercely guarded. Governments spend vast resources to keep this kind of knowledge from the public. They also attack whistleblowers as traitors and subversives, especially when their revelations embarrass the state and destroy carefully constructed dissembling narratives. Those who attempt to pull back veils of ignorance, as in the cases of Julian Assange and Edward Snowden, find few friends in government (or the general public), even when such whistleblowing leads to fundamental reforms. When it comes to the production of new knowledge, governments enact laws to prevent researchers and innovators from making discoveries that may harm the public or conflict with social and cultural norms. Sweden, Finland, Belgium, Greece, Britain, Denmark, and the Netherlands allow experimentation using the human embryo. Germany, Austria, Ireland, Italy, and Portugal do not. There are over one thousand laws in over one hundred countries that regulate, control, and decide what we can know from human subjects research and what must remain in the dark corners of our collective ignorance.[2]

Countries around the world, meanwhile, take dramatically different approaches to combating ignorance. The social democracies of postwar northern Europe have beat back ignorance by making schools, colleges, and universities accessible to their citizens. They've provided high-quality, tuition-free education from as early as preschool to as late as graduate school. Outcomes including and well beyond international test scores suggest that their evolving and expanding commitment to education since the mid-twentieth century has helped significantly reduce ignorance, crime, disease, and poverty in the region. In the United States, President Lyndon B. Johnson, a former schoolteacher, had a similar postwar vision. He saw in education a solution to some of the country's most pressing problems. Johnson signed into law elementary, secondary, and postsecondary legislation that, as he put it, represented a broader "revolution of the spirit against the tyranny of ignorance." Since then, other movements, like the Reagan Revolution, have moved the United States toward a pay-for-use approach to public and private educational services. With retrenchment and declining state support, public institutions of higher education have relied on breathtaking increases in tuition, fees, and student debt to maintain their place in the global academic community. As a result, the student debt crisis is now being compared to the subprime mortgage bubble that threatened to bring down the global economy, raising new questions about the economic consequences of ignorance.[3]

Despite its significance to everyday life, scholars have ignored ignorance for too long. On occasion the subject captured the attention of great minds, but ignorance itself has rarely come under systematic scrutiny. Socrates, through Plato, told us we're all enslaved ignorant cave dwellers who need to unshackle our minds and seek the light of true knowledge. His dialogues left us with some clues about how to deal with the problem, but they're rooted in ancient norms and conditions. Medieval scholars were too busy swimming in mass ignorance and superstition to pay careful attention to the riptides taking them away from the shores of knowledge. The modern philosophical project, beginning with René Descartes, questioned all knowledge and, later, all language used to understand knowledge, but the headlights guiding this adventure didn't shine on the contours and qualities of ignorance itself. Noticing ignorance on the way to getting a handle on knowledge failed to produce much in the way of systematic study. Those with a more historical, rather than analytical, bent have brought us closer to the mark. Karl Marx's study of history led to new ideas about the relevance of ignorance in terms of political economy. His works inspired schools of thought about the role of religion as the "opiate of the masses" and the denial of class conflict through false consciousness, both of which highlighted how one social or economic group has potentially induced ignorance by manipulating the perceptions of another. Nevertheless, before the twentieth century, ignorance hadn't made a debut in the titles of major analytic or historical projects, nor had it earned a place in authoritative reference works on Western and non-Western thought.[4]

At the start of the twentieth century, however, ignorance began to receive sustained attention from historian and educator Carter G. Woodson. His reflections on the topic in *Miseducation of the Negro*, a work often and undeservedly overlooked, appeared after Booker T. Washington and W. E. B. Du Bois staked out their positions on the role of education in uplifting the African American community in postbellum America. Washington championed industrial education as the fastest way to integrate newly freed slaves into the workforce and, ultimately, to establish an economic foundation from which to gradually assert social and political equality. Du Bois, meanwhile, had little patience for the gradualist approach. At the time, he fought to give a "talented-tenth" of the black community the same kind of traditional liberal arts education received by elite whites in order to train a leadership class that would run for office or conduct research to prevent a second slavery.[5]

Woodson sidestepped both positions and focused on what we would now call the problem of ignorance-making. He had no sympathy for Washington's vision of industrial education, which would train African Americans to use yesterday's technology. As Woodson saw it, the machinery used at Tuskegee and other schools run by Washington's contemporaries was largely dated and irrelevant. By the time students graduated from these schools, their skill sets would be useless in the modern factory or on the technologically advanced farm. He also opposed Du Bois's easy acceptance of traditional liberal arts education. Of what value, asked Woodson, was a degree from a prestigious university if at the end of the day such a degree failed to help the African American gain employment in a segregated world? He also wondered why Du Bois supported a tradition of education that tended to ignore the African American's past, present, and potential future. To Woodson's mind, neither Washington nor Du Bois advocated for much more than miseducation and ignorance-making. Washington ignored the kind of education that would yield competent journalists, physicians, lawyers, and bankers. Du Bois wanted to prepare the talented-tenth for such roles, but with a training that omitted and occluded the particular needs, aspirations, and contexts of African Americans. Woodson believed they both left the black community worse off than before—miseducated, ignored, and prepared for failure.[6]

Woodson's study broke new ground in thinking through patterns and processes of ignorance-making within institutions expected to advance knowledge and dispel myths. He offered a solution, or "new program," as he dubbed it, that called for the creation of schools, colleges, and universities designed specifically to serve the needs and interests of African Americans. Included in this program were courses focused on exploring the past and present experiences of black America, as well as the professional means for building a thriving, self-sufficient future within a segregated society. At the same time, his work left wide open fields of unanswered questions about how the patterns and processes of miseducation extended beyond his community. As such, Woodson's contribution wasn't only in identifying a problem affecting a specific segment of the population. His study highlighted how ignorant we all were and, until recently, continued to be about ignorance.

This book is a response to our lack of knowledge about ignorance-making in America and beyond. It offers a historical exploration of ignorance in the context of schools, colleges, universities, and a variety of other public and private settings.

The essays contained in this volume take as a given the longstanding claim that education occurs within and outside the schoolhouse gates. They assume that education encompasses, in the words of historian Bernard Bailyn many decades ago, "the entire process by which a culture transmits itself across the generations." The newspaper, museum, and radio, in this light, have as much claim to cultural transmission as the textbook and seminar room.[7]

But the book also moves beyond this view. It strikes up a dialogue with current research on the social construction of ignorance coming from the field of history of science, which has opened up new avenues of inquiry for historians, educational researchers, and policymakers. Historians of science have targeted ignorance-making in cases like tobacco, asbestos, and climate change. We now know much more, as a result, about how cigarette manufacturers spent a fortune in the mid-twentieth century to create doubt about epidemiological research that linked smoking to cancer. These scholars have been interested in such settings as corporate board rooms, government agencies, and private laboratories. Drawing on historian Robert Proctor's approach to the systematic study of ignorance, or what he calls *agnotology*, the goal of this book is to extend the discourse about cultural transmission into the realm of ignorance maintenance, regulation, and perpetuation through processes of miseducation. Put another way, this book expands on Woodson's ideas about miseducation, revises Bailyn's views about cultural transmission, and integrates Proctor's work on agnotology to chart new pathways in the exploration of the educational past.[8]

This book was written with a common understanding of Proctor's three main categories for defining the parameters of agnotology: native, passive, and active. *Native* ignorance comes largely from "naïvety of youth" or "improper education." The absence of knowledge in this case is less by design than by default. Children, for instance, are born ignorant of reading and writing. The knowledge deficit will continue to exist until that native ignorance is addressed. *Passive* ignorance occurs when, individually or collectively, we select and choose to limit what we know and understand. Since it's not possible to learn about everything that interests us, we must inevitably make decisions about the range of interests we pursue. College students face these kinds of decisions when selecting their majors. Some might have a passion for history but ultimately pick medicine or accounting for career-based reasons. Others might have the passion and resources for all three but must prioritize because of limits imposed by time. These basic constraints dictate research topics or places to investigate to the exclusion of others that might serve equally well. The knowledge gaps that form become lost realms created by choice or passivity, both of which may occur without a deliberate decision to ignore or destroy sources of knowledge.[9]

The final and most provocative of the three—*active* ignorance—has the potential to bring about a fundamental shift in how scholars view, understand, and uncover the educational past. This kind of ignorance moves beyond naïveté and passivity and into the territory of active construction, maintenance, regulation, and diffusion. The stakes are much higher with these strategic or active constructs. Consider, for example, recent cases in which the fracking industry has refused to release information about chemicals injected into drinking water resources because of proprietary knowledge or "trade secrets." Attorneys used the same veil of secrecy and ignorance-making strategies to defend Freedom Industries after the contamination of water reserves in Charleston, West Virginia. Historians of science have explored numerous examples of this kind of socially produced ignorance, including military secrets, classified research, and moral or religious resistance to certain forms of scientific and medical knowledge.[10]

The task of this book is to secure a place for education in the vibrant historical discussion about ignorance by describing the role of education and educational materials in the social construction of ignorance. Contributors to this volume, recognized nationally or internationally in their areas of expertise, approached this assignment with the following questions in mind: What role does ignorance play in your particular specialty? How might the native-passive-active framework produce new insights in your area? And how might your area, when studied from an agnotological perspective, advance our understanding of the study of ignorance? The goal, in short, was to produce new historical perspectives fueled by a desire to understand the substantive and methodological concerns raised by ignorance and its social construction in educational contexts. The answers offered in the following chapters suggest that education and ignorance-making have stood at the center of some of the most important developments in the United States and around the world. Miseducation, in its diverse forms, has served as a prime agent—one as powerful as social movements, economic crises, political conflicts, and intellectual advances—in shaping the course of the past and present.

The answers and insights generated by chapter contributors inspired this book's three-part organization. Part I, "Legalizing Ignorance," offers a history of legal, legislative, and regulatory measures used to prevent or suppress content considered too controversial for public consumption. Kim Tolley opens the volume with a chapter on antiliteracy laws targeting slaves and those wanting to teach literacy skills to slaves during the antebellum period. Jennifer Burek Pierce and Matt Pierce follow with a stillborn attempt at a sex-education radio broadcast in the 1930s that led to an overhaul of the nation's federal communications laws and regulations. Karen Graves focuses on mid- to late twentieth-century state laws enacted to erase diversity of sexual orientation in school materials and settings. And Adam Shapiro and Kevin Elliott offer chapters on laws, public policies, and information campaign strategies that have distorted the public understanding of evolution (Shapiro) and anthropogenic climate change (Elliott).

Part II, "Mythologizing Ignorance," targets myths produced in educational settings and materials or about specific populations in educational contexts. Daniel Perlstein unpacks traditional assumptions about the nobility of progressive education and active learning in the context of the Great Depression, suggesting that romantic ideas about the frontier—a staple for progressive educators—obscured more pressing, immediate, and relevant problems. Eileen Tamura offers an internalist and externalist account of how the myth of the model minority (and associated disinformation) has impacted the Asian American community since the phrase was coined in the sixties and later used to define, expand, and limit educational opportunity. Adam Laats examines mythmaking about American history through Christian textbook publishing houses of the late twentieth century. And Donald Warren closes the section by probing how the field of history of education itself might face its own construction of ignorance about the educational past by highlighting provocative omissions and silences in the historical literature.

The third and final part, "Nationalizing and Globalizing Ignorance," takes readers abroad to Latin America, Europe, the Middle East, and Asia in a study of how political and economic ideologies intersect with education and agnotology. Lisa Jarvinen analyzes the use of "schooling" as a rhetorical and policy device to educate the US public about the need for colonization in Cuba, Puerto Rico, and the Philippines during and after the War of 1898. Lisa Pine revisits and recasts what we know about the use of propaganda and schools as vehicles of misinformation in Nazi Germany. E. Thomas Ewing seeks to complicate our understanding of the Soviet Communist Party's Central Committee control over content in the teaching of history in Stalinist schools during the 1930s and 1940s. Soli Vered and Daniel Bar-Tal examine the role of Israel's Ministry of Education in the miseducation of Israeli children about the conflict with Palestinians from the 1940s to the 1970s. And Dongping Han and Stephen Smith raise a series of agnotology-inspired research questions about approaches to understanding China since Mao and the Cultural Revolution.

With this structure, the book's central purpose is to highlight three kinds of miseducation. The first section analyzes forms of ignorance-making imposed by or intersecting with laws and regulations (de jure). The second focuses on the production of ignorance emerging from social, cultural, and intellectual processes (de facto). And the final section globalizes elements drawn from the first two (de jure and de facto). All three sections aim to offer fresh insights and case studies for reassessing what we know about the educational past and its relationship to the social construction of ignorance. It's an aim that underscores a struggle between competing ways of knowing and, in many cases, between those who want access to certain kinds of literacies—basic, sexual, and scientific, among others—and those who want to prevent or alter the diffusion of those literacies. It's a history that begins with slaves, fire, and the most elementary literacy of all in antebellum Georgia.

NOTES

1. McKinsey and Company, *Global Media Report 2013: Global Industry Overview*, http://www.mckinsey.com/~/media/mckinsey/dotcom/client_service/Media%20and%20Entertainment/PDFs/Global_Media_Report_2013.ashx (accessed October 26, 2014); Neilsen, "Global Ad Spend Grows 3.2% in 2012," 11 April 2013, http://www.nielsen.com/us/en/insights/news/2013/global-ad-spend-grows-3.2-percent-in-2012.html; Gabriel Beltrone, "A By-the-Numbers Look at Hollywood's Marketing Machine: Media Spend and Agencies of the Biggest Studios," *Adweek*, 24 February 2014; David S. Hilzenrath, "Goldman Fined $22M for 'Willfully' Violating Law on Information-Sharing, SEC Says," *Washington Post*, 12 April 2012; consent order, *Massachusetts v. Goldman Sachs*, no. E2009-0079, MA, 9 June 2011; Matt Williams, "'Legitimate Rape' Rarely Leads to Pregnancy, Claims US Senate Candidate: Todd Akin, the Republican Nominee for Missouri, Causes Furor with Explanation of His No-Exceptions Policy on Abortion," *Guardian*, 19 August 2012; Sam Harris, *End of Faith: Religion, Terror, and the Future of*

Reason (New York: Norton, 2005); Nancy Qian and David Yanagizawa-Drott, "Government Distortion in Independently Owned Media: Evidence from U.S. Cold War News Coverage of Human Rights," Harvard University, 4 June 2013, 1–56, http://www.hks.harvard.edu/fs/dyanagi/Research/HRMedia_20130604.pdf; Steven Kull, Clay Ramsay, and Evan Lewis, "Misperceptions, the Media, and the Iraq War," PIPA/Knowledge Networks: The American Public on International Issues, 2 October 2003, 1–21, http://www.worldpublicopinion.org/pipa/pdf/oct03/IraqMedia_Oct03_rpt.pdf.

2. The United States spends more than $11 billion annually on information security and classification. Information Security Oversight Office, *2013 Report to the President*, National Archives and Records Administration, 2 June 2014, http://fas.org/sgp/isoo/2013rpt.pdf; Sean Wilentz, "Would You Feel Differently about Snowden, Greenwald, and Assange If You Knew What They Really Thought?," *New Republic*, 19 January 2014; "International Compilation of Human Research Standards," Office for Human Research Protections, US Department of Health and Human Services, 2014, http://www.hhs.gov/ohrp/international/intlcompilation/2014intlcomp.pdf.pdf. Restrictions on research, as presented in this paragraph, do not imply an inherently good or bad quality to either the pursuit of knowledge or the construction of ignorance. Social norms dictate, justifiably, that some destructive pursuits of knowledge—like those of Joseph Mengele in Nazi Germany or Oliver Wenger with the Tuskegee syphilis study—should not be permitted. Consider, also, the public response to rDNA research in Cambridge, Massachusetts, in Everett Mendelsohn's "'Frankenstein at Harvard': The Public Politics of Recombinant DNA Research," in *Tradition and Transformation in the Sciences: Essays in Honor of I. Bernard Cohen*, ed., Everett Mendelsohn (1984; reprint, New York: Cambridge University Press, 2003), 317–335. The point here is that neither knowledge pursuits nor ignorance construction should receive praise or condemnation without considered judgment.

3. Tom Hertz et al., "The Inheritance of Educational Inequality: International Comparisons and Fifty Year-Trends," *B.E. Journal of Economic Analysis & Policy* 7 (2007): 1–46; Miles Corak, "Income Inequality, Equality of Opportunity, and Intergenerational Mobility," *Journal of Economic Perspectives* 27 (Summer 2013): 79–102; Jordan Weissmann, "A $150 Billion Burden: What to Do about Subprime Student Loans," *Atlantic*, 20 July 2012; Glenn Harlan Reynolds, *The Higher Education Bubble* (New York: Encounter Books, 2012); Thomas G. Donlan, "College Loans: The New Debt Bubble," *Barron's*, 29 March 2014.

4. As of 2015, the *Stanford Encyclopedia of Philosophy* (http://plato.stanford.edu/) does not have an entry for *ignorance* in its table of contents nor its index, despite recent philosophical work on the topic. See, e.g., Shannon Sullivan and Nancy Tuana, eds., *Race and Epistemologies of Ignorance* (Albany: SUNY Press, 2007). The nineteenth-century Scottish philosopher James Frederick Ferrier coined the term *agnoiology* in his *Institutes of Metaphysics* (1854) to describe the study of ignorance, but his term and proposed subfield failed to inspire much interest. The term was briefly revived as an aside in Keith Leherer, "Why Not Skepticism?," *Philosophical Forum* 2 (1971): 283–98, but there too the term gained little traction. For more on the evolution of ignorance studies, see Nancy Tuana, "Coming to Understand: Orgasm and the Epistemology of Ignorance," in *Agnotology: The Making and Unmaking of Ignorance*, ed. Robert N. Proctor and Londa Schiebinger (Stanford, CA: Stanford University Press, 2008), 141.

5. Carter G. Woodson, *Miseducation of the Negro* (Washington, DC: Associated Publishers, 1933); Booker T. Washington, *Up from Slavery: An Autobiography* (Garden City, NY: Doubleday, 1901); idem, "Industrial Education for the Negro," in *The Negro*

Problem, by Booker T. Washington et al. (New York: J. Patt, 1903), 7–30; W. E. B. Du Bois, *The Souls of Black Folk: Essays and Sketches* (Chicago: A. C. McClurg, 1903). Recent scholarship has shown that the positions held by Washington and Du Bois were more complex and nuanced than traditional interpretations have suggested. See, e.g., Derrick Alridge, *The Educational Thought of W.E.B. Du Bois: An Intellectual History* (New York: Teacher's College Press, 2008); K. Anthony Appiah, *Lines of Descent: W. E. B. Du Bois and the Emergence of Identity* (Cambridge, MA: Harvard University Press, 2014); and Michael Scott Bieze and Marybeth Gasman, eds., *Booker T. Washington Rediscovered* (Baltimore: Johns Hopkins University Press, 2012).

6. For an elaboration on Woodson's own educational proposals, see "The New Program," in Woodson, *Miseducation*, 144–56. See also, A. J. Angulo, "Agnotology: Game Changer for History of Education?," Annual Meeting of the American Educational Research Association, Chicago, IL, 20 April 2015.

7. Bernard Bailyn's oft-cited definition of education can be found in his *Education in the Forming of American Society* (New York: Vintage Books, 1960), 14. For a brief survey of Bailyn's work, as well as that of Lawrence Cremin, Michael Katz, and others who have contributed to the development of educational history, see William Reese and John Rury, "Introduction: An Evolving and Expanding Field of Study," in *Rethinking the History of American Education*, ed. Reese and Rury (New York: Palgrave, 2008), 1–16.

8. Robert N. Proctor, "Agnotology: A Missing Term to Describe the Cultural Production of Ignorance (and Its Study)," in Proctor and Schiebinger, *Agnotology*, 1–36; Paul Erker, "A Comparative Perspective on Asbestos as an Industrial and Environmental Hazard," *Global Environment* 7 (2014): 72–105; Naomi Oreskes and Erik M. M. Conway, *Merchants of Doubt: How a Handful of Scientists Obscured the Truth on Issues from Tobacco Smoke to Global Warming* (New York: Bloomsbury, 2011). Revisionist and postrevisionist historians of education might wonder whether their work already engages the problem of ignorance production, even if it's not described as such, through studies on segregation, tracking, and many other topics. These works have certainly made important contributions in identifying which groups have received and which have been denied access to educational opportunities. Agnotology, however, differs in approach by highlighting the way ignorance production appears in all settings, even those sites valorized by historians of education (e.g., high-performing schools, well-resourced institutions, and AP and honors courses).

9. Proctor and Schiebinger, *Agnotology*, 4.

10. Kate Sheppard, "For Pennsylvania Doctors, A Gag Order on Fracking Chemicals," *Atlantic*, 27 March 2012; Paul M. Barrett, "A Second Chemical Spilled in West Virginia, and the Company Said Nothing until Now," *BusinessWeek*, 23 January 2014. See Peter Galison on classified research and Londa Schiebinger on abortifacients in Proctor and Schiebinger, *Agnotology*, 37–54, 149–62.

PART I / Legalizing Ignorance

1

Slavery

KIM TOLLEY

> Sec. 11. *And be it further enacted*, That if any slave, negro, or free person of colour, or any white person, shall teach any other slave, negro, or free person of colour, to read or write either written or printed characters, the said free person of colour or slave shall be punished by fine and whipping, or fine or whipping at the discretion of the court; and if a white person so offending, he, she, or they shall be punished with fine, not exceeding five hundred dollars, and imprisonment in the common jail at the discretion of the court before whom said offender is tried.
>
> —ASSENTED TO, DECEMBER 22, 1829, GEORGE R. GILMER, GOVERNOR

In the twenty-first century, Americans often equate literacy with personal empowerment, economic opportunity, and political freedom. Nevertheless, governments have not always promoted literacy with these outcomes in mind. As scholars Robert Arnove and Harvey Graff have suggested, governments have often sought to promote literacy as a means of advancing their own political agendas and ideologies. In the past, particular triggering events—such as the sixteenth-century Protestant Reformation, the 1861 abolition of serfdom in Russia, or the Cuban Revolution of 1959—have motivated political literacy campaigns. Moreover, as literacy historian E. Jennifer Monaghan has noted, some triggering events in the past have instigated campaigns *against* literacy.[1]

Nowhere is the role of state governments in suppressing knowledge more evident than in the enactment of antiliteracy laws in the antebellum South. These laws fall squarely into the category of ignorance-making characterized by its active constructs—ignorance as a deliberately engineered and strategic ploy—as described by recent research on agnotology. Through these laws, which prohibited the teaching of slaves to read and write, governments purposely fostered, maintained, and regulated the structural production of ignorance.[2]

Table 1.1. Antiliteracy Legislation, 1740–1834

States of the Confederacy	Year Antiliteracy Law Enacted	Year(s) Antiliteracy Law Revised
South Carolina	1740	1800, 1834
Georgia	1770	1829
North Carolina	1830	1835
Louisiana	1830	—
Virginia	1819	1830
Alabama	1831	1833
Arkansas	—	—
Tennessee	—	—
Mississippi	1823	1831
Texas	—	—
Florida	—	—

Sources: McCord, ed., *Statutes at Large of South Carolina*, 364–65, 460; South Carolina, *Acts and Regulations of the General Assembly of the State of South Carolina . . . 1834*, 13–14; Cushing, *Earliest Printed Laws of the Province of Georgia, 1755–1770*, 2:31; Georgia, *Acts of the General Assembly of the State of Georgia . . . 1829*, 168; Alabama, *Acts Passed at the State of Alabama . . . One Thousand Eight Hundred and Thirty-One*, 16; Louisiana, *Acts Passed at the Second Session of the Ninth Legislature of the State of Louisiana*, 96; Mississippi, *Laws of the State of Mississippi, Passed at the Fourteenth Session of the General Assembly*, 86–87; North Carolina, *Acts Passed by the General Assembly of the State of North Carolina . . . 1830–31*, 11; Virginia, *Acts Passed at a General Assembly of the Commonwealth of Virginia . . . One Thousand Eight Hundred and Thirty*, 107–8; Cornelius, *When I Can Read My Title Clear.*

Of the eleven states that eventually joined the Confederacy, seven passed such legislation before the Civil War, as shown in table 1.1. Historians agree that the legal prohibitions against writing aimed to prevent slaves from forging passes, running away, and communicating with one another. The laws against reading reinforced the suppression of communication among slaves and between slaves and freemen and also prevented slaves from access to antislavery literature.[3]

Nevertheless, there has been virtually no discussion among scholars about why some southern states passed antiliteracy laws and others did not. Nor have historians considered such developments from an agnotological perspective. For more than half a century, scholars have simply followed in the footsteps of twentieth-century historian Henry W. Farnam, who reported that “a good many States thought to protect themselves against rebellion and conspiracy by keeping the Negroes in ignorance, and laws intended to prevent their instruction are found in about half of the southern states.” In fact, as historian Peter Kolchin has pointed out, only four states maintained laws from the 1830s to the Civil War totally prohibiting teaching slaves to read and write. Apart from Georgia, the Carolinas, and Virginia, other states enforced such laws for shorter periods or banned the teaching of assembled slaves rather than individuals. These statistics raise a number of questions: If white southerners believed antiliteracy laws prevented insurrection and conspiracy, why didn't every slave state pass such ignorance-producing laws? What prompted some states to pass such laws, while others did not? What can we learn by using an agnotological perspective to investigate the triggering events that prompted the passage of antiliteracy laws in a state like Georgia?[4]

The timing of the 1829 antiliteracy laws in Georgia presents a puzzle. Given the public safety benefits a slaveholding society could reap from suppressing literacy among its slave population, it is surprising that Georgia passed legislation during the late colonial period penalizing anyone teaching a slave to write, while the state did not prohibit the teaching of reading until 1829. Some scholars have concluded that southern state legislatures strengthened existing antiliteracy laws in response to Nat Turner's rebellion, an insurrection that took place in Southampton County, Virginia. As one historian put it, "After Nat Turner's insurrection, in all slave states except Maryland, Kentucky, Tennessee and Arkansas, it was against the law to teach a slave to read or write." However, Turner's rebellion, which took place in August 1831, could not have been the triggering event for Georgia's antiliteracy legislation, for it had gone into effect more than a year and a half earlier. Other scholars have identified the circulation of David Walker's abolitionist *Appeal to the Colored Citizens of the World*, published in the fall of 1829, as an important triggering event. For example, Monaghan has argued that although slave owners had always considered writing a dangerous skill for slaves to possess, since they could use it to forge passes and communicate with one another, Protestant Christians generally had sanctioned reading because they believed that every converted soul should be able to read the Bible. Monaghan claims that this view of reading changed with the rise of the abolition movement in the North and the appearance of Walker's *Appeal*—an eighty-eight-page booklet that condemned slavery and urged slaves to rise up against oppression—in the South. In response, slaveholders came to characterize reading as potentially seditious. Monaghan's argument is compelling, but as this chapter will demonstrate, Georgia's State Assembly passed its antiliteracy legislation several weeks before Walker's booklet appeared in the state.[5]

This chapter aims to identify the triggering event that led the Georgia legislature to pass new antiliteracy legislation in fall of 1829 that actively and strategically sought the production of ignorance. Using state records, newspaper accounts, and the journals of Susan Nye Hutchison and James Stuart, this chapter explores the events leading to that legislation and analyzes some of the contradictory perspectives that preceded its passage. It all began with a series of fires in Augusta.

The 1829 Fires in Augusta

Every community experienced one or more fires a year during the early national period. The practice of lighting candles and cooking fires in wooden structures made the outbreak of fire inevitable. Southerners were well aware of the danger; newspapers commonly advertised the sale of fire insurance. Fire was a fairly common occurrence for the residents of Augusta in the 1820s. The town was an important financial and trade center located on the Savannah River, and the state's cotton exchange, banks, and merchant houses lined its broad streets.[6]

Before the events of 1829, Augusta residents did not usually attribute outbreaks of fire to arson. In 1827 Susan Nye Hutchison, a married northern woman who lived in the town, recorded in her journal three outbreaks of fire. In May, the house of a neighbor burned down, and on a separate occasion town citizens extinguished a fire near the counting room where her husband worked. In July, a fire consumed a well-known local tavern. Local newspapers concluded that the tavern fire had probably been caused by negligence. In 1828, Hutchison reported another outbreak of fire. In none of these instances did the authorities suspect arson.[7]

But in 1829 the number of fires in Augusta increased exponentially (figure 1.1). Hutchison described thirteen separate incidents that year. In February, fires broke out two days in a row. Toward the end of the month, she wrote, "I heard the cry of fire and soon afterward a whole square was in flames . . . for a while flashes of fire passed over our roof but at length the wind turned." According to the *Augusta Chronicle and Georgia Advertiser*, more than thirty buildings went up in flames.[8]

Augusta residents probably would have interpreted the February fires as the result of commonplace carelessness had the authorities not discovered evidence of arson. According to the *Augusta Chronicle and Georgia Advertiser*, some individuals had found "some fodder, paper, cotton, &c. . . . placed against

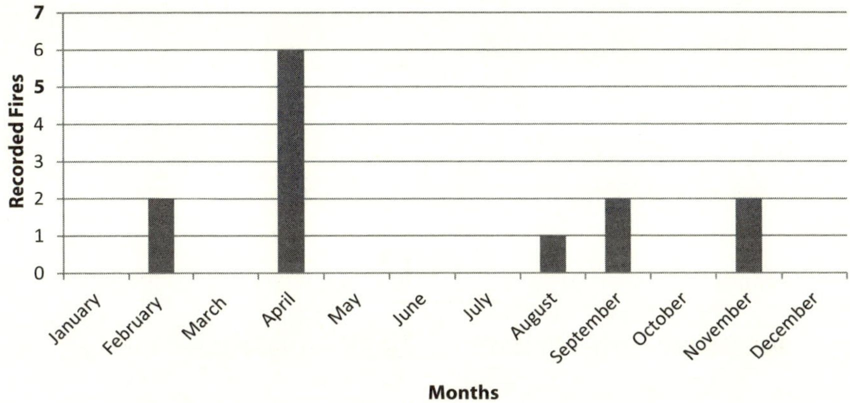

Figure 1.1. Thirteen Fires Reported in Augusta, Georgia, 1829. Susan Nye Hutchison Journal, January 3, 1829, to January 2, 1830. Southern Historical Collection, Wilson Library, University of North Carolina, Chapel Hill.

a pile of small pieces of lightwood, under a shed" and concluded that these represented an aborted attempt to burn down some buildings on the corner of Reynolds and Macintosh Streets. The authorities urged citizens to take "such measures as circumstances may suggest, for the protection of their lives and property and the discovery of the inhuman wretches who are prowling about them." In early March the newspaper reported that a Volunteer City Guard, numbering "nearly three hundred of our most respectable citizens," had been created to patrol the town.[9]

In April, six separate fires erupted across the town. On April 3 Hutchison reported, "Today a most awful conflagration occurred . . . at first a dark smoke arose but soon the blaze burst out and rising toward heaven swelled and spread till all the southeastern part of the city was one wide conflagration—It continued till near sunset destroying all in its reach except two or three lonely habitations, from 400 to 600 habitations are supposed to have been consumed besides a vast amount of other property." According to the *Augusta Chronicle and Georgia Advertiser*, this fire raged for about five hours, destroying "half a million dollars" worth of property, an enormous sum at the time. Fortunately, no lives were lost. On April 6 Hutchison wrote, "Another alarm of fire—It seems that some incendiary is determined to effect our destruction—the whole town is a scene of alarm." Fire broke out again the following day. The newspaper proclaimed that the fires were the premeditated work of a group of arsonists:

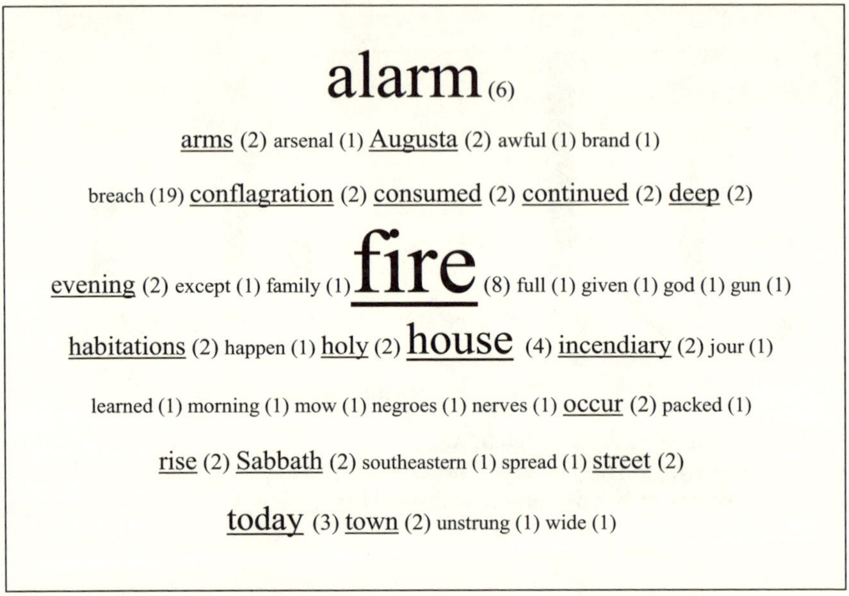

Figure 1.2. Tag Cloud Displaying 41 Words Most Often Appearing in the Susan Nye Hutchison Journal, 3—23 April 1829, with Frequencies. Susan Nye Hutchison Journal, Southern Historical Collection, Wilson Library, University of North Carolina, Chapel Hill.

"[There] can be no doubt, that we have among us . . . desperate incendiaries, who contemplate the entire destruction of the City." On April 7 Hutchison noted, "Great fear begins to be prevalent that the negroes are about to rise." The newspaper exhorted residents to be vigilant: "Let the most rigid police be instituted—let the guards be doubled—let every citizen be on the alert, and watch with a wary eye the movements of *strange* and suspected individuals." Hutchison's journal entries for the month of April reflected the level of panic in town, as illustrated in figure 1.2.[10]

In response to the cluster of fires in April 1829, Hutchison's husband and other white males in Augusta began to arm themselves. On April 9, town residents found "a brand of fire thrown in a hay mow," confirming suspicions of arson. Ten days later, Hutchison reported another fire alarm and described "Mr. H busy in loading his gun and pistols in expectation of an alarm of insurrection." By late April the military had occupied the town. "Today the street[s] are full of armed men," she wrote. "500 stands of arms came from the arsenal and all capable of doing military duty are to be prepared for the evil which I trust in God may never happen—Col. Cumming is Commandant."[11]

The authorities imposed martial law in the city because of a deep-seated fear of insurrection and massacre. Fears of a slave uprising were often associated with arson in the South. News of the bloody Haitian Revolution (1791–1804), which had erupted when slaves, armed with machetes and pruning hooks, torched plantation buildings in and around the French colony of St. Domingue and hacked slave owners to death in their beds, haunted southerners for decades afterward. Thwarted slave revolts in Virginia in 1800 and 1805 raised fears that a slave revolution on the same scale could occur in the United States. Reports of a slave uprising in New Orleans Territory in 1811, in which scores of slaves marched from the German Coast to the city of New Orleans, burning several plantation houses, sugarhouses, and crop fields, reinforced these fears.[12]

When the Englishman Adam Hodgson traveled through Charleston, South Carolina, in 1820, he reported that the town's residents were so fearful of a slave uprising that "a person is stationed every night on the steeple of one of the churches to watch and give the alarm in case of fire, as the inhabitants are never free from the apprehension of an insurrection of the slaves in the confusion of a premeditated or accidental conflagration." Anxiety in Charleston turned to panic with the arrest of Denmark Vesey in 1822. Vesey, a freedman, purportedly had developed an elaborate plan for a massive slave revolt involving thousands of slaves on plantations around Charleston. According to some sources, the plan involved starting a major fire at night and then killing the slave owners and their families under cover of the ensuing chaos. Vesey and thirty-four of his presumed co-conspirators were hanged, and others involved were deported, but officials never identified many of the other leaders believed to have plotted with him.[13]

Over the years, Augusta had experienced several thwarted slave uprisings of its own. In 1809 authorities had learned that a slave in town had written to a slave in North Carolina planning an armed insurrection to occur on April 22, 1810. Nine years later a slave named Mills, in retaliation for being whipped, had burned down his owner's stable, killing twelve horses. Shortly afterward, a slave named Coco or Coot reportedly had developed a plan to break into local stores for guns and ammunition, set fire to the town perimeter, and slaughter the white residents as they mobilized to put out the fire. The authorities had learned of this plan before it was put into action and arrested

Coot and several of his co-conspirators. The men were tried, found guilty, and publicly executed in 1819.[14]

It is not surprising that the fires in 1829 led the authorities to suspect another conspiracy like the one they had narrowly averted ten years earlier. No one was sure who was starting the fires, but suspicion fell squarely on the slaves. From April on, the Augusta authorities rounded up slave and free black suspects on the basis of their proximity to the fires. For instance, when a second fire broke out in the buildings above some stables in town in early April, suspicion fell on the black residents who lived next-door. According to the newspaper, "The fire was soon discovered in a building on Ellis Street, above [the] street's stables, immediately contiguous [to that] occupied by a negro blacksmith named Joe Rockburn, and other persons of color. They have all been taken into custody for examination. It was evidently, as in the former instance, an incendiary attempt."[15]

Susan Nye Hutchison's journal indicates that the authorities arrested only black individuals, not whites, who resided in buildings where fires originated. For example, although she reported that combustible materials had been found behind her own house, unlike the black residents on Ellis Street, neither Susan nor her husband, Adam, faced any scrutiny from authorities. The sort of racial profiling that Susan Nye Hutchison described was never questioned in the antebellum era, not only because free blacks did not have citizenship in most of the South but also because white Americans generally approved the right of authorities to question, arrest, and imprison individuals on the basis of race. In the atmosphere of heightened fear that existed in Augusta, Savannah, and other southern cities that year, the color of one's skin was a mark of innocence or guilt.[16]

A climate of fear and hysteria could give rise to a witch-hunt. *Witch-hunt* is defined in *Webster's New Universal Unabridged Dictionary* as a public campaign "supposedly to uncover subversive political activity, disloyalty, etc., but really to harass and weaken the entire political opposition." In Georgia, this kind of political persecution sometimes led the authorities to arrest innocent individuals. For example, after arsonists tried to set fire to Savannah in 1818, authorities rounded up nine suspects, but none was convicted at trial. Two years later a fire that began in a livery stable swept through the town and quickly reduced it to ashes, destroying 463 buildings worth $4 million. Although the exact cause of the fire was unknown, the mayor of Savannah ordered the city watch to round up and confine all vagrants and "especially black boys and those of colour." However, once again there was insufficient evidence for a conviction. A year later, a slave named Susan was brought to trial for allegedly setting a house on fire in Savannah, but the evidence was so flimsy that she was exonerated. In fact, after all the arrests in Savannah in this period, only one slave, a female slave named Molly who was found guilty of burning down her owner's house, was convicted and executed.[17]

Fears of slave conspiracy and insurrection led politicians in some areas of the South to pass new restrictions on slaves and free blacks. For instance, on the heels of the Vesey Conspiracy, a group of Charleston residents sent a memorial to the state legislature in 1822 recommending the "principle of fear" as the only means of safely maintaining slavery: "We should always act as if we had an enemy in the very bosom of the state, prepared to rise upon and surprise the whites, whenever an opportunity afforded." They asked the legislature to expel every free person of color from the state so that "we may extinguish at once every gleam of hope which the slaves may indulge of ever being free—and . . . govern them on the only principle that can maintain slavery, the 'principle of fear.' "[18]

The excitement and panic in Augusta subsided over the summer of 1829, but anxiety resurfaced when fires started up again in August and September. On August 13 a fire destroyed Augusta's Eagle Tavern, a famous local landmark. The *Georgia Constitutionalist* described Augusta as besieged by enemies of another race: "Fellow citizens, be watchful!—We are in the midst of a race of beings who are determined upon the destruction of our city. . . . Let the number of our Marshals be increased—let us be more vigilant—let us not give sleep to our eyes, nor slumber to our eye lids, until we have defeated their machinations, and averted the evil they have plotted." The increased vigilance did not appear to help matters, because in late September flames again swept the town, burning a number of homes to the ground.[19]

After a second fire in late September, authorities arrived on September 29 to interrogate a hired slave who worked in the Hutchison's household. Susan wrote, "At ten or after, two gentlemen called to examine Harriet, a servant of ours who with some others alleged that a black man carried smoking cotton into an alley yesterday where a little after a barn was found on fire." However, despite the rumors of slave insurrection circulating in the community from February to November 1829, authorities had insufficient evidence to convict any of the individuals who had been arrested. None could identify the alleged arsonists.[20]

Fire broke out again in early November. On November 6 Hutchison wrote, "About two o'clock an alarm of fire aroused us—we got up [and] found the fire—was only distant two squares being just above Col. Cumming's new stores. Amanda and I packed up and waited to remove as the danger became more imminent—the fire fell thick around us sometimes landing upon our roof—In a little time a gentleman passing under the window informed me that the danger was arrested." Four days later she reported, "A fire was discovered under Mrs. Walker's stable just as it had got under way." Among those arrested was Jenny, a slave belonging to the widow Elizabeth Jones.[21]

Conviction, Execution, and Unresolved Questions in Augusta

Susan Nye Hutchison knew Jenny through her friendship with Elizabeth Jones. A wealthy woman who owned slaves and property in town, Jones was a devout Presbyterian and the widow of Seabold Jones, a former mayor and member of the Augusta city council. She had hired Jenny out to a Mrs. Dunn.[22]

Hiring out, or "jobbing," slaves was a common practice in the antebellum South. Studies have shown that a wide range of employers hired slaves, from middle-class business owners who provided some training and hired individuals for several years to working-class individuals who hired slaves for short-term jobs lasting only several days, weeks, or months. In Susan and Adam Hutchison's household, two or more hired slaves worked as cooks, housekeepers, nurses, and handymen when the family could afford it. The practice of hiring out slaves made many members of the community beneficiaries of slavery, whether or not they owned slaves. Antebellum critics viewed jobbing slavery as particularly inhumane because the frequent turnover in their assignments allowed slaves little contact with their own family members and friends. Under this system, slaves often worked for relatively short periods of time in different homes and businesses. Masters began hiring out slaves as young as eight or nine years to work for wages. Infants accompanied their mothers, who were hired out more cheaply than single women. Very young children could be hired out to work for room and board only, saving their masters the cost of supporting them as they grew older. The economic self-interest of the slave owner often conflicted with humanitarian concerns, but most owners had at least some vested interest in maintaining the long-term health and earning potential of their slaves. In contrast, those who employed jobbing slaves on a temporary basis had no longstanding personal connection to the slaves they hired and little financial incentive to treat them particularly well.[23]

After the outbreak of fire at Walker's stable, rumor circulated throughout Augusta that a slave named Cinda had confessed to the crime of arson and implicated Jenny. On November 12 Hutchison wrote, "Jenny Mrs. Jones['s] servant living with Mrs. Dunn has confessed she set fire to Mrs. Walker's stable—she was under punishment for her faults and was excited to take this dreadful revenge by Cinda living with Mrs. Arnold who has by her confession established the fact—both are in confinement." The *Georgia Constitutionalist* described the convicted women as "the misguided and foolish blacks engaged in a recent attempt to set fire to the city." The papers described Cinda as a woman and Jenny as "a young negro girl." Given this description, it is likely that Jenny was twelve to sixteen years old.[24]

Jenny and Cinda were quickly brought to trial, found guilty, and convicted. Nine days after their arrest, the *Augusta Chronicle and Georgia Advertiser* offered the following account: "In the *Inferior Court* of this County convened in the City Hall yesterday, for the trial of several negroes, for *Arson* present their Honors, V. Walker, A. Rhodes, and H. McTurk. Verdicts of *Guilty* were found against a negro woman, Cinda, belonging to Mrs. Lubbeck, and a young negro girl Jinney [*sic*], belonging to Mrs. Jones, and a verdict of *Not Guilty* in favor of Mr. L. Roll's negro man Reuben. The Court adjourned to to-day, when it will again convene to hear the trial of Mr. Parson's negro Lewis, and pass sentence on Cinda and Jinney." Four days later, "a verdict of *Not Guilty* was returned in favor of Mr. Parson's negro Lewis, indicted for Arson; and a sentence of death was passed on the two negroes Jinney and Cinda." The court found Cinda "the instigator, and Jinney, the immediate actor, in the attempt to fire the city on R[ey]nold Street, on the 10th instant." The judge set the date of June 4 for Cinda's execution, a delay of six months because she was pregnant. Jenny's execution was scheduled to take place on December 4.[25]

Hutchison wrote two entries in her journal on the day of Jenny's hanging. The first is in present tense: "Jenny is as calm as it is possible for a living thing to be. She desires death." She must have written this after another visit with Jenny that morning. Later, Hutchison and her husband walked together to the Augusta jail to witness Jenny's execution. A local minister preached a sermon. Jenny herself prayed aloud and then addressed the crowd with her "last dying words." When she returned home, Hutchison added the following entry: "Poor

Jenny was today launched into eternity—on the gallows she prayed after the Reverend Mr. Sinclair and spoke in a collected and impressive manner to the spectators."[26]

Was Jenny actually guilty of arson, or was she the innocent victim of a political witch-hunt? Susan Nye Hutchison's journal offers no clues. Certainly Hutchison was sympathetic to Jenny's plight; she referred to her as "poor Jenny" more than once in her journal, but she never wrote that Jenny might have been innocent. Nevertheless, the question of Jenny's guilt persisted in Augusta long after her hanging. According to the Scotsman James Stuart, who visited Augusta four months after her death, both Jenny and Cinda denied any guilt, and both were convicted on insufficient evidence. Stuart reported, "The fire was believed to be the work of incendiaries among the people of colour. One slave, a female, was convicted, executed, dissected, and exposed, but she died denying the crime. Another, now with child, is sentenced to be executed in June, but she still denies her guilt. I fear these unhappy creatures are convicted on what we should consider very insufficient evidence."[27]

Like Stuart, some Augusta residents were convinced the women had not received justice. When Cinda was executed on June 6, three months after Stuart's visit, rumors apparently circulated through Augusta that she had died proclaiming her innocence. To counter these reports, the editor of the *Augusta Chronicle and Georgia Advertiser* printed the following notice: "The negro woman, *Cinda*, convicted in November last of Arson, was executed on Friday the 6th instant, agreeably to the sentence passed on her. The execution was deferred thus long on account of her pregnancy. As it has been publicly stated that she persisted to the last in asserting her innocence of the crime for which she suffered, it may be proper to state, for the purpose of preventing misapprehension on the subject, that her guilt had been fully developed."[28]

Jenny and Cinda's incarceration, however, did not put an end to the fires in Augusta. Two weeks after their arrest, a fire broke out in Macon Academy, and the entire school burned to the ground. Perhaps because the authorities believed they had captured the town's two arsonists, the newspaper attributed the fire to a stove accident. In February 1830, two months after Jenny's execution, a small fire broke out below the wooden floor of the dining room in the house Susan Nye Hutchison and her husband rented. The Hutchisons smelled the smoke and woke up in time to extinguish the fire. They concluded that a burning coal that had fallen through a crevice in their fireplace and rolled beneath the floorboards had started this fire.[29]

Despite the frequency of accidental fires throughout the antebellum era, southern whites believed that slaves deliberately set most of the fires that destroyed major portions of some towns and cities, though not everyone agreed. Some northerners suspected that the greater number of fires in the South had more to do with poor fire prevention than with slave insurrection. Cadwallader David Colden, mayor of New York City, stated as much in a letter to the mayor of Savannah in 1820: "Certainly, [southerners] have comparatively suffered much more by conflagration than the other states—in part, perhaps, from the general construction of their buildings, but more from the want of proper apparatus and *free labor* to extinguish fires." Colden's letter, which was replete with references to the superior equipment and social institutions of the North, probably did nothing to sway southern readers. Yet perhaps he was right.[30]

The thirteen outbreaks of fire in Augusta in 1829 may simply have been a random cluster of events. As cognitive psychologists have pointed out, humans often believe that statistical significance exists in cases where clusters of random events occur. Of course, in the case of the Augusta fires it is impossible to know with certainty whether all thirteen outbreaks were random, a mix of random events and instances of arson, or the result of acts by individual arsonists or a group of slaves bent on insurrection.[31]

In Jenny's case, there was never any evidence of an orchestrated conspiracy to promote a slave uprising. Not only the newspaper accounts but also Susan Nye Hutchison's journal identified retaliation for punishment as the women's motivation for burning down Walker's stable. Nevertheless, the two women may have deliberately turned to arson as a means of striking back at the whites who had abused them. As historians have pointed out, arson as a form of retaliation against oppression is a form of slave resistance. Based on an analysis of executions of female slaves, David V. Baker has identified poisoning and arson as the most common methods slave women used to strike back at their masters.[32]

Women commonly turned to arson as a means of retaliation. According to Philip J. Schwartz, during the years 1785–1831 the rise in the number of convictions for arson in Virginia outstripped population growth, and 23 percent of all executed as well as banished slave arsonists during those years were women. Arson clearly appealed to women as a weapon against southern whites, and by the end of 1829 the Georgia state legislature had come to view individual acts of arson committed by slaves as dangerous forms of resistance virtually indistinguishable from instances of organized insurrection.[33]

The Timing and Context of Georgia's 1829 Antiliteracy Act

Young Jenny was the first slave convicted of arson to receive a mandatory death sentence in the city of Augusta. Before that, a convicted arsonist might receive a whipping and some jail time. For example, in 1819, when the slave known as Mills was convicted for burning down his owner's stable, his punishment consisted of a flogging and banishment. However, during the weeks leading up to Jenny's trial, the Georgia legislature proposed and passed new laws to prevent future episodes of slave insurrection. In particular, the legislature amended "several laws of this State for the trial and punishment of slaves and free persons of color." Henceforth, "the willful and malicious burning, or setting fire to, or attempting to burn a house in a city, town or village, when committed by a slave or free person of colour, shall be punished with death."[34]

The revised punishment for arson accompanied a number of acts designed to strengthen the state's control over its slave and free black populations, including an act prohibiting the teaching of slaves to read and write. What prompted Georgia's legislature to pass such punitive and restrictive laws? The timing of Georgia's antiliteracy laws prohibits consideration of Nat Turner's 1831 insurrection as a triggering event, but it is possible that advance knowledge of David Walker's *Appeal . . . to the Coloured Citizens of the World* played a role in galvanizing Georgia politicians to suppress slave literacy.

Walker's booklet, published in Boston on September 28, 1829, appeared in some areas of the South at about the time of Jenny's arrest and execution. Walker, a free black man from North Carolina who had moved north to Boston and become active in the antislavery movement, published his document to persuade readers of the injustice and evil of slavery. He urged slaves to deliver themselves from the "abject ignorance and wretchedness" in which slaveholders aimed to keep them, seek education, and resist oppression. White southerners characterized Walker's *Appeal* as a brazen effort to promote insurrection, pointing to passages like the one in which he urged slaves to "fight under our Lord and Master Jesus Christ, in the glorious and heavenly cause of freedom and of God—to be delivered from the most wretched, abject and servile slavery."[35]

Although whites across the South reacted to Walker's booklet with horror and outrage, the timeline of events raises some doubt about the direct influence of Walker's document on the passage of Georgia's antiliteracy law. It is likely that no politician living in Georgia had read the *Appeal* when the legislation was passed. Jenny was executed on December 4, three weeks before copies of Walker's *Appeal* first surfaced in the state. On December 26, William T. Williams, mayor of Savannah, wrote to the governor, George Gilmer, to inform him that sixty copies of the *Appeal* had been found and turned over to the local police. The new antiliteracy law therefore preceded its appearance in Georgia.[36]

Nevertheless, Georgia legislators may well have been aware of this document weeks, if not months, before it reached their borders. Recent work by historians has documented the active social, religious, and business networks that bound northerners and southerners together. News traveled relatively quickly from North to South and back again through personal letters, business dealings, religious conference meetings, newspapers, and other outlets. Moreover, Georgia's antiliteracy law was part and parcel of an act that clearly aimed to keep slaves in a state of ignorance, especially when it came to ideas then circulating among the abolitionists in the North. The act was designed to "prevent the circulation of written or printed papers . . . calculated to excite disaffection among the coloured people of this state, and to prevent said people from being taught to read or write." The act also mandated quarantine on ships carrying free black sailors in order "to prevent such persons of colour from coming into this State or from communicating with the coloured people of this State." Even though the act was signed into law on December 22, four days before Savannah's mayor wrote to the governor to alert him to the dangers of Walker's *Appeal*, the legislation makes clear that the authorities suppressed literacy not only to extinguish the communication of insurrection plans among southern slaves but also to prevent the communication of abolitionist ideas between southern slaves and free black outsiders.[37]

Yet while Walker's *Appeal* may have played a part in the passage of Georgia's antiliteracy law, the fires in Augusta that year also played a significant role. It is important to remember that the timing of antiliteracy laws in the South varied state by state. Virginia passed its first legislation in 1819 in response to insurrection fears and strengthened those laws in 1830, whereas Alabama passed its first legislation in 1831 and strengthened its laws two years later. In other words, the passage of antiliteracy legislation depended on the social and political context within each state.

The most important trigger for Georgia's law against teaching slaves to read was a growing fear that just one slave bent on retaliation—even a female youth—could wreak havoc in the urban centers. The antiliteracy law was not solely the result of abolitionist literature arriving in the state's ports. After all, contemporary accounts suggest that copies of *Walker's Appeal* were easily collected at the point of entry in Savannah and suppressed. Nor did the law result from the initial outbreak of fires in Augusta; some newspaper editors and some Georgia residents initially believed that many, if not most, of the fires besetting their cities and towns were caused by negligence. The most important trigger for Georgia's antiliteracy laws was growing fear in the urban areas—fear not only of planned insurrection but also of the substantial damage that could be inflicted by individual jobbing slaves seeking retaliation, not only against their owners but also against the middling and working-class whites who hired their labor. That fear was widespread, and it transcended class lines.

The fires in Augusta revealed a glaring security weakness in southern society: cities and towns were vulnerable to attack, and not just from large gatherings of male insurgent slaves. Southerners realized they had a great deal to fear from a twelve- or fourteen-year-old girl like Jenny. Such a female could move without suspicion throughout the town, communicate with other jobbing slaves like Cinda, and perhaps read and write notes encouraging others to start a fire. Even if she acted on her own, a young girl could potentially burn down half the town, destroying banks, markets, and important trade centers. She could easily create as much destruction as an armed, organized group of adult male slaves.

Up until 1829, Georgia had refrained from following South Carolina's example in prohibiting the teaching of slaves to read, but the thirteen fires in Augusta that year, coupled with several small fires in Savannah and elsewhere, changed everything. The Georgia General Assembly did not publish legislative debates, so what remains is the record of the law itself, along with the comments of the governor. In his legislative report at the end of the year, Governor George Gilmer stated, "The plots devised some years ago in Charleston, and very lately in Georgetown, South Carolina, the late fires in Augusta and Savannah, have shewn us the danger to be apprehended in the cities from the negroes." This rhetoric supported the legislature's effort to pass antiliteracy legislation in December 1829. The legislation, which revised and expanded the state's antiliteracy law to include the prohibition against teaching slaves to read, represented but one part of the slave codes, the overarching set of laws restricting and controlling the interactions of slaves and free blacks.[38]

Through its antiliteracy legislation, Georgia deliberately fostered the structural production of ignorance. The state's justification for suppressing knowledge among its slave population rested on rhetoric about the importance of enhanced safety and security. As Governor Gilmer explained, the slave codes passed in 1829 had "become highly necessary and essential to the welfare and safety of the good people of this state." Although some Augusta residents were shocked at the execution of Jenny and Cinda, and a few continued to raise questions about their innocence for months afterward, within a relatively short time those voices fell silent as white residents conformed to the new state of affairs.[39]

In later years, advocates of slavery frequently pointed to the illiteracy and ignorance of the enslaved population as evidence of the African's "natural" inferiority. For example, in a journal entry written in 1837, a former student of Susan Nye Hutchison's named Sarah Frew Davidson described a conversation with a wealthy mine owner from Charlotte, North Carolina, in which he persuaded her of the inherent inferiority of Africans. She wrote, "He spoke of the intellectual capacity of the human race—as evinced in national character & convinced me that the Africans are inferior whether it is their ignorant degraded condition—climate—or the curse entailed on the progeny of Canaan or Ham it is not easy to determine." Davidson had never heard this argument, but it was hardly new. Messages about the intellectual inferiority of Africans had supported the slave trade from the very beginning, and southerners continued to advance them to justify slavery.[40]

It is important to understand that despite the rhetoric about the intellectual inferiority of slaves, the South's antiliteracy laws exposed a fear of the intellectual equality of blacks and whites. As Brown University president Francis Wayland pointed out in *Elements of Moral Science*, a text that was highly popular in northern schools by 1835, "Such laws suppose the capacity of negroes for intellectual culture, and are an implicit confession that it is necessary to degrade their minds in order to keep their bodies in slavery." Nevertheless, by midcentury, proslavery Americans increasingly emphasized the inherent intellectual inferiority of slaves. The old awareness and fear of slaves' and free blacks' potential to achieve literacy on a par with whites became knowledge forgotten, overlooked, or relinquished. On the eve of the Civil War, proslavery arguments continued to demonstrate a willful disregard, both active and strategic, of past political actions that had served to maintain a large portion of the South's population in a state of ignorance.[41]

NOTES

1. Georgia, *Acts of the General Assembly of the State of Georgia, Passed in Milledgeville at an Annual Session in November and December, 1829* (Milledgeville, GA: Camak & Ragland, 1830), 170–71; Robert A. Arnove and Harvey J. Graff, eds., *National Literacy Campaigns and Comparative Perspectives* (New York: Plenum, 1987); Harvey J. Graff, *The Literacy Myth: Literacy and Social Structure in the Nineteenth-Century City* (New York: Academic, 1979); E. Jennifer Monaghan, "Reading for the Enslaved, Writing for the Free: Reflections on Liberty and Literacy," *Proceedings of the American Antiquarian Society* 108 (1998): 309–41.

2. Robert N. Proctor, "Agnotology: A Missing Term to Describe the Cultural Production of Ignorance (and Its Study)," in *Agnotology: The Making and Unmaking of Ignorance*, ed. Robert N. Proctor and Londa Schiebinger (Stanford, CA: Stanford University Press, 2008), 1–36.

3. See Janet Duitsman Cornelius, *When I Can Read My Title Clear: Literacy, Slavery, and Religion in the Antebellum South* (Columbia: University of South Carolina Press, 1991); Monaghan, "Reading for the Enslaved, Writing for the Free"; Heather Andrea Williams, *Self-Taught: African American Education in Slavery and Freedom* (Chapel Hill: University of North Carolina Press, 2005), 7–29. On the passage of antiliteracy laws in the South, see David J. McCord, ed., *The Statutes at Large of South Carolina; Edited under Authority of the Legislature, Volume Seventh* (Columbia, SC: A. S. Johnston, 1840), 364–65, 460; South Carolina, *Acts and Regulations of the General Assembly of the State of South Carolina . . . December, 1834* (Columbia, SC: E. F. Branthwaite, [1834]), 13–14; John D. Cushing, ed., *The Earliest Printed Laws of the Province of Georgia, 1755–1770*, 2 vols. (Wilmington, DE: Michael Glazier, 1978), 2:31; Georgia, *Acts of the General Assembly of the State of Georgia . . . 1829*, 168; Alabama, *Acts Passed at the State of Alabama . . . One Thousand Eight Hundred and Thirty-One* (Tuscaloosa, AL: Wiley, M'Guire & Henry, 1832), 16; Louisiana, *Acts Passed at the Second Session of the Ninth Legislature of the State of Louisiana* (Donaldsonville, LA: U. W. Huby, 1830), 96; Mississippi, *Laws of the State of Mississippi, Passed at the Fourteenth Session of the General Assembly* (Jackson, MS: Peter Isler, 1830), 86–87; North Carolina, *Acts Passed by the General Assembly of the State of North Carolina, at the Session of 1830–31* (Raleigh, NC: Lawrence & Lemay, 1831), 11; Virginia, *Acts Passed at a General Assembly of the Commonwealth of Virginia . . . in the Year of Our Lord, One Thousand Eight Hundred and Thirty* (Richmond, VA: Thomas Ritchie, 1831), 107–8.

4. Henry Walcott Farnam, *Chapters in the History of Social Legislation in the United States to 1860*, ed. Clive Day (Union, NJ: Lawbook Exchange, 2000), 193; Peter Kolchin, *American Slavery: 1619–1877* (New York: Macmillan, 2003), 129.

5. Amy Reynolds, "The Impact of *Walker's Appeal* on Northern and Southern Conceptions of Free Speech in the Nineteenth Century," 319, in *Communication and Law: Multidisciplinary Approaches to Research*, ed. Amy Reynolds and Brooke Barnett (New York: Routledge, 2006). For other depictions of Turner's insurrection as the triggering event for antiliteracy legislation, see Stephanie Foote, "Education and Polemic," in *A Companion to American Fiction, 1780–1865*, ed. Shirley Samuels (New York: John Wiley & Sons, 2008), 105; Scott E. Casper, "Antebellum Reading Described and Prescribed," in *Perspectives on American Book History: Artifacts and Commentary*, by Scott E. Casper, Joanne D. Chaison, and Jeffrey D. Groves (Amherst: University of Massachusetts Press, 2002), 158; Clement Eaton, *The Growth of Southern Civilization, 1790–1860* (New York: Harper & Row, 1961), 77; Monaghan, "Reading for the Enslaved, Writing for the Free"; Cornelius, *When I Can Read My Title Clear.*

6. For an example of an ad for fire insurance, see "Insurance Against Fire," in *Georgia Courier*, 11 June 1827. For descriptions of nineteenth-century Augusta, see Edward J. Cashin, *The Story of Augusta, Georgia* (Augusta, GA: Springfield Village Park Foundation, 1995).

7. "Augusta—*FIRE!*," *Augusta Chronicle and Georgia Advertiser*, 4 July 1827; Susan Nye Hutchison Journal, Southern Historical Collection, Wilson Library, University of North Carolina, Chapel Hill (hereafter SNHJ), 4 May, 2 July, 27 December 1827, 18 February 1828.

8. SNHJ, 22, 23, 24 February 1829; "Augusta—Incendiaries," *Augusta Chronicle and Georgia Advertiser*, 25 February 1829.

9. "Augusta—Incendiaries"; "Augusta—Volunteer City Guard," *Augusta Chronicle and Georgia Advertiser*, 4 March 1829.

10. SNHJ, 3, 6, 7 April 1829; "Terrible and awful conflagration," *Augusta Chronicle and Georgia Advertiser*, 4 April 1829; "The Fire," ibid., 8 April 1829; "On Tuesday Last," ibid., 11 April 1829.

11. SNHJ, 8–11, 13, 22 April 1829.

12. David Patrick Geggus, ed., *The Impact of the Haitian Revolution in the Atlantic World* (Columbia: University of South Carolina Press, 2001); Adam Rothman, *Slave Country: American Expansion and the Origins of the Deep South* (Cambridge, MA: Harvard University Press, 2005), 100–116.

13. Adam Hodgson, *Remarks During a Journey through North America in the Years 1819, 1820, and 1821* (New York: Samuel Whiting, 1823), 134; Walter J. Fraser, *Savannah in the Old South* (Athens: University of Georgia Press, 2003), 197–200. Some historians have questioned whether the so-called Vesey Conspiracy was an orchestrated plot. Michael Johnson argues that although white Carolinians believed in the existence of a slave conspiracy, there is little documentary evidence to support this belief. Johnson is not the first scholar to question the existence of a planned slave revolt; in 1964 Richard C. Wade claimed that the conspiracy amounted to little more than "angry talk." However, other scholars disagree, citing evidence of contemporaries who spoke or wrote about Vesey's plans to lead a revolt. See Wade, "The Vesey Plot: A Reconsideration," *Journal of Southern History* 30 (May 1964): 143–61; Michael P. Johnson, "Denmark Vesey and His Co-Conspirators," *William and Mary Quarterly* 58 (October 2001): 915–76; Michael P. Johnson et al., in "Forum," ibid., 59 (January 2002); Robert L. Paquette, "From Rebellion to Revisionism: The Continuing Debate about the Denmark Vesey Affair," *Journal of the Historical Society* 4 (Fall 2004): 291–334.

14. Edward J. Cashin, *Old Springfield: Race and Religion in Augusta, Georgia* (Augusta, GA: Springfield Village Park Foundation, 1995), 28–29.

15. "On Tuesday Last."

16. SNHJ, 8–11, 13, 22 April 1829; "Destructive Fire," *Augusta Chronicle and Georgia Advertiser*, 14 April 1829.

17. *Webster's New Universal Unabridged Dictionary* (New York: Simon & Schuster, 1979), 2100; Fraser, *Savannah in the Old South*, 197–200, quotation on 198. On white hysteria in the face of potential insurrection, see Bertram Wyatt-Brown, *Southern Honor: Ethics and Behavior in the Old South* (New York: Oxford University Press, 1981), 402–34.

18. Ulrich Bonnell Phillips, ed., *Plantation and Frontier Documents, 1649–1863; Illustrative of Industrial History in the Colonial and Antebellum South* (Cleveland, OH: Arthur H. Clark, 1909), quotations on 111 and 114.

19. SNHJ, 13 August, 24 September 1829; "FIRE!," *Georgia Constitutionalist*, 18 August 1829.

20. SNHJ, 24, 26, 27, 29 September 1829; "FIRE!"

21. SNHJ, 6, 10 November 1829.

22. Ibid., 16 November 1827. Seabold Jones is mentioned in Cashin, *Story of Augusta*, 58. According to the will of Seabold Jones, "Estate to be divided equally between my wife Elizabeth and children, she to have slaves, etc., which I got by her." Georgia State DAR, "Richmond County, Georgia Wills," in *Historical Collections of the Georgia Chapters Daughters of the American Revolution: Records of Richmond County, Georgia, Formerly Saint Paul's Parish* (Baltimore: Clearfield, 1929), 45.

23. See Jennifer Oast, "'The Worst Kind of Slavery': Slave-Owning Presbyterian Churches in Prince Edward County, Virginia," *Journal of Southern History* 76 (November 2010): 867–92.

24. SNHJ, 12 November 1829; "Augusta," *Georgia Constitutionalist*, 24 November 1829.

25. *Augusta Chronicle and Georgia Advertiser*, 21, 25 November 1829. The newspapers spelled Jenny's name as "Jinney." I've used Susan Nye Hutchison's spelling, "Jenny," in this account.

26. SNHJ, 4 December 1829.

27. Georgia, *Acts of the General Assembly of the State of Georgia . . . 1829*, 119; James Stuart, *Three Years in North America* (Edinburgh: Robert Cadell, 1833), 2:123.

28. "Augusta," *Augusta Chronicle and Georgia Advertiser*, 9 June 1830.

29. "Fire!," ibid., 27 November 1829; SNHJ, 8, 9 February 1830.

30. *Niles' Weekly Register* 18 (March–September 1820): 89.

31. See Thomas Gilovich, *How We Know What Isn't So: The Fallibility of Human Reason in Everyday Life* (New York: Free Press, 1991).

32. Eugene D. Genovese, *Roll, Jordan, Roll: The World the Slaves Made* (New York: Vintage Books, 1976), 587–97; Herbert Aptheker, "American Negro Slave Revolts," *Science & Society* 1 (Summer 1937): 512–38; David V. Baker, "Black Female Executions in Historical Context," *Criminal Justice Review* 33 (March 2008): 64–88. On slave resistance during this era, see Steven Hahn, *A Nation Under Our Feet: Black Political Struggles in the Rural South from Slavery to the Great Migration* (Cambridge, MA: Harvard University Press, 2003); and Susan Eva O'Donovan, *Becoming Free in the Cotton South* (Cambridge, MA: Harvard University Press, 2010). On female resistance, see Stephanie M. H. Camp, *Closer to Freedom: Enslaved Women and Everyday Resistance in the Plantation South* (Chapel Hill: University of North Carolina Press, 2004).

33. Philip J. Schwartz, *Twice Condemned: Slaves and the Criminal Laws of Virginia, 1705–1865* (Union, NJ: Lawbook Exchange, 1998), 211–14.

34. Cashin, *Old Springfield*, 28; Georgia, *Acts of the General Assembly of the State of Georgia . . . 1829*, 172.

35. David Walker, *Walker's Appeal, in Four Articles; Together with a Preamble, to the Coloured Citizens of the World, but in Particular, and Very Expressly, to Those of the United States of America, Written in Boston, State of Massachusetts, September 28, 1829* (Boston: Revised and Published by David Walker, 1830), 4, 15, http://www.docsouth.unc.edu.

36. W. T. Williams to George M. Gilmer, 26 December 1829, in *David Walker's Appeal to the Coloured Citizens of the World*, ed. Peter P. Hinks (University Park: Pennsylvania State University Press, 2000), 93–94.

37. On the close bonds between northern and southern merchants and politicians during this era, see Daniel Kilbride, *An American Aristocracy: Southern Planters in Antebellum Philadelphia* (Columbia: University of South Carolina Press, 2006). For the religious and educational bonds between the two regions during the Second Great

Awakening, see Kim Tolley, *Heading South to Teach: The World of Susan Nye Hutchison, 1815–1845* (Chapel Hill: University of North Carolina Press, 2015). For the development of the press, see Jeffrey L. Pasley, *"The Tyranny of Printers": Newspaper Politics in the Early American Republic* (Charlottesville: University Press of Virginia, 2002). On the exchange of ideas through religious and reform organizations, see Ronald G. Walters, *American Reformers, 1815–1860* (New York: Hill & Wang, 1978). Georgia, *Acts of the General Assembly of the State of Georgia . . . 1829*, 168.

38. Georgia, *Acts of the General Assembly of the State of Georgia . . . 1829*, 119.

39. Ibid., 168.

40. Sarah Frew Davidson, *The Private Journal of Sarah F. Davidson, 1837*, ed. Karen M. McConnell, Janet S. Dyer, and Ann Williams (Charleston, SC: History Press, 2005), 107. On proslavery writers, see Larry E. Tise, *Proslavery: A History of the Defense of Slavery in America, 1701–1840* (Athens: University of Georgia Press, 1990); and Drew Gilpin Faust, ed., *The Ideology of Slavery: Proslavery Thought in the Antebellum South, 1830–1860* (Baton Rouge: Louisiana State University Press, 1981).

41. Francis Wayland, *The Elements of Moral Science* (New York: Cooke, 1835), 198.

2

Sex

JENNIFER BUREK PIERCE AND MATT PIERCE

On the evening of November 19, 1934, radio listeners across America tuned in to hear Dr. Thomas Parran, New York State commissioner of health, deliver a short lecture on the educational program *Dollars, Doctors, and Disease.* William F. Snow, a physician and head of the American Social Hygiene Association, envisioned "many thousands of people throughout the United States" waiting to hear a talk that he characterized as "one of the most important and informative in the series . . . conducted under the auspices of the National Advisory Council on Radio in Education." Instead of Parran's words, however, piano music filled the airwaves.[1]

Columbia Broadcasting System executives and Parran, who would be named the sixth US surgeon general two years later, were embroiled in a dispute over program content that resulted in the last-minute cancellation. Parran's prepared remarks discussed sexually transmitted infections as a public health threat. He wanted to dispel myths and provide medical information to the general public at a time when this kind of educational talk was regarded as an important disease prevention strategy. CBS, meanwhile, questioned the appropriateness of Parran's content and decided against allowing the doctor to utter words like *syphilis* on air. Network executives saw that sort of detail as incompatible with the company's responsibility to ensure the decency of material disseminated under its auspices. Parran insisted that failing to name a medical condition injured rather than aided listeners' understanding. Neither would concede.[2]

This case of sex education censorship proved to be a pivotal point in the evolution of broadcast regulation. What's inescapably significant about the incident is that the resulting conversations created conventions that would govern what the public might hear on the radio, including educational broadcasts deemed integral to a federal requirement to serve the public interest. The station prevented the doctor from airing his lecture, but CBS officials could not keep him from publicizing the standoff, which quickly made its way into newspapers and other sources that documented the controversy and its resolution. Those in the medical community who supported the diffusion of knowledge and reduction of ignorance about sexually transmitted infection clashed with industry executives who were concerned about social and cultural norms of propriety; it was the latest in a series of similar episodes that had taken place since the turn of the century. Although writers called attention to the problem of individuals' ignorance about sex in sources ranging from medical journals to the *Ladies' Home Journal*, advocates continued to face controversy. Numerous strategies, such as couching sex education in terms of religious and moral imperatives or teaching parents how to talk to their children about reproduction, broke through long-standing silence on these subjects but never quite eliminated the stigma from the subject. Earlier conflicts over how much information young people ought to have about their bodies, now encapsulated in the commonly used phrase *sexual health information*, had been the province of reformers concerned with eugenics, women's rights, and other causes.[3]

The Parran case was different. It involved a new technology many saw as having tremendous power because of the audiences it could reach. As Henry Field stated in his 1933 proposal for a new educational program, "Within the past decade a powerful and unprecedented agent for the dissemination of knowledge has been put into our hands—the Radio. Today the radio rivals the press in being the strongest and most irresistible influence on the minds of the people. To what are young Americans, still in their formative years, listening each day in their own homes?"[4] In the same year, experts estimated that two-thirds of US homes had a radio, creating millions of listeners. As such, Parran's case marks an important yet often overlooked episode in US educational, medical, media, and public policy history. It is also a story that stands at the intersection of science and the social construction of ignorance, which has recently drawn much attention from historians of science. This now little-known episode in broadcast history centered on the idea that information provided through a broadcast medium that was new and rapidly gaining popularity should be suitable for a child's ears. Such a restriction on educational programming had the potential to stymie attempts to educate the American public about sexually transmitted infections, including the importance of prevention and treatment.

Social health campaigners often defended their work in terms of protecting "innocence, not ignorance," but executives within the emerging broadcast industry were not willing to differentiate between the two.[5]

Arguments about the age appropriateness of content had already figured in earlier professional discussions about print media in public libraries. Most librarians, at least in the printed professional literature of the era, advocated conservative limits on sensitive materials, and some subjects, such as contraception, were regarded as too indelicate to even consider in this context. These older arguments reappeared in debates over broadcast decency at a time when many viewed radio as an instrument that could support education and the public good. In these debates, the "child" became an abstraction, a symbol, rather than an actual example of negative consequences resulting from mature content reaching younger listeners. Broadcast companies insisted on limitations based on the possibility that children might listen to radio broadcasts, but they did not simply rehearse old arguments. Parran's case set the stage for a series of closed-door discussions about how to interpret the few government-issued guidelines on content, which included statements of interest in how media should serve the public good.

Advocates for child-friendly broadcasting would characterize the programs they wanted to counter as "dangerous education." While broadcast companies were caught up in developing federal regulation for new media, the Motion Picture Producers and Distributors of America (MPPDA) opted for self-regulation via a set of "in-house" rules popularly known as the Hays Code, beginning in 1934. By 1948, when television became a technically and financially viable enterprise, these earlier, private decisions revolving around the trope of the young listener had effectively become public policy, and they have continued to inform public policy to this day. Although the cases of print, radio, and television all hinged on questions of decency and appropriateness, their agnotological implications in terms of information access constitute a larger, and largely unaddressed, theme in the historiography.[6]

Parran's efforts to challenge CBS's restrictions on sexual health programming emerged out of broader trends in broadcast regulation and historic, procensorship concerns about the impact of print on children. As government regulators and industry officials discussed how to use new media to inform and educate the public in the early twentieth century, ideas about libraries as public institutions figured in their deliberations about how to manage radio as a public resource. Both federal laws and prior controversies involving indecency and censorship demonstrate print history's influence on radio. The confrontation between Parran and CBS, particularly as organizations interested in public education to reduce rates of sexually transmitted infection became involved, foreshadows the way previous arguments were refashioned for the purposes of shaping subsequent policy. Even though national norms for broadcast decency were lacking, concerns about media's effects on children created a code of operations based on the idea that electronic media was "a guest in your home."[7]

Broadcast Regulation, Self-Censorship, and the Child

The Parran sex education case highlights a strategic, ignorance-promoting anthropomorphism absent from the earliest federal laws and policy statements governing the public airways. Because of radio's limited capacity and potential to serve public safety and military interests, the federal government enacted and revised a series of laws between 1910 and 1934. Initially, regulation governed the point-to-point communications used in military operations and wireless telegraph operation, and existing laws were extended and adapted to suit new purposes; a handful of court decisions that followed augmented these statutes. Even as radio became sufficiently widespread to justify federal regulation, these laws were more concerned with technical specifications than with defining the tripartite requirement that radio operate "consistent with the public interest, convenience, and necessity," the primary legal provision governing content. These limited references to content meant that for many years the evaluation of radio programs for appropriateness was neither standardized nor a government priority.[8]

The lack of attention to broadcast appropriateness reflects pragmatic concerns in the developing industry. After the first officially licensed broadcast station went on the air in 1920, broadcasters spent most of the decade struggling to improve technical aspects of radio and searching for a profitable business model. Simultaneously, Congress grappled with how to regulate the new medium to reduce interference among stations and ensure it served the public interest. Once basic aspects of operation were addressed, broadcasters focused on developing content to attract the largest possible audience while avoiding offensive content, which was characterized in the Radio Act of 1927 as "obscene, indecent, or profane language." Children were assumed to be among the potential listeners who should be protected, as the entity now known as the Federal Communications Commission (FCC) had received complaints since its inception about whether broadcast content was appropriate for young people. While the majority of official documents concentrated on technical specifications, other outlets and venues discussed what the new medium might do to and for listeners. Issues familiar to librarians, who had long provided print media to the public, would also return in debates over how radio should serve the public.[9]

Among these issues was how to balance responsible regulation against restrictions that might verge on censorship. Herbert Hoover, in his capacity as secretary of commerce in the 1920s, avoided making judgments about program content, believing that such decision making too closely resembled censorship. Aside from ensuring that obscene or suggestive content was not aired, Hoover favored deferring to the public's sensibilities, believing that appropriate limitations could be achieved without undue government influence. His argument that communities of listeners would have their own standards, which would have power because "the stations naturally are endeavoring to please their listeners," prevailed for years.[10]

Some shared Hoover's confidence that broadcast media could function without explicit federal control of content. While exploring regulatory issues in 1924, the House of Representatives Committee on Merchant Marine and Fisheries heard David Sarnoff, then vice president and general manager of the Radio Corporation of America (RCA), argued that libraries and broadcast media alike were aspects of an expansive, engaged, and educated community. He told Congress: "I believe that the expression of educators and statesmen should reach [the public] uncensored and uncontrolled. The air belongs to the people. . . . Its main highways should be maintained for the main travel. To collect a tax from the radio audience would be a reversion to the days of toll roads and bridges; to the days when schools were not public or free, and when public libraries were unknown."[11] The public library, however, represented challenges as well as an ideal for those involved in the development of radio as a public information resource.

Popular ideals that defined childhood and family spaces as pure and protected forged connections between the evolving norms of public library service and the political regulation of the airwaves in the first half of the twentieth century. For both, questions about the decency of content and the innocence of the child provoked scrutiny of professional conduct. While the outcomes eventually differed, their Progressive Era origins were pronounced and interrelated. A prolific rhetoric about childhood innocence motivated censorship in public libraries and over the public airwaves, and librarians' prior professional discussions about protecting children by prohibiting access to questionable material informed broadcast regulations. Broadcasters alternately relied on and resisted these notions. The potential for a young person to be harmed by inappropriate ideas or overly adult language prompted restrictions on content in both milieus. Just as moral entertainment had been associated with the public library, many believed that radio should provide positive, educational material suited to the enjoyment of the entire family, a notion reflected in industry advertisements and governmental deliberations alike. In the years following the Parran controversy, an executive with another network concluded, "This American institution of home life has driven people back to their 'fire-sides' . . . to listen to radio."[12]

Most accepted that broadcast regulation was inevitable, and their hope was that limitations would be thoughtfully crafted. Albert Pfeiffer, who worked for the Social Hygiene Division of the New York Department of Health, advised those who worked in broadcasting to resist broad censorship rules, a longtime library concern. Pfeiffer railed against the idea of "radio and the unmentionable." He believed that it was "tragic to find a great implement of potential educational value controlled by men with mid Victorian attitudes." Despite these criticisms, he acknowledged that "a broadcasting system must censor its programs but censoring must be done in an intelligent manner. . . . Proper censoring should be left in the hands of an intelligent person whose judgment the broadcasting system can depend upon and this person should have the advice of experts on the various subjects . . . discussed over the system." While this sort of peer-review system had been proposed as an alternative to censorship in 1924, in part because of the "danger of presenting any sort of medical subject" on the air, deliberations continued about how to ensure that radio programs avoided inappropriate content. Libraries, too, had long been concerned with patrons' access to the specifics of medical knowledge. In the debates over decency in the decades ahead, the library, the book, and the child would continue to be used to evaluate the ways broadcast stations fulfilled their role.[13]

As the number of stations and programs proliferated, so did questions about how decency should be understood, and new federal guidance emerged in the 1930s. Just five months before the CBS-Parran incident, Congress reaffirmed its prohibition against obscene, indecent, or profane language over the airwaves when it included the same provision in the Communications Act of 1934. This new law, however, did little to render the language and concepts it barred more clear or precise, opening the door to further controversy.[14]

"Radio and the Unmentionable": Parran's Address and the Regulation of Radio Programming

Some scholars have asserted that Parran and the health issues he wished to publicize in 1934 received more attention through the controversy his case caused than they would have if the fifteen-minute talk had been aired over sixty CBS stations as planned. It is in any case evident that Parran and his colleagues worked to direct attention to both the canceled broadcast and his resignation from the National Advisory Council on Radio in Education (NACRE). NACRE was an independent organization created to advocate for the educational use of radio, a purpose congruent with the federal requirement to operate in the public interest. Parran's resignation from the organization may have resulted in part from the council president's waffling over how to handle CBS's censorship case.[15]

Word about the sex education censorship fracas quickly spread across the country, and almost immediately Parran began receiving recognition and support from far-flung places. New Mexico's director of the Bureau of Public Health "missed the broadcast . . . [but] saw this item in the Albuquerque paper." Others lambasted "the calm assumption by radio powers that they have the right to sacrifice a vital movement for public health because it happens to conflict with their peculiar, personal, and outworn concept of 'purity.'" Still others observed that Parran controlled a budget that represented multiples of his annual salary, a significant portion of which had gone to creating "prophylactic stations, clinics, moving pictures, lectures and pamphlets to teach New Yorkers the ravages of syphilis and gonorrhea, and to cure as many of those infected as could be induced to apply for treatment." Newspapers and magazines presented Parran as a man of considerable responsibility and authority whose multimedia informational campaign had reached and aided countless individuals. Given the doctor's reputation for public health activism, his sexual health advocacy could have been anticipated by those who agreed to put him on the air. Furthermore, it would not have been lost on Parran and his circle that these publications deliberately printed the very terms prohibited on the CBS network.[16]

CBS responded to Parran supporters by arguing that print and broadcast media reached different audiences and had different effects. In a press release on the incident, the station resorted to the age appropriateness argument: "In deciding what is proper for us to broadcast, we must always bear in mind that broadcasting reaches persons of widely varying age levels and reaches them in family and social groups of almost every conceivable assortment." The child and the possible diversity of audiences meant that "some things may more properly be discussed in print, where each person may individually and privately concern himself with the subject." Some fought this position and advocated for using radio to provide information to those who lacked either literacy or access to printed materials. Henry Field, of the Field Museum of Natural History in Chicago, proposed radio programs for young listeners and urged NBC to be the library of the airwaves. He said that research information that "should be made available to every man, woman, and child" was located in "the great libraries of the world," and he argued that "a radio program, dealing with these and many other highly significant subjects, broadcast each day to homes of America, will stimulate the minds and broaden the outlook of young citizens." The wholesome, if idealized, image of radio in the family setting was integral to the way the industry understood its future.[17]

The censorship of Parran's proposed address in 1934 was not an isolated instance. Broadcasters had cut off New York City's health commissioner in the middle of a similar address. Following the Parran incident, however, public health activists like Snow challenged CBS and its competitors to prove that their concerns were justified and began collecting information from social hygiene societies around the nation about local broadcasts on the subject. Snow aggressively pursued "the question of what may or may not be said over the radio." Meanwhile, John L. Rice, the prominent and respected New York City health commissioner, began writing letters of protest. Both men soon received responses, including a letter from Stanford University president Ray Lyman Wilbur, that promised federal-level inquiries and continued national media attention. These men hoped that the collection of information and scrutiny of broadcast practices would result in fewer restrictions on educational messages in the future.[18]

Snow's aims were manifested in a letter sent to private state health associations and public state boards of health alike, asking that they report whether they or their colleagues had been able to speak about the prevention of sexually transmitted infection and whether they faced restriction when they did so. He asked for "any information from you regarding your radio experience that will bear on this point" but hoped to demonstrate that "the public views accurate and frank social hygiene education with approval rather than protest," indicating that he intended to use the data gathered in an early January meeting with representatives from NACRE, NBC, and CBS. It seems significant that the government was not present during this conversation about how to interpret federal rules regulating broadcast content.[19]

Responses from around the nation soon demonstrated that hygiene activists and health professionals encountered divergent standards. At times, commentary came from local broadcasters themselves, as when the assistant manager of the *Milwaukee Journal* station insisted that the subject of syphilis could not possibly be broached on air. Of the more than three dozen responses, some of which accounted for multiple broadcasts or stations, twenty-five writers reported some ability to discuss sexually transmitted infection on local or regional radio programs.[20]

Those who supported Parran did not ignore the trope of the child. A letter of complaint to a CBS official recited the statistics of infection rates, noting the percentage of infants born with syphilis and the infection rates for teens of each sex: "Most of the infections, as you see, occur in young people." Another physician was sympathetic to concerns about the questions children might ask "uniformed parents" about the subject during a daytime program but insisted that the late-night broadcast should have been permitted.[21]

Beyond the idea that a preponderance of stations facilitated access to sexual and reproductive health information, specific instances in which health educators were permitted to talk to the public about sexually transmitted infection reveal themes invoked in other contemporary conversations. G. W. Grignon of WISN in Milwaukee wrote that while he would not prohibit the use of the word *syphilis* on air, he would recommend that a program host use *social disease* instead: "A radio station has as its' [*sic*] audience many children of tender years. . . . Adults would have a perfect understanding of the phrase 'social disease' while my experience with children has shown that use of the words . . . causes no question in their young minds." Grignon was one of many local broadcasters who did not feel obligated to censor based on existing indecency law and made decisions based on his personal sense of what was appropriate. Others in public health would make the point as well, though differently, by asking, for example, "What is there in such a statement of facts that should be taboo in New York and accepted without question in Boston?" It seems that few were unaware of the potential for the sort of dispute that had arisen when Parran was asked to address CBS audiences.[22]

Meanwhile, NBC president M. H. Aylesworth, who had been appointed head of the network in 1926 and was charged with seeing that its "operations . . . reflect enlightened public opinion," urged those who protested CBS's decision to think of younger listeners: "I know you will agree with me that the handling of the radio is a serious problem, particularly where people in the home are concerned and especially children. Most people look at radio as an invited guest and expect us to do something quite different than the writer of a book." Despite his position, Aylesworth offered to meet with Rice and Snow, "to sit down and talk with you about this and other matters of similar character," as he put it. He continued, "A frank discussion of an important subject, affecting the lives of our friends, their children and future children, seems imperative."[23]

In response to the *New York Post*'s advice that "someone ought to take the radio executives by the hand and gently break the news to them that the dear, sweet smirking Victorian days are dead," Aylesworth's "frank discussion" took place in 1935. With their collection of broadcast data from across the nation in hand, health experts hoped to influence the policies and practices that would emerge from this meeting. The outcome was not, however, favorable to those in Parran's corner. The discussion took the form of closed-door meetings between broadcasters and public health advocates and resulted in an industry policy statement that articulated the same conservative, child-conscious standards that had kept Parran off the air. Asserting an "ever greater responsibility" to the public, the document focused specifically and immediately on programs "designed for children, and those involving unpleasant discussions of bodily functions, bodily symptoms, or other matters which similarly infringe upon good taste." In addition to barring such programs in the interest of protecting the public, the document expressed the idea that "the personal and often intimate quality of the human voice, radio's medium of presenting its sponsors' messages, forbids discussing subjects on the air in a manner which might be wholly acceptable when read in type." Prominent individuals within the public health community had expected substantive change; instead, broadcasters' interest in commercial viability held sway. In short, profits trumped education. The resulting network statement on what the major broadcast companies would and would not air made its way into later modifications to federal policy.[24]

Four years later, H. L. Mencken's *American Mercury* ran an article titled "Syphilis: Mrs. Grundy's Disease," which recalled the 1934 radio censorship incident and lambasted media standards. Mocking the refusal to air Parran's speech as puritanical, the author echoed NACRE president Levering Tyson's account of the event in noting a plethora of ads for products like Ex-Lax: "If the moral difference between one biological function and another is any greater than the anatomical distance, the explanation is that radio stations, like the newspapers, take their cues from advertisers: Salvarsan [a drug used to treat syphilis] does not happen to be a patent medicine."[25] The article reinforced the views of observers who speculated that when it came to radio broadcast censorship policy, financial success loomed larger than decency in deciding what the public might hear on the air.

Indecency Revisited

It would be easy to observe the ironies inherent in the Parran-CBS conflict, which silenced a leading public health authority and replaced an educational program with instrumental music. Although the program was designed to comply with minimal federal regulation of content, CBS decided against airing Parran's sexual health broadcast for fear that it provided too much information or provided that information to an unintended audience. The forces at work against Parran represented more than reactionary interests. The controversy triggered the creation of new broadcast guidelines aimed at ensuring child-appropriate programming, along with other limits designed to prevent youthful exposure to mature content.

The new guidelines foregrounded a Progressive Era viewpoint that radio had great potential to support education and the public good. Sexual health campaigners imagined the airwaves as powerful educational outlets, but unlike libraries, public airwaves were regulated by the federal government. Concurrently with evolving government restrictions, controversy over broadcast decency unfolded; its resolution centered on constructing standards for acceptable information. The question of age appropriateness and the fear that a child might encounter unsavory ideas and words like *syphilis* during a radio program was a significant factor in subsequent decency regulations that still govern the industry.[26]

Early twentieth-century broadcast decency standards are enacted in federal laws that have not changed significantly since the 1930s. Even in 1973, when the FCC was confronted with the broadcast of humorist George Carlin's "Filthy Words" over Pacifica Foundation's WBAI station in New York, the child listener was once again the focus. The case was ultimately decided by the US Supreme Court, which held that although an earlier case involving printed matter, *Cohen v. California*, 403 US 15 (1971), had resulted in a different conclusion, children's sensibilities had to be a factor in broadcast policymaking: "Broadcasting is uniquely accessible to children, even those too young to read. Although Cohen's written message might have been incomprehensible to a first grader, Pacifica's broadcast could have enlarged a child's vocabulary in an instant. Other forms of offensive expression may be withheld from the young without restricting the expression at its source."[27]

The Court also cited another case, *Ginsberg v. New York*, 390 US 629 (1968), noting that "the government's interest in the 'well-being of its youth' and in supporting 'parents' claim to authority in their own household' justified the regulation of otherwise protected expression" as a precedent for reiterating restrictions on indecent content. Here the Court's reasoning for upholding the FCC's censure echoed broadcasters' justification for preventing Dr. Parran's use of the word *syphilis* over the airwaves forty-four years earlier. The 1978 ruling signaled the lasting significance of the 1934 Parran incident and further illustrates the importance of libraries' willingness to champion unfettered access to information. In 2012 the Supreme Court, in its *FCC v. Fox Television Stations* decision, essentially affirmed standards last revised in the 1970s, which in turn bear the hallmarks of those formulated in the 1930s.[28]

The Parran-CBS incident shows that Parran and his contemporaries viewed the controversy as a matter of policy versus prudery but it meant more than that to those who lived without access to basic information resources in the 1930s. Some stations were willing to air programs like the one CBS prohibited, and one physician shared an excerpt from a letter that arrived following a radio talk he delivered responding to Snow's survey of broadcasting practices: "By chance I tuned in while you were on the air one week ago Thursday and heard your talk on a disease called syphilis, Dr. I live in constant fear of just that one word, and when you stated there was a cure it helped me to feel a great deal easier in my mind. . . . What I would like to know Dr. is, [is] it possible for a local Dr. to treat my 13 yr old daughter and really cure her?"[29]

Although the Parran controversy is now known as an effort to test the limits of educational programming and broadcast decency, reaching this sort of isolated and concerned listener, who suffered because of ignorance, motivated the challenge Parran and his peers presented to CBS. Those who provided education about sexual and reproductive health knew that the child whose innocence was much talked about by their opponents was not the only child who needed and deserved professional attention. Rather than allow an idealized notion of childhood innocence to serve as a rationale for the perpetuation of ignorance, Parran and his colleagues sought to ensure that individuals' literacy would not prevent their access to contemporary, scientific health information.

APPENDIX

Selected Local Station Practices Related to Programs with Sexual Health Content, 1930–1936

Station/ Place	Date[a]	Broadcasts on Sexual Health Subjects Permitted?	Terms or Subjects Permitted	Regular Program or Special Broadcast	Source Reporting to ASHA
KOB, Albuquerque	1934	Yes	Syphilis		J. R. Earp, Director, Bureau of Public Health
WBZ Boston	1930–33, 1934	Yes, until WBZ became NBC affiliate	Gonorrhea, syphilis		H. D. Chadwick, Commissioner of Public Health
WEEI, Boston	1930+	Yes			H.D. Chadwick, Commissioner of Public Health
Washington, DC	1935	No			"R.H.E."
Harrisburg, PA	1935	No			T. B. Appel, Secretary of Health
Virginia	1935				D. C. Smith, University of Virginia Hospital
Little Rock, AR	1935				W. B. Grayson, State Health Officer
KMOX, St. Louis	1933, 1934	Yes	Syphilis, gonorrhea, venereal disease	Both	H. S. Cory, Executive Secretary, Missouri Social Hygiene Assoc.
KWK, St. Louis	1933	Yes	Syphilis, venereal disease, prostitution	Both	H. S. Cory, Executive Secretary, Missouri Social Hygiene Assoc.

Station/ Place	Date[a]	Broadcasts on Sexual Health Subjects Permitted?	Terms or Subjects Permitted	Regular Program or Special Broadcast	Source Reporting to ASHA
KSD, St. Louis	1934	Y	Syphilis, venereal disease	Regular program	H. S. Cory, Executive Secretary, Missouri Social Hygiene Assoc.
Grand Rapids, MI	1935	Yes/limited	"General problems pertaining to family-life"		M. A. Gizon, Director, Welfare Guidance Bureau
WEEU, Reading, PA	1934	No/Yes	Social diseases, delinquent girls' home, infection	Special broadcast	J. Bamford, Director, Community Chest Campaign
WLW, Cincinnati	1935+	Yes/limited	"Syphilis in connection with other diseases" but not as an independent subject	Regular program	C. A. Wilzbach, Executive Secretary, Cincinnati Social Hygiene Society
WSAI, Cincinnati	1935+	No			C. A. Wilzbach, Executive Secretary, Cincinnati Social Hygiene Society
WKRC, Cincinnati	1935+	No			C. A. Wilzbach, Executive Secretary, Cincinnati Social Hygiene Society
WCKY, Cincinnati	1935+	No			C. A. Wilzbach, Executive Secretary, Cincinnati Social Hygiene Society
KSO, Des Moines, IA	1935	Yes/limited	Social diseases, not specific ones	Special broadcast	F. J. Swift, Deputy Commissioner in Charge of V.D. Control, Iowa State Dept. of Health
WGY, Albany, NY	1925+	Yes	Syphilis	Regular program	Albert Pfeiffer, Director, Division of Social Hygiene, NY Dept. of Health

(continued)

Station/ Place	Date[a]	Broadcasts on Sexual Health Subjects Permitted?	Terms or Subjects Permitted	Regular Program or Special Broadcast	Source Reporting to ASHA
Seattle	1935	No			E. R. Coffey, Director, Dept. of Health
WJDX, Jackson, MS	1930+	Yes		Regular program	F. J. Underwood, Executive Officer, Mississippi State Board of Health
WTMJ, Milwaukee	1935	No			H. I. McCarthy, Instructor, City of Milwaukee Health Dept.
WINS	1935	Yes/limited	Social diseases, not syphilis		H. I. McCarthy, Instructor, City of Milwaukee Health Dept.
WTIC, Hartford, CT	1933, 1934	Yes/reversal of earlier refusal	Venereal diseases, syphilis, gonorrhea	Regular program	S. H. Osborn, Commissioner of Health; 11/23/34 news article
San Francisco	1935	Yes/limited	No "sex"	Both	G. A. McGraw, Executive Secretary, Sex Education Society
WSUI, Iowa City	1935	Yes	Venereal disease prevention, syphilis, gonorrhea		J. H. Kinnaman, State Dept. of Health
WOI, Ames, IA	1935	Yes	Venereal disease prevention, syphilis, gonorrhea		J. H. Kinnaman, State Dept. of Health
WHAM, Rochester, NY	1934	Yes	Syphilis or gonorrhea		W. W. Bauer, AMA Bureau of Health and Public Instruction
WGY, Schenectady, NY	1934	Yes	Syphilis or gonorrhea		W. W. Bauer, AMA Bureau of Health and Public Instruction
WJJD, Chicago	1934	Yes	Syphilis or gonorrhea		W. W. Bauer, AMA Bureau of Health and Public Instruction

Station/ Place	Date[a]	Broadcasts on Sexual Health Subjects Permitted?	Terms or Subjects Permitted	Regular Program or Special Broadcast	Source Reporting to ASHA
WBBM, Chicago	1934	Yes	Syphilis or gonorrhea		W. W. Bauer, AMA Bureau of Health and Public Instruction
WGN, Chicago	1934	Yes	Syphilis or gonorrhea		W. W. Bauer, AMA Bureau of Health and Public Instruction
WCAE, Pittsburgh	1934	Yes	Syphilis or gonorrhea		W. W. Bauer, AMA Bureau of Health and Public Instruction
WPTF, Raleigh, NC	1935	Yes	"Correct terms"		J. C. Knox, Medical Director, NC State Board of Health
WAPI, [Birmingham, AL]	1936	Yes	Syphilis, syphilitic, venereal		J. D. Dowling, Health Officer, Jefferson County Board of Health
Erie, PA	1932	Yes	Venereal disease, prostitution, treatments for venereal disease	Special program	V. H. Parker

[a]When a correspondent's letter provided dates for broadcasts, those numbers are given. When the dates were not specified, the date of the letter is used.

NOTES

1. William F. Snow to William S. Paley, 20 November 1934, American Social Hygiene Association Collection, Elmer L. Andersen Library, University of Minnesota, Box 70, Folder 4. A portion of this chapter was delivered as a coauthored conference paper titled "Censorship in Print and over Public Airwaves: Youthful Audiences as Impediments to Information Access, 1900–1953" at Library History Seminar XII: Libraries in the History of Print Culture, sponsored by the Center for the History of Print Culture in Modern America, in Madison, Wisconsin, on 10–12 September 2010. The authors also wish to acknowledge the assistance of the archivists at the University of Minnesota's Andersen Library.

2. While this confrontation has been written about in terms of the history of public health and Parran's professional life, based on his writings and interviews, the episode has been little explored in relation to the development of broadcast regulation, in which lasting regulatory pronouncements followed debates over media use, as documented in ASHA. See, e.g., Lynne Page Snyder, "New York, the Nation, the World: The Career of Surgeon General

Thomas J. Parran, Jr., MD (1892–1968)," *Public Health Reports* 100 (1995): 630–32; and Mike Stobbe, *Surgeon General's Warning: How Politics Crippled the Nation's Doctor* (Oakland, CA: University of California Press, 2014). On the state of sex education in US schools, see Julian Carter, "Birds, Bees, and Venereal Disease: Toward an Intellectual History of Sex Education," *Journal of the History of Sexuality* 10 (2001): 213–49. As Carter concluded, "Movement for sex education in the public schools began in the second decade of the twentieth century, at the cultural moment when vocal public criticism of many social ills was paired with an equally articulate optimism about education's ability to effect their cure" (213).

3. See, e.g., "Syphilis & Radio," 3 December 1934, ASHA, Box 70, Folder 5. See also Jennifer Burek Pierce, *What Adolescents Ought to Know: Sexual Health Texts in Early Twentieth Century America* (Amherst: University of Massachusetts Press, 2011).

4. Henry Field, "Radio Program for Young America," unpublished manuscript, Winter 1933–34 (October 1933): 4, RG 173: Federal Communications Commission, Office of the Executive Director, General Correspondence, 1927–46, National Broadcasting Co.–National Citizens Political Committee, Box 404, Envelope 89-6 NBC, National Archives and Records Administration, College Park, Maryland.

5. Christopher H. Sterling and John Michael Kittross, *Stay Tuned: A History of American Broadcasting*, 3rd ed. (Mahwah, NJ: Lawrence Erlbaum Associates, 2002), 139. See also Pierce, *What Adolescents Ought to Know*.

6. On the relationship between the media and morality at this time, see Thomas Doherty, *Pre-Code Hollywood: Sex, Immorality, and Insurrection in American Cinema, 1930–1934* (New York: Columbia University Press, 1999); Field, "Radio Program for Young America." For research on the debate, see Snyder, "New York, the Nation, the World," 630–32; and Stobbe, *Surgeon General's Warning.*

7. "A Welcome Guest in the House" aired in 1957. *Television Broadcast History Films: The Creation, Innovation & Expansion of the Televisual Medium* ([Bayonne, N.J.]: Earthstation1.com, 2007), DVD.

8. "The Communications Act of 1934," in *Documents of American Broadcasting*, ed. Frank J. Kahn, rev. ed. (New York: Meredith, 1973), 64; Hugh R. Slotten, *Radio and Television Regulation: Broadcast Technology in the United States, 1920–1960* (Baltimore: Johns Hopkins University Press, 2000), 49–50. Slotten labeled this stage of regulation a "technocratic" one.

9. Radio Act of 1927, Pub. L. No. 69-632, §29, 44 Stat. 1162 (1927).

10. Slotten, *Radio and Television Regulation*, 30–31; Herbert Hoover, quoted in *Radio's Hidden Voice: The Origins of Public Broadcasting in the United States*, by Hugh R. Slotten (Urbana: University of Illinois Press, 2009), 102.

11. *Hearings on H.R. 7357 before the H. Comm. on the Merchant Marine and Fisheries*, 68th Cong. 160 (1924).

12. Eveready Radio Batteries playing cards, produced by the National Carbon Company, date unknown; Slotten, *Radio's Hidden Voice*, 205.

13. Albert Pfeiffer to Snow, 4 January 1935, ASHA, Box 70, Folder 6; *Hearings on H.R. 5589 before the H. Comm. on the Merchant Marine and Fisheries*, 69th Cong. 56–57 (1926); Jennifer Burek Pierce and Debora Shaw, "Looking for Love in All the Wrong Places: Accessing Sexual and Reproductive Health Information via the *Readers' Guide to Periodical Literature*," *Library & Information Science Research* 27 (2005): 467–84; Martha Cornog and Timothy Perper, *For Sex Education, See Librarian: A Guide to Issues and Resources* (Westport, CT: Greenwood, 1996).

14. Federal Communications Commission Act, Pub. L. No. 73-416, §326, 48 Stat. 1064 (1934).

15. Bess Furman, *A Profile of the United States Public Health Service, 1798–1948* (Bethesda, MD: US Department of Health, Education, and Welfare, 1973), 391–401; "Syphilis & Radio"; "Talk Censored, Dr. Parran Quits Radio Council," *Herald Tribune*, 12 November 1934, ASHA, Box 70, Folder 5. NACRE president Levering Tyson's explanation of his participation in the pre-airtime discussion of the lecture is also located in ASHA. "Memorandum Covering the Director's Participation in the Incident Resulting in the Refusal of the Columbia Broadcasting System to Broadcast the Address in the 'Doctors, Dollars & Disease' Series on November 19, 1934," 24 December 1934, ASHA, Box 70, Folder 6. While some historians report that Parran resigned from an NBC advisory group, NACRE's own materials describe itself as independent. See National Advisory Council on Radio in Education, finding aid and records at http://archives.nypl.org/mss/2096; and "National Advisory Council on Radio in Education (NACRE)," in *The Evolution of American Educational Technology*, ed. Paul Saettler (Charlotte, NC: Information Age, 2004), 207.

16. J. Rosalyn Earp to Thomas Parran, 23 November 1934, ASHA, Box 70, Folder 6; "Nice Nellies of the Air," *New York Post*, 4 December 1934, ASHA, Box 70, Folder 5; "Syphilis & Radio."

17. "Syphilis & Radio"; Field, "Radio Program for Young America," 2–3.

18. Snow to Parran, 21 November 1934; Snow to Thomas C. Edwards, 22 November 1934; John L. Rice to Snow, 22 November 1934; and Ray Lyman Wilbur to Snow, 27 November 1934, all in ASHA, Box 70, Folder 4.

19. Snow to Social Hygiene Societies, 29 December 1934, ASHA, Box 70, Folder 6. Slotten has observed that "complex negotiations among different groups and individuals" have influenced the development of present-day broadcast technology in the US and that "a sharp dichotomy between public and private sectors" has not always existed in these interactions.

20. Russ Winnie to Hazel I. McCarthy, 15 January 1935, ASHA, Box 70, Folder 4; "Iowa Is No Ostrich," 24 May 1935, ASHA, Box 70, Folder 6, among other multireport responses. See also appendix 2.1.

21. Henry D. Chadwick to Frederic A. Willis, 31 December 1934, and Raymond H. Greenman to Snow, 8 January 1935, both in ASHA, Box 70, Folder 6.

22. G. W. Grignon to Hazel I. McCarthy, 10 January 1935, ASHA, Box 70, Folder 4; Chadwick to Willis, 31 December 1934.

23. "Announcing the National Broadcasting Company, Inc.," September 1926, http://earlyradiohistory.us/1926nbc.htm; M. H. Aylesworth to Rice, 22 November 1934, ASHA, Box 70, Folder 4.

24. "Nice Nellies of the Air"; Parran to Snow, 26 November 1934, ASHA, Box 70, Folder 4 ("Tyson says that he believes this incident may be used as a basis for a conference with the Columbia broadcasting authorities in an effort to bring about a change in policy"); E. R. Coffey to Snow, 8 January 1934, ASHA, Box 70, Folder 6; "New Policies: A Statement to the Public, To Advertisers and to Advertising Agencies," 15 May 1935, 3, 8–9, ASHA, Box 70, Folder 4.

25. Anthony M. Turano, "Syphilis: Mrs. Grundy's Disease," *American Mercury*, April 1937, 403.

26. "Talk Censored, Dr. Parran Quits Radio Council."

27. *FCC v. Pacifica Foundation*, 438 US 726, 748–49 (1978).

28. *FCC v. Fox Television Stations*, 567 US ___ (2012), http://www.supremecourt.gov/opinions/11pdf/10-1293f3e5.pdf.

29. Pfeiffer to Jean B. Pinney, 17 January 1935, ASHA, Box 70, Folder 6.

3

Sexuality

KAREN GRAVES

> What keeps ignorance in one place, while it evaporates in some other? And which among our myriad ignorances will be tolerated or combated?
>
> —ROBERT N. PROCTOR

> I really was a very strong believer in civil liberties and equal rights. . . . And I just could not see any way that I couldn't do what I had taught [students] we should all do—stand up for our rights.
>
> —JAMES GAYLORD

In 2006 the American Library Association's Office for Intellectual Freedom reported that the children's book *And Tango Makes Three* topped the list of censorship challenges filed with libraries and schools. The award-winning book, a "true story of two male penguins who hatch and parent a baby chick at New York's Central Park Zoo," was challenged on the issues of homosexuality, appearing to be antifamily, and being unsuitable for children; it remained among the most challenged books for many years to follow. A 2013 analysis of the continuing censorship challenges to *And Tango Makes Three* revealed that parental rights, claimed by both censors and defenders of the book, was the most common theme to surface in the *Tango* imbroglio. Third-person effect, claims based on perceived harmful impact on others, also characterized many of the objections. For instance, radio host Michael Medved complained that children (and penguins) were being used as "pawns of propaganda." At times the objections took turns toward the sensational, as when one parent objecting to homosexual content clarified, "I'd feel the same way if the penguins were murderers or rapists."[1]

Barbara Jones, director of the American Library Association Office for Intellectual Freedom, responded to the *Tango* censors with a basic argument:

> While we firmly support the right of every reader to choose or reject a book for themselves or their families, those objecting to a particular book should not be given the power to restrict other readers' right to access and read that book. As

> members of a pluralistic and complex society, we must have free access to a diverse range of viewpoints on the human condition in order to foster critical thinking and understanding. We must protect one of the most precious of our fundamental rights—the freedom to read.[2]

Recent high-profile censorship cases have not been limited to children's books. In 2014 the South Carolina House of Representatives approved Representative Garry Smith's bill to cut funding from two state colleges that had adopted gay-themed books for first-year student orientation programs. The measure proposed eliminating $52,000 from the College of Charleston's budget and $17,000 from the budget of the University of South Carolina–Upstate. Legislators were upset that students at the College of Charleston were reading Alison Bechdel's acclaimed *Fun Home: A Family Tragicomic* and that USC-Upstate faculty had assigned *Out Loud: The Best of Rainbow Radio*, a history of South Carolina's first gay and lesbian radio show. Some lawmakers tried to restore the funds, arguing that the legislation was designed to "limit the views that college students can be exposed to," but the proposed amendments failed by a considerable margin. Ironically, the state senate voted to require that the two institutions spend an amount equal to the funds the legislators had cut on teaching principles of the US Constitution, the Declaration of Independence, and *The Federalist Papers*, presumably including lessons on the First Amendment and freedom of expression. Governor Nikki Haley let stand this "compromise," including a provision that students be allowed to avoid educational material they found "objectionable based on a sincerely held religious, moral, or cultural belief." Ten organizations, including the National Coalition Against Censorship, the American Civil Liberties Union of South Carolina, the American Association of University Professors, the Modern Language Association, and the National Council of Teachers of English, signed a statement protesting the state's action. "Such leveraging of public funds with the goal of micromanaging curriculum and excluding disfavored ideas is a destructive assault on academic freedom," they noted, adding that it was likely unconstitutional as well.[3]

As scholarship over the last two decades shows, school administrators, board members, government officials, and citizens have tried to remove evidence of the existence of gays and lesbians from the school environment, from primary school to university, by censoring curricula, library holdings, student groups, and even the names of schools. Concentrated efforts to censor began in the

1970s in parallel with the gay rights movement and are still likely to emerge, in spite of significant advances toward securing civil rights for LGBTQ citizens. This resistance to the reality of sexual diversity poses an important question for the generation that has witnessed Supreme Court decisions that decriminalized homosexuality and supported marriage equality. To paraphrase Robert Proctor, why does ignorance concerning sexuality remain in the schools as it evaporates elsewhere in the culture?[4]

This chapter approaches that question through a historical analysis of how school and other government officials have tried to repudiate the existence of gay and lesbian people by purging gay and lesbian teachers. The act of removing gay and lesbian bodies from schools disrupts the work environment, obviously, but also inhibits teaching about human sexuality. The battles grew more contested over time, as more teachers stepped out of the closet to challenge the strictures against them. Today, teachers in twenty-nine states have no recourse if they are fired for being gay, and there is no federal law to protect against job discrimination on the basis of sexual orientation or gender identity. In many schools, the threat of removal hangs like the sword of Damocles over gay and lesbian educators, destabilizing the educational environment in such a way that many teachers are reluctant to address LGBTQ issues in the curriculum.[5]

The geography of ignorance is a useful cognate in studying the history of sexuality in twentieth-century education in the United States. The theoretical ground for mapping how school officials facilitated dominant misperceptions concerning sexual orientation is the third type of ignorance discussed by Robert Proctor in his volume on agnotology, "ignorance as a deliberatively engineered and *strategic ploy* (or active construct)." For decades, officials aimed this line of attack directly and unambiguously at teachers, purging gay and lesbian educators, as well as those presumed to deviate from the heterosexual norm, from schools and universities. The witch-hunts that characterized the Cold War period eventually subsided in the wake of the gay rights movement, but in the 1970s and 1980s lesbian and gay teachers were still being dismissed as a matter of course in the absence of laws to guard against employment discrimination.[6]

It is important to acknowledge a distinction between school officials' construction of ignorance regarding gay and lesbian people and its passive production prior to midcentury. Until about 1930, school officials were not much concerned with the sexual orientation of teachers or students, although gender norms remained intensely guarded. The animus that came to be directed toward lesbian and gay educators reflected a growing awareness of diverse sexualities in the broader culture that eventually found expression in widespread condemnation and repression, severe employment prohibitions, and harsh laws and punishments. What had once not even been noticed was now inciting public controversy. In order to gain a fuller understanding of this history, then, we need to consider Proctor's second type of ignorance as well, the one grounded in political geography. As he put it, "It is less like a vacuum than a solid or shifting body—which travels through time and occupies space, runs roughshod over people or things, and often leaves a shadow." The early twentieth-century cultural ignorance of sexuality, which paid little mind to lesbian and gay teachers, was displaced by Cold War partisans, who took brutal steps to erase gay bodies not only from educational institutions but also from society at large. The mid-twentieth-century construction of the deviant, homosexual teacher has cast a long shadow over schools, eclipsing and rendering absent fundamental lessons on the diversity of human sexuality.[7]

The Florida Teacher Purge, 1956–1965

> We hope that many citizen organizations in Florida will use this report as a jumping off point for a serious and meaningful consideration of the problem of homosexuality and as a source of information with which to prepare their children to meet the temptations of homosexuality lurking today in the vicinity of nearly every institution of learning.
>
> —*HOMOSEXUALITY AND CITIZENSHIP IN FLORIDA*, 1964

Cold War ideologues constructed the deviant homosexual profile and located its most reviled figure in the schoolhouse. After World War II, politicians pursued a domestic containment agenda that paralleled American foreign policy. Marshaling economic, political, and military pressure against the Soviet Union was only part of the postwar strategy; social engineers tried to strike a counterbalance by stressing stability at home through uniformity of thought and behavior. School and family became key institutions for enforcing boundaries and restrictions. Meanwhile, Republicans in Congress seeking to discredit the Truman administration latched onto the Lavender Scare, "a fear that homosexuals posed a threat to national security and needed to be systematically removed from the federal government," connecting homosexuality to communism. Linking communism to homosexuality for political gain triggered moral panic and repressive laws at all levels of government, ruining untold numbers of lives. Gay and lesbian teachers, cast as predators of youth, became a particularly vulnerable target in this climate. The most aggressive state offensive against teachers occurred in Florida, part of an unrestrained legislative investigation initially designed to stall desegregation.[8]

The history of the Florida Legislative Investigation Committee is complex and has intersected with such topics as the Cold War, the black freedom struggle, lesbian and gay history, and education history. Established in 1956, the committee, more generally known as the Johns Committee (for state senator Charley Johns), was a reactionary response to the Supreme Court's *Brown v. Board of Education* decisions. Similar to legislative committees that sprang up in other states, the Johns Committee intended to delay desegregation by investigating the NAACP. During its first two years the committee focused on campaigns to desegregate the University of Florida Law School, public schools in Miami, and the bus system in Tallahassee. By 1958 the Johns Committee had moved onto the campus of the University of Florida to conduct an undercover investigation of homosexuality. The NAACP investigation stalled later that year when the civil rights organization balked at turning over membership records. That legal issue would not be resolved until 1963, when the US Supreme Court sided with the NAACP. In the meantime the investigation at the University of Florida set off a period of furtive activity, as investigators tracked and interrogated gay and lesbian professors, teachers, and students throughout the state. In the years that followed, scores of schoolteachers, professors, and university students were fired or expelled on the basis of assumptions concerning their sexuality.[9]

The Johns Committee left an explicit record of how the state constructed ignorance regarding sexuality at midcentury. Working in secrecy as much as possible, the group surfaced at key points to deliver "authoritative" findings concerning homosexuality, education, and citizenship. Marred by contradictions that were apparent at the time, the committee's reports nevertheless established assumptions that linger in some quarters to this day.[10]

From the beginning, the Johns Committee understood the power inherent in controlling information. Departing from their practice of holding public hearings as it did with the NAACP investigations, interrogators grilled educators suspected of engaging in homosexual activity in private sessions, usually without the benefit of legal counsel. The committee argued that the "better interests of the State and the educational institutions could not have been served by public hearings." It was not unusual for investigators to appear on a campus or at a school and summon a professor or teacher directly from class. In private correspondence committee members acknowledged the strategic edge to be gained by surprising witnesses, controlling media access to proceedings, and lessening the chance that a witness might appear with a lawyer. Beyond the reach of public accountability, Johns Committee investigators took free rein in violating witnesses' legal rights and, as a matter of course, held up the possibility of public exposure as a form of blackmail to get people to talk. The committee was just as calculating when it came to releasing information. Noting that public disclosure of testimony during the course of the closed hearings would disrupt future investigations, the Johns Committee relied upon biennial reports to the legislature as the primary means of communicating with the public. Even then, it offered a distilled version of events, claiming that "it would serve no useful purpose to enumerate the full extent and details of the findings."[11]

The Johns Committee pressed the inflammatory myth that homosexuals posed a great danger to society because they tended to recruit others. Believing that one's sexuality was largely determined by environmental factors, that "homosexuals are made by training, rather than born," the committee worried that gay and lesbian educators would apply their professional teaching skills to inappropriate ends. Asserting that educators and other professionals accounted for a higher percentage of the homosexual population than other workers, the Johns Committee concluded that homosexuals were infiltrating the Florida educational system like a pack of Russian dolls. The major "finding" in the Johns Committee's 1959 legislative report was that "some of the State's instructional personnel at the higher educational level have been and are recruiting young people into homosexual practices and these young people have been and are becoming teachers in the public school system of Florida, and some of them are recruiting teen-age students into homosexual practices." The committee informed the legislature that there was no effective means of "treating" homosexuality (except psychotherapy) and that the "prognosis for cure" varied. Therefore, it urged the state to devise a centralized records system for tracking teachers so that investigators could "ferret out concrete proof on the vast majority of all practicing homosexuals" in the schools.[12]

The Johns Committee intensified its rhetoric in the next biennial report. In the two-year period beginning in July 1959, thirty-nine teachers' licenses were revoked, fourteen cases were pending at the board of education, and the testimonies of seventy-five teachers were being processed. The wide sweep of the Florida campaign led the Johns Committee to conclude that the "problem" of homosexuality was more serious than previously believed. It claimed, "Practically all children are susceptible to being recruited into homosexual practices." In the 1961 report the Johns Committee went a step further, blaming school administrators for ignoring the purported crisis and thereby contributing to its severity: "Most administrators, because of lack of knowledge and appreciation of the extent of this problem, prefer to think it does not exist in their institution and consequently close their eyes to it. With few exceptions, there is an almost uniform inability or unwillingness on the part of the responsible administrators to cope actively, aggressively and effectively with the problem." The committee made a few recommendations about how the educational system might police itself, including some suggestions for tidying up the process for revoking teachers' credentials and threatened legislative action in 1963 should the school system fail to enact an effective plan.[13]

In 1964 the Johns Committee released what it considered a comprehensive report on homosexuality. *Homosexuality and Citizenship in Florida* was directed primarily to state administrators and personnel officers, although the preface added that "it can be of value to all citizens; for every parent and every individual concerned with the moral climate of the state." Once citizens got word of the booklet, with its provocative photographs and glossary of crude sexual terms, demand for the twenty-five-cent copies was high. This turn of events prompted the committee to reframe its intent: the "purple pamphlet," as some were now calling the publication, really was intended just for state officials on a "need to know" basis.[14]

The report's authors presented major conclusions they had drawn from multiple sources, including the testimony of witnesses who appeared before the Johns Committee, materials provided by the Florida Children's Commission and the Sheriffs Bureau, courts of law, educators, mental health officials, and scientific and popular publications. The writers took pains to acknowledge the absence of a clear, agreed-upon understanding of homosexuality, as well as the limits of their own knowledge of the subject, citing "many conflicting opinions, contradictory statistics, and a serious lack of responsible research." Then they got down to business, describing a "homosexual world" of pornography, blackmail, disease, mental imbalance, crime, and, of course, recruiting: "The plain fact of the matter is that a great many homosexuals have an insatiable appetite for sexual activities and find special gratification in the recruitment to their ranks of youth." The publication cast the Johns Committee's eight-year pursuit of gay and lesbian educators as part of an ongoing battle for morality. An unnamed investigator urged citizens to commit themselves to the fight, vaguely alluding to the gay rights sea change that was just beneath the surface: "Since the homosexual has seen fit to come out into the open and try to get himself accepted by society, I think it is about time the thinking members of society, the persons in positions of responsibility, get up off their duffs and realize that if we don't stand up and start fighting, we are going to lose these battles in a very real war of morality." The Johns Committee, however, miscalculated the impact of its tax-supported salvo. The release of *Homosexuality and Citizenship in Florida* offended many citizens with its sordid discussion of sexuality and its photographic layout. The Dade County state attorney threatened legal action against the committee for producing such "obscene and pornographic" material, and the state found itself in a most unfavorable light.[15]

During the 1965 session of the Florida legislature no attempt was made to extend the committee's charter, and its nine-year assault on civil liberties came to an end. Politically humiliated, the Johns Committee was not so much defeated, however. Its 1964 report included updated statistics on the numbers of teachers whose licenses had been revoked since 1959. The number stood then at sixty-four, with an additional eighty-three cases pending. And as historian Stacy Braukman shows, the ideology that powered the committee's crusade simply went underground. Its 1977 reprisal in Miami pushed Florida to the center of the national spotlight on the gay rights movement.[16]

Teachers' Struggle for Civil Rights, 1977–1985

> It is simply not true that all human beings have the same rights.
>
> —ANITA BRYANT

Educational institutions have been complicit in the construction of ignorance of sexuality for decades. For most of the twentieth century, school officials simply eliminated gay and lesbian staff and those assumed to be gay, making dismissal policies the primary mechanism for enforcing heterosexist norms. In terms of intensity and unrelenting focus, the Johns Committee episode stands as the most extreme example of the harsh purges that characterized the immediate postwar years. These became less frequent as the 1970s gay rights movement advanced, but gay and lesbian teachers were still dismissed when awareness of their sexuality surfaced. As states slowly began to decriminalize homosexuality, starting with Illinois in 1961, public school employees gained new leverage to fight for their jobs. No longer criminals, gay and lesbian teachers could challenge job discrimination on the basis of the Fourteenth Amendment. Antigay backlashes in jurisdictions that included sexual minorities in antidiscriminatory legislation, however, honed in on teachers. They were both criticized for seeking "special rights" and at the same time set aside as special exceptions to the emerging gay rights legislation. Midway through the four decades between passage of the first state law that decriminalized sodomy and the US Supreme Court's 2003 finding that sodomy laws were unconstitutional (*Lawrence v. Texas*), legal battles over the employment rights of gay and lesbian teachers reached a climax in Florida, California, and Oklahoma. Outlandish rhetoric by antigay activists, combined with legal victory by the narrowest of margins, seems to have staved off national campaigns to ban lesbian and gay teachers from the profession, although individual LGBTQ educators continue to lose their jobs. Antigay activists eventually trained their sights on curricula, a transition that was becoming apparent in the 1970s.

A couple dozen cities and counties across the United States had enacted legislation to prohibit antigay discrimination before the Miami-Dade County Metropolitan Commission passed its bill in January 1977. But when this wave of civil rights action reached South Florida, it collided with the rise of the Christian Right. Commissioner Ruth Shack proposed the Miami bill to ban discrimination on the basis of sexuality in employment, housing, and public accommodations. The bill did not exempt religious institutions, particularly schools, from discriminatory hiring practices, a tactical error that drew the attention of local religious leaders. Opponents to the measure enlisted singer and former Miss Oklahoma Anita Bryant to lead the campaign to overturn the ordinance. She and her colleagues asserted the myth of predatory homosexual schoolteachers in the months leading up to the referendum, and, in June voters recalled the ordinance by a two-to-one margin. The defeat resounded across the nation. Fred Fejes captured the impact: "The feeling of inclusion and strength many lesbians and gay men felt in January 1977—that they were becoming accepted as part of American society—now turned into fear and vulnerability."[17]

Astute political players began to appreciate the strategic value of addressing "the homosexual menace" in campaigns of all stripes. They exploited negative images of "the homosexual teacher" as a wedge issue in the effort to win support for conservative issues and advance political careers. California state senator John Briggs, who joined Bryant in Florida to witness the referendum returns on the night of the election, launched his own campaign to ban gay and lesbian teachers. The Briggs Initiative, listed as Proposition 6 on a 1978 California ballot, ensured that the national focus on gay rights would stay firmly fixed on teachers. However, Briggs overreached, writing his bill in such a way that any public school teacher who supported gay rights could be dismissed. This had the effect of widening the field of opponents to Proposition 6 to include those concerned with free speech issues and collective bargaining rights. The Briggs Initiative was defeated with a 58 percent majority.[18]

Campaign materials from the Miami referendum and the Briggs Initiative, as well as information in Bryant's and Briggs's correspondence, interviews, and writings, suggest another answer to the fundamental question in this chapter, why ignorance concerning sexuality remains in schools as it evaporates elsewhere in US culture. Perhaps it takes longer for cultural change to permeate schools because they are firmly embedded within a network of other institutions. Public schools are, by design, responsible to a diverse group of stakeholders. As long as sexual politics are still at play in the institutions of law, the family, and religion, schools are likely to trail, rather than reflect, the cultural edge. Bryant and Briggs argued that the very presence of lesbian and gay teachers in schools endangered children by providing access to role models who would advocate for gay and lesbian students in class, counseling sessions, and student clubs. Allowing gay and lesbian people to teach, they feared, would set a legal precedent for "special privileges" in terms of employment that would reach other occupations and, in addition, encourage "more flagrant sexual behavior by these disturbed teachers." As they saw it, legal protection of this magnitude would end up destroying the family and debasing religion.[19]

In spinning this argument, Bryant charged that lesbian and gay teachers were not deserving of "so-called civil rights for homosexuals who are not a legitimate oppressed minority." She repeatedly referred to the Miami ordinance as a "special-privileges" ordinance and rebuffed the notion that gay men and lesbians faced discrimination. At least as long as they "keep their perversions in the privacy of their homes. They can hold any job, transact any business, join any organization—so long as they do not flaunt their homosexuality and try to establish role models for the impressionable young people—our children." Bryant was advancing an early form of an argument that legal scholar William Eskridge later described as "no promo homo" reasoning, "rhetoric justifying the traditionally degraded social and legal status of GLBT people." No promo homo arguments arose most often in cases like the Miami referendum, when opponents charged that antidiscrimination laws actually were designed to *promote* homosexuality rather than to protect homosexuals. This strand of Bryant's argument was meant to drive home the point that the laws were to benefit only "normal" people. She considered any departure from that standard offensive. "The laws of the land have always been to protect the normal, not the abnormal," she argued. "If you're going to have a preferential legislative piece for everyone in the whole world, it becomes ridiculous."[20]

While Bryant charged that nondiscrimination laws amounted to special privileges for gay and lesbian teachers, Briggs argued that his proposal appropriately targeted lesbian and gay teachers *because* they presented a special case. In a 1977 press release, Briggs clarified that Proposition 6 was aimed at schoolteachers only. He offered two points in support of his position. First, Briggs defined homosexuals as "anti-life" persons (because, he said, they couldn't procreate) and therefore pegged them as immoral. He claimed that since immoral people could not teach morality, and since teachers were charged with teaching morality, homosexual teachers could not teach morality. Therefore, according to Briggs's syllogism, they couldn't *do* the job. Briggs also relied on the commonly cited argument that since sexuality was learned and children were mandated by the state to attend school through compulsory education laws, the mere presence of a homosexual teacher could lead students to identify as homosexuals. Therefore, Briggs concluded, homosexual teachers *couldn't be allowed* to do the job.[21]

Given the rhetoric surrounding the Miami referendum and the Briggs Initiative, the role-model argument seems not to have lost any force in the two decades since the Johns Committee warned citizens about homosexual teachers recruiting students. With his hallmark flair, Briggs summed up the charge: "How else can they enlarge their flock? They don't have any children of their own. If they don't recruit children or very young people, they'd all die away. They have no means of replenishing. That's why they want to be teachers and be equal status and have those people serve as role models and encourage people to join them."[22]

Departing momentarily from the perspective that homosexuals could not be allowed to teach children because sexuality was learned, Briggs clarified that it really didn't matter how a person developed sexual identity: "Whether you learn it or you're born that way doesn't make any difference . . . you sure in hell don't have to give him a [teaching] license. What they want to do is to license immorality." Any teacher who deviated from the norm, according to Briggs, would be a bad role model. He proceeded to associate homosexuals with prostitutes, dope addicts, alcoholics, perjurers, and criminals—bad teachers all. And yet both Briggs and Bryant confused the issue of whether a closeted homosexual should be allowed to teach. Having just argued that any homosexual would be a bad role model for students, Briggs confirmed, "My law says you can be a homosexual and even be a teacher and be a homosexual if you want. You just can't be a public homosexual and go out and brag about it and recruit children." This logic applied to prostitutes as well, evidently, although in making the case Briggs contradicted himself on whether teachers should teach morality.[23] Briggs opined, "Well, I think a teacher should go in and teach, not go in and teach values. You can be a prostitute in California and go in and teach geography on Monday and Tuesday and Wednesday. You're not going to be in trouble. But if you stand up and say, 'I'm a whore on weekends,' or 'I write articles for Coyote and I'm a member of it,' then that person is going to be removed from the classroom because it's illegal to be a prostitute."[24]

It was no longer illegal in California to be gay when Briggs initiated his campaign, a fact that deprived him of a powerful element the Johns Committee had at its disposal in driving gay men and lesbians from the schools. By the late 1970s antigay activists leaned more heavily on the morality clauses in teachers' contracts.[25]

Bryant and Briggs were also concerned that if lesbian and gay teachers gained a foothold in schools, they would work to integrate gay and lesbian issues into the curriculum and establish student groups and counseling services for gay and lesbian students. Bryant targeted Jean O'Leary's education proposals, warning readers to "be aware of these demands and to guard carefully your own local educational institutions against inroads by flaunting and militant Lesbians and homosexuals whose motives are subject to question." Campaign materials for the Briggs Initiative denounced O'Leary's efforts to support lesbian and gay teachers, counselors, and students. They also criticized an educational outreach program in California schools run by the Gay Community Services Center in Hollywood, as well as the San Francisco School Board's policy to eliminate negative references to homosexuality in its sex education curriculum. These programs were offered as evidence of what "may happen in your school" if Proposition 6 failed. Bryant and Briggs were correct in noting that the employment status of gay and lesbian teachers could have a tremendous impact on the overall school climate and that educators were on the threshold of incorporating gay and lesbian issues into the curriculum. It would take time, of course, but a powerful movement had been unleashed by the Miami and Briggs campaigns. Fejes observes, "While the Stonewall riots of 1969 may have marked the beginning of the modern lesbian and gay movement, the campaigns of 1977 and 1978 marked the emergence of a nationally politically self-conscious lesbian and gay community."[26]

In January 1978 Oklahoma state senator Mary Helm introduced a bill similar to the Briggs Initiative, then under way in California. Helm invited Anita Bryant to address the senate, while gay activists were left outside, on the capitol steps. Helm's bill allowed public schools to dismiss or refuse to hire teachers who engaged in "public homosexual activity" or "public homosexual conduct." The activity clause referred to violations of the state's sodomy law, while the conduct clause prohibited "advocating, soliciting, imposing, encouraging or promoting public or private homosexual activity in a manner that creates a substantial risk that such conduct will come to the attention of schoolchildren or school employees." As would become clear when the case reached the US Supreme Court, the broad wording in the law was problematic from a legal standpoint. Oklahoma City University law professor Dennis Arrow would have to admit to the Court that the statute did not provide an adequate definition of "private homosexual activity." Nonetheless, the Oklahoma house passed the measure, 88-2, and the state senate passed the bill unanimously.[27]

In 1980 the National Gay Task Force (NGTF) brought suit against the Oklahoma City Board of Education in federal court, challenging the constitutionality of Helm's law on the basis of multiple issues: freedom of speech, freedom of association, freedom of religion, the right to privacy, the Equal Protection Clause of the Fourteenth Amendment, and the fact that the law was too vague and too broad. Judge Luther Eubanks ruled in favor of the school district, stating that the right of privacy did not extend to same-sex couples and rejecting the notion that teachers would be fired for advocating gay rights unless that activity rendered them unfit to perform their jobs. Although gay advocacy groups disagreed about whether to appeal, weighing the risk that the appellate court might issue a more stringent ruling, the case did make its way to the Tenth Circuit Court in 1984. That court ruled for the NGTF in a 2-1 decision. Judge James Barrett issued a raging dissent, contending that advocacy of the "*unnatural and detestable act of sodomy*" by a teacher was worse than "advocacy of violence, sabotage, and terrorism." When the case reached the Supreme Court docket in September 1984, Justices Brennan and

Marshall voted to affirm the appellate court ruling, Justice Stevens did not vote, and the majority voted to hear the school district's appeal.[28]

Board of Education of Oklahoma City v. National Gay Task Force was the first US Supreme Court case to address the rights of lesbian and gay teachers, but it was not the first opportunity the court had to weigh in on these issues roiling the nation. In 1974, 1977, and just before hearing arguments in the Oklahoma case in January 1985, the justices rejected the petitions of three gay and lesbian educators who had been fired on the basis of their sexuality. In *Acanfora v. Board of Education of Montgomery County* the Fourth Circuit Court of Appeals ruled that schools could not fire teachers on the basis of sexuality or force teachers to remain closeted. But it upheld the dismissal of Joseph Acanfora on a technicality. In *Gaylord v. Tacoma School District No. 10 et al.* the Supreme Court of Washington upheld a school district's decision to fire a teacher on the basis of his gay identity, claiming that public knowledge of his alleged "immorality" damaged his ability to teach—in spite of excellent evaluations over the course of a twelve-year teaching career and in the absence of any charges of misconduct. James Gaylord was fired simply because people in his community had discovered he was gay. The US Supreme Court denied *cert* in both cases, showing an unwillingness to resolve contradictions regarding the constitutionality of dismissing public school teachers on the basis of sexual identity. Days before they would hear the appeal of the Oklahoma City school board, the justices denied *cert* yet again, letting the dismissal of a bisexual guidance counselor stand in *Rowland v. Mad River Local School District, Montgomery County, Ohio*. This time, however, Justice Brennan took the unusual step of issuing a dissent, one that would become a landmark document in due course. The former clerk Charles Curtis recalled that Brennan's thinking was, "We don't have a prayer now. Let's lay it all out and build a record for 50 to 100 years from now."[29]

The Tenth Circuit Court's ruling struck down only the part of the Oklahoma law that prohibited "public homosexual conduct," finding that it violated the First Amendment right to freedom of speech. It upheld the ban on "public homosexual activity," so the Supreme Court did not address that issue. Oklahoma could still fire teachers simply for being gay.[30] Dennis Arrow, who argued the school board's case before the Supreme Court, suggested that the decision would set a precedent concerning the limits of public school teachers' advocacy. He highlighted five issues for the court's consideration. First, he pointed to the multifaceted interests that government, parents, and students held in public education, noting in particular the goals of "undisrupted academic education," citizenship training, and "providing an appropriate introduction to traditional, fundamental cultural values." Arrow indicated that "promoting respect and authority for law" was among the most important lessons in citizenship training. He claimed that the *Pickering* balancing test rather than the less restrictive alternative was the appropriate standard in cases involving teachers' freedom of speech. *Pickering v. Board of Education* established that courts must balance the interests of a teacher commenting on matters of public concern, as a citizen, against the interests of the school in promoting efficient public service. Arrow made the case that the law in question advanced the interests of government, parents, and students, arguing that "teacher interests in advocating the specified crime of homosexual sodomy" was outweighed by these other interests, "effective public education and the preservation of student morality." Finally, Arrow argued that no other law was available to meet the "threat presented by the advocacy of criminal homosexual sodomy by public school teachers."[31]

When Harvard law professor Laurence Tribe stepped up to deliver his argument on behalf of the NGTF, he became the first Supreme Court litigator to explicitly address a gay civil rights issue. He later remarked that the experience engendered an unprecedented level of hostility from the bench. Tribe designed his case narrowly, as a First Amendment, free speech issue. This was not about the power of Oklahoma to prescribe a curriculum. And he wasn't challenging the law on firing teachers for violating the Oklahoma sodomy statutes. When pressed about the role of teachers in modeling appropriate values for students, Tribe reminded the court that when Ronald Reagan was governor of California he had spoken against the Briggs Initiative on that very point. Tribe persistently pointed to the free speech rights of teachers, arguing that "public school teachers have First Amendment rights every bit as broad as private citizens when they talk on matters of public interest, whether they are talking politics or philosophy or ethics or social change." Tribe returned to this point later in his argument, citing Justice Felix Frankfurter's depiction of teachers as the "high priests of our democracy," called upon to model the "value of open inquiry, the value of free speech and an informed citizenry."[32]

Only eight justices heard *Board of Education v. National Gay Task Force*. Justice Lewis F. Powell was absent owing to surgery and did not participate in deciding the case. The decision resulted in a tie, with four justices voting to affirm and four voting to reverse. In split decisions the lower court ruling is affirmed, which in this case resulted in a victory for the NGTF. The part of Oklahoma's law that prohibited teachers from engaging in "public homosexual conduct" was struck down. It was a significant victory but not much cause for celebration. Oklahoma school psychologist Kathleen McKean explained: "We knew that the Supreme Court hadn't affirmed our rights to be. It was a fluke. We didn't feel that it was any grand victory. Actually, we had a much bigger celebration when we threw Mary Helm out of office."[33]

Essential Participants in the Debate

The Miami-Dade referendum, with teachers at the core of the debate, was a stumbling block for the nascent gay rights movement in the United States. Bryant's campaign revived the panic of an earlier generation that had lived through the Johns Committee's purge. When California voters weighed in on the matter of banning lesbians and gay men from the teaching profession, the issue turned, too, on an intersecting question regarding the free speech rights of all teachers. When the matter reached the US Supreme Court, by way of Oklahoma, the body-speech split was neatly divided. In the decades since, countless efforts on both fronts—to shore up the presence of gay and lesbian teachers and to protect all teachers' right to free expression—have brought schools closer to acknowledging the diversity of human sexuality. As a result, the cloud of ignorance that has hovered over gay and lesbian teachers since the Cold War has been dissipating.

Ironically, the 1977–78 campaigns against lesbian and gay teachers may have altered the topography of ignorance by opening up new possibilities for reflection. According to one observer, "Christian anti-gay movement activists successfully drew the debate about gay rights into the public sphere at a point when the lesbian and gay movement could not."[34] This juxtaposition that characterizes the gay rights battles of the 1970s and 1980s—opening dialogue by repressing gay bodies—is striking. In a 1977 interview, Dr. John Deady, the superintendent of schools in Springfield, Massachusetts, spoke about the climate at the time for gay teachers: "Any teacher publicly stating that he is gay would be attempting to change a social condition and would have to face the possibility of a storm. There would be negative talk and some further action might be taken. Such a person would become a cause, the focus of debates and publicity. We would have to consider the effect of this notoriety on the school system and the community."[35]

Indeed, the mere acknowledgment of the presence of gay and lesbian teachers incited debate in the 1970s through public referendum and in court. But discourse on sexuality, not surprisingly, did not keep pace in schools; as long as gay and lesbian teachers could be fired for being out in their schools, they and their colleagues would not be very likely to honestly address issues of sexuality with students.

Schools have had, and continue to have, great difficulty moving beyond a state of ignorance pertaining to the fundamental human experience of sexuality. Ignoring the rights and dignity of lesbian and gay teachers only exacerbates the problem. Justice Brennan's "stirring and historic plea" on behalf of gay Americans, penned thirty years ago, attempted to move the country in a different direction. The importance of his dissent from the Court's decision to deny *cert* in *Rowland v. Mad River Local School District* rested not so much on the eloquence of his rhetoric as on the fact of a Supreme Court justice acknowledging "the immediate and severe opprobrium often manifested against homosexuals," who had "historically been the object of pernicious and sustained hostility." Brennan, joined by Justice Thurgood Marshall, laid the groundwork for protecting the rights of gay and lesbian educators to work in public schools. He staked this claim both on the Equal Protection Clause of the Fourteenth Amendment and on the First Amendment, protecting both body and soul, as it were.[36]

In 1985 Brennan considered it impossible to remain ignorant of the public debate swirling around the rights of homosexuals. Urging his colleagues to consider the serious questions raised by Rowland's case, he noted, "The fact of the petitioner's bisexuality, once spoken, necessarily and ineluctably involved her in that debate. Speech that 'touches upon' this explosive issue is no less deserving of constitutional attention than speech relating to more widely condemned forms of discrimination." His plea anticipated ongoing efforts to call on schools to confront cultural ignorance concerning sexuality and recognize lesbian and gay teachers as participants in the conversation.[37]

NOTES

Epigraphs: Robert N. Proctor, "Agnotology: A Missing Term to Describe the Cultural Production of Ignorance (and Its Study)," in *Agnotology: The Making and Unmaking of Ignorance*, ed. Robert N. Proctor and Londa Schiebinger (Stanford, CA: Stanford University Press, 2008), 26; James Gaylord, quoted in *Courting Justice: Gay Men and Lesbians v. the Supreme Court*, by Joyce Murdoch and Deb Price (New York: Basic Books, 2001), 197.

1. Justin Richardson and Peter Parnell won the following awards for *And Tango Makes Three*: American Library Association Notable Children's Book, ASPCA's Henry Bergh Award, Gustavus Myer Outstanding Book Award, Nick Jr. *Family Magazine* Best Book of the Year, Bank Street Best Book of the Year, Cooperative Children's Book Council Choice, and CBC/NCSS Notable Social Studies Trade Book. "The Awards," *And Tango Makes Three*, accessed 8 May 2014, http://andtangomakesthree.weebly.com/the-awards.html; "*And Tango Makes Three* Tops Most-Challenged List," *American Libraries* 38 (April 2007): 10; "'And Tango Makes Three' Tops Most Challenged List, Again," *School Library Journal*, 12 April 2011, http://www.slj.com/2011/04/collection-development/and-tango-makes-three-tops-most-challenged-list-again/#_; Marta L. Magnuson, "Perceptions of Self and the 'Other': An Analysis of Challenges to *And Tango Makes Three*," *School Library Media Research* 14 (January 2011): 4; *Education Research Complete*, accessed 16 June 2014, http://www.ala.org/aasl/sites/ala.org.aasl/files/content/aaslpubsandjournals/slr/vol14/SLR_PerceptionsofSelf_V14.pdf.

2. Barbara Jones, quoted in "*And Tango Makes Three* Heads ALA's 2010 'Most Challenged' List," *Newsletter on Intellectual Freedom* 60 (July 2011): 127.

3. Charles Huckabee, "S.C. House Keeps Penalties on Colleges That Assigned Gay-Themed Books," *Chronicle of Higher Education*, 11 March 2014, http://chronicle.com/blogs/ticker/s-c-house-keeps-penalties-on-colleges-that-assigned-gay-themed-books/74099; J. Bryan Lowder, "South Carolina Champions Academic Freedom . . . by Punishing Colleges for Assigning Gay Books," *Slate*, 20 February 2014, http://www.slate.com/blogs/outward/2014/02/20/gay_book_punishment_south_carolina_legislators_fine_college_of_charleston.html; Nick DeSantis, "U. of South Carolina–Upstate Cuts Center That Held Gay-Culture Event," *Chronicle of Higher Education*, 14 May 2014, http://chronicle.com/blogs/ticker/u-of-south-carolina-upstate-cuts-center-that-sponsored-gay-culture-event/77635?cid=at&utm_source=at&utm_medium=en; Andy Thomason, "S.C. Governor Upholds Penalties for Gay-Themed Books in State Budget," ibid., 16 June 2014, http://chronicle.com/blogs/ticker/s-c-governor-upholds-penalties-for-gay-themed-books-in-state-budget/79865?cid=at&utm_source=at&utm_medium=en; "This Compromise Is Not Acceptable: Constitutionally Suspect South Carolina Budget Measure Is an Assault on Academic Freedom," National Coalition Against Censorship, 13 June 2014, http://ncac.org/update/this-compromise-is-not-acceptable-constitutionally-suspect-south-carolina-budget-measure-is-an-assault-on-academic-freedom/.

4. Proctor, "Agnotology," 26. The most important US Supreme Court ruling on homosexuality was the 2003 *Lawrence v. Texas* decision, which struck down sodomy laws as unconstitutional. In 2013 the Supreme Court declared the federal Defense of Marriage Act unconstitutional in *United States v. Windsor*. Two years later a Supreme Court majority determined that same-sex couples had a fundamental right to marry under the Due Process Clause and the Equal Protection Clause of the Fourteenth Amendment. See *Lawrence v. Texas*, 539 US 558 (2003), *United States v. Windsor*, 570 US 12 (2013), and *Obergefell v. Hodges*, 576 US ___ (2015). For an overview of education scholarship on lesbian, gay, bisexual, transgender, queer issues, see Cris Mayo, *LGBTQ Youth and Education: Policies and Practices* (New York: Teachers College Press, 2014); and George L. Wimberly, ed., *LGBTQ Issues in Education: Advancing a Research Agenda* (Washington, DC: American Educational Research Association, 2015).

5. The US military ended discrimination against LGBTQ citizens on 20 September 2011. At that time, California, Colorado, Connecticut, the District of Columbia, Hawaii, Illinois, Iowa, Maine, Minnesota, Nevada, New Jersey, New Mexico, Oregon,

Rhode Island, Vermont, and Washington had laws prohibiting employment discrimination on the basis of sexual orientation or gender identity. Delaware, Maryland, Massachusetts, New Hampshire, New York, and Wisconsin prohibit job discrimination on the basis of sexual orientation.

6. Proctor, "Agnotology," 3, emphasis in the original.

7. See Jackie M. Blount, *Fit To Teach: Same-Sex Desire, Gender, and School Work in the Twentieth Century* (Albany: State University of New York Press, 2005); John D'Emilio, "Capitalism and Gay Identity," in *Making Trouble: Essays on Gay History, Politics, and the University* (New York: Routledge, 1992), 3–16, essay first published 1983; D'Emilio, "The Homosexual Menace: The Politics of Sexuality in Cold War America," in ibid., 57–73, essay first published 1989; D'Emilio and Estelle B. Freedman, *Intimate Matters: A History of Sexuality in America*, rev. ed. (Chicago: University of Chicago Press, 1997); Lillian Faderman, *Odd Girls and Twilight Lovers: A History of Lesbian Life in Twentieth-Century America* (New York: Penguin, 1991); Proctor, "Agnotology," 6–7. See also David Bruton, "Insisting on Ignorance: The Paradox of Withholding Knowledge in Our Schools," in *One Teacher in 10: Gay and Lesbian Educators Tell their Stories*, ed. Kevin Jennings (Boston: Alyson, 1994), 177–92.

8. Stephen Preskill, "Contradictions of Domestic Containment: Forestalling Human Development during the Cold War," in *Inexcusable Omissions: Clarence Karier and the Critical Tradition in History of Education Scholarship*, ed. Karen Graves, Timothy Glander, and Christine Shea (New York: Peter Lang, 2001), 181–94; David K. Johnson, *The Lavender Scare: The Cold War Persecution of Gays and Lesbians in the Federal Government* (Chicago: University of Chicago Press, 2004), 9.

9. See Karen L. Graves, *And They Were Wonderful Teachers: Florida's Purge of Gay and Lesbian Teachers* (Urbana: University of Illinois Press, 2009); Stacy Braukman, *Communists and Perverts under the Palms: The Johns Committee in Florida, 1956–1965* (Gainesville: University Press of Florida, 2012); and Judith Poucher, *State of Defiance: Challenging the Johns Committee Assault on Civil Liberties* (Gainesville: University Press of Florida, 2014).

10. Traces of discredited myths (that lesbian and gay teachers recruit students to a "homosexual lifestyle"; that lesbian and gay teachers are immoral, therefore unfit to teach) are apparent in legal cases brought by teachers who have been harassed at work and in their local communities, denied promotions, or dismissed. In some cases, administrators transferred students from gay and lesbian teachers' classes at parents' requests; in another documented case a lesbian teacher's classroom was vandalized. It is not unusual for teachers to draw scrutiny for their work with gay-straight alliances or for including LGBTQ themes in the curriculum. In addition, "no promo homo" policies such as those adopted by the Anoka-Hennepin School District in Minnesota rest upon the assumption that teachers can impose sexuality on students. See, for example, Madelaine Adelman and Catherine A. Lugg, "Public Schools as Workplaces: The Queer Gap Between 'Workplace Equality' and 'Safe Schools,'" *Law Journal for Social Justice* 3 (September 2012): 27–46; Stuart Biegel, *The Right to Be Out: Sexual Orientation and Gender Identity in America's Public Schools* (Minneapolis: University of Minnesota Press, 2010); Mayo, *LGBTQ Youth & Education*; and the following from The Lambda Legal website, all accessed 31 August 2014: "Lambda Legal Sues School District in California on Behalf of Lesbian Teacher," http://www.lambdalegal.org/blog/20131119-lambda-legal-sues-ca-usd; "Lambda Legal Senior Staff Attorney Joins Plymouth-Canton Teachers for Safe School Forum," http://www.lambdalegal.org/news/il_20020403_lambda-senior-staff-attorney-joins-plymouth-canton-teachers; "News Conference: Thursday,

March 18, 2 p.m. PST: Gay Teacher Wins Agreement That Bakersfield School Won't Discriminate," http://www.lambdalegal.org/news/ca_19990318_news-conference-gay-teacher-wins-agreement; "California Department of Industrial Relations Vindicates Lesbian Teacher,"http://www.lambdalegal.org/news/ca_20000620_ca-dept-of-industrial-relations-vindicates-lesbian-teacher; and "Lesbian Teacher, Under Siege for Years, Wins Her Day in Court," http://www.lambdalegal.org/news/ca_20000420_lesbian-teacher-under-siege-wins-day-in-court.

11. "Report of Florida Legislative Investigation Committee to 1959 Session of the Legislature," 13 April 1959, 1, File 21, Box 1, S1486 Florida Legislative Investigation Committee, State Archives of Florida; Graves, *And They Were Wonderful Teachers*, 3, 31–33; "Report of Florida Legislative Investigation Committee to 1959 Session of the Legislature," 1, 3.

12. "Report of Florida Legislative Investigation Committee to 1959 Session of the Legislature," 4–5, 6–8.

13. "Report of the Florida Legislative Investigation Committee to the 1961 Session of the Legislature," 19–20, File 21, Box 1, S1486 Florida Legislative Investigation Committee, State Archives of Florida, 20, 23–24. The battle between the Johns Committee and the State Department of Education following the release of the 1961 report played a significant role in the Florida purge. See Graves, *And They Were Wonderful Teachers*, 105–13.

14. *Homosexuality and Citizenship in Florida: A Report of the Florida Legislative Investigation Committee* (Tallahassee, 1964), preface, File 21, Box 1, S1486 Florida Legislative Investigation Committee, State Archives of Florida, available online at the University of Florida Digital Collections, Florida History and Heritage, accessed 12 May 2014, http://ufdc.ufl.edu/UF00004805/00001; Braukman, *Communists and Perverts under the Palms*, 172.

15. *Homosexuality and Citizenship in Florida: A Report of the Florida Legislative Investigation Committee* (Tallahassee, 1964), 1, 8–10, 11. The backlash against the Johns Committee began to accelerate during its 1962 investigation at the University of South Florida. See Graves, *And They Were Wonderful Teachers*, 72–82, 122; and Braukman, *Communists and Perverts under the Palms*, 169–77.

16. See Braukman, *Communists and Perverts under the Palms.*

17. Kevin K. Kumashiro, *The Seduction of Common Sense: How the Right Has Framed the Debate on America's Schools* (New York: Teachers College Press, 2008), 9–10; Fred Fejes, *Gay Rights and Moral Panic: The Origins of America's Debate on Homosexuality* (New York: Palgrave Macmillan, 2008), 69–83, 150.

18. See Blount, *Fit To Teach*, 135–55; and Karen M. Harbeck, *Gay and Lesbian Educators: Personal Freedoms, Public Constraints* (Malden, MA: Amethyst, 1997), 61–82.

19. California Defend Our Children campaign flyer, Harvey Milk Archives—Scott Smith Collection (GLC 35), Gay and Lesbian Center, San Francisco Public Library.

20. Anita Bryant, *The Anita Bryant Story: The Survival of Our Nation's Families and the Threat of Militant Homosexuality* (Old Tappan, NJ: Fleming H. Revell, 1977), 34; see also 21–39, 62. William N. Eskridge Jr., "No Promo Homo: The Sedimentation of Antigay Discourse and the Channeling Effect of Judicial Review" (2000), *Faculty Scholarship Series*, Paper 3793, http://digitalcommons.law.yale.edu/fss_papers/3793, 1327; Bryant, *Anita Bryant Story*, 73.

21. Statement by Sen. John Briggs, Los Angeles Press Club, 7 September 1977, Harvey Milk Archives–Scott Smith Collection (GLC 35), Gay and Lesbian Center, San Francisco Public Library.

22. John Briggs, quoted in Robert Levering, "The Guardian Interviews John Briggs," *San Francisco Bay Guardian* 12 (5–13 October 1978): 6, Harvey Milk Archives–Scott Smith Collection (GLC 35), Gay and Lesbian Center, San Francisco Public Library.

23. Ibid., 7–8. In many statements Bryant refers to "known," "flaunting," "militant," "avowed," or "self-proclaimed" homosexuals. See Bryant, *Anita Bryant Story*, 14, 62, 68, 113–20. Bob Green claimed that students often know when a teacher is homosexual, whether or not the teacher is out. See Anita Bryant and Bob Green, *At Any Cost* (Old Tappan, NJ: Fleming H. Revell, 1978), 136.

24. Briggs, quoted in Levering, "Guardian Interviews John Briggs," 7.

25. Harbeck, *Gay and Lesbian Educators*, 101–48.

26. Bryant and Green, *At Any Cost*, 126; California Defend Our Children campaign flyer; Fejes, *Gay Rights and Moral Panic*, 214. O'Leary was co-executive director of the National Gay Task Force (later named the National Gay and Lesbian Task Force) at the time. See Briggs's reference to O'Leary in Levering, "Guardian Interviews John Briggs," 8.

27. Text of the law quoted in Murdoch and Price, *Courting Justice*, 252. See also Bryant and Green, *At Any Cost*, 145–46.

28. Judge James Barrett, quoted in Murdoch and Price, *Courting Justice*, 255, emphasis in the original; for analysis of the case, see 251–60.

29. Blount, *Fit to Teach*, 116–19; Charles Curtis, quoted in Murdoch and Price, *Courting Justice*, 246.

30. Linda Greenhouse, "Supreme Court Roundup; 4-to-4 Vote Upholds Teachers on Homosexual Rights Issue," *New York Times*, 27 March 1985, LexisNexis. Oklahoma was one of eighteen states that still had sodomy laws on the books after the landmark *Lawrence v. Texas* decision. Even though *Lawrence* declared sodomy laws unconstitutional, the stress of living in states that declare sodomy a criminal offense is significant. These laws technically attach criminal status to gay and lesbian citizens, giving laws banning gay and lesbian teachers the patina of legal cover. See Carlos Maza, "State Sodomy Laws Continue to Target LGBT Americans," *Equality Matters: A Campaign for Full LGBT Equality* (blog), 8 August 2011, http://equalitymatters.org/blog/201108080012; Lou Chibbaro Jr., "Sodomy Laws Remain on Books in 17 States, including Md. and Va.," *Washington Blade*, 17 April 2013, http://www.washingtonblade.com/2013/04/17/sodomy-laws-remain-on-books-in-17-states-including-md-and-va/.

31. "*Pickering v. Board of Education*," Oyez Project at IIT Chicago-Kent College of Law, accessed 14 May 2014, http://www.oyez.org/cases/1960-1969/1967/1967_510; "*Board of Ed. v. National Gay Task Force*," Oyez Project at IIT Chicago-Kent College of Law, 4 April 2014, http://www.oyez.org/cases/1980-1989/1984/1984_83_2030; "*Board of Ed. v. National Gay Task Force*."

32. Murdoch and Price, *Courting Justice*, 257–58; "*Board of Ed. v. National Gay Task Force*."

33. Kathleen McKean, quoted in Murdoch and Price, *Courting Justice*, 260.

34. Tina Fetner, "Working Anita Bryant: The Impact of Christian Anti-Gay Activism on Lesbian and Gay Movement Claims," *Social Problems* 48 (August 2001): 424.

35. John Deady, quoted in Cash Lockhart Clay, "Gays: The Colossal Closet," *Springfield (MA) Union*, 21 March 1977, 9.

36. Murdoch and Price, *Courting Justice*, 246; "Rowland v. Mad River Local School District, Montgomery County, Ohio," Justia US Supreme Court, accessed 15 May 2014, http://supreme.justia.com/cases/federal/us/470/1009/case.html.

37. Ibid.

4

Evolution

ADAM R. SHAPIRO

Antievolution is a loaded term. It denotes an inherently negative claim, one defined as opposing a theory rather than affirming another one. But it's also a term with a historical trajectory. Before the late eighteenth century, when some of the first evolutionary hypotheses about the diversification of life were put forward, there was no *anti*evolution. People may have been Bible believers, they may have accepted that Adam and Eve were real living human beings created directly by God and that all human beings were descended from this pair. Some of them may even have believed the world was only a few thousand years old. But these people were not in any real sense antievolutionists, because there was no well-articulated theory of evolution to oppose. If we imagine that ignorance in the first instance is a sort of primordial chaos out of which knowledge emerges—that first-order ignorance is some sort of naïve natural state of unknowing—then that innocent ignorance of evolution is not the same as antievolution. Evolution and antievolution were born together, a theory and its negation.[1]

Antievolution is more than just the affirmation of certainty in a position other than an evolutionary one. Many people in prior centuries had no doubt about Adam and Eve's real historical existence and their separate creation; more do today, but many of these people are also aware that evolution exists as an explanation irreconcilable with that certainty, and so they further avow that evolution is false. Antievolution is an inherently agnotological position. In particular it is engaged in the kinds of ignorance-making that Robert Proctor has described as an "active construct." There are populations today who have never heard of evolution, who may inhabit a cultural space or a "political geography" in which evolution is not taught, not a part of the popular consciousness, without being the result of intentional concealment by any party. Yet with rare exceptions these populations are also absent any intentional antievolutionism.[2]

Even though it is no longer an anachronism, *antievolution* remains a loaded term today. Evolution is a biological theory about the development of diverse and complex life through natural processes of heredity, variation, and selection. The term *evolution* refers to the theory of changes in biological populations that continue to occur today and to the application of that theory to establish the fact that life on Earth was the result of evolutionary processes. Most people who oppose evolution do not simply deny the validity of this scientific theory; they advocate alternative explanations for the diversity and complexity of life. There are several antievolutionary positions, which are greatly different from one another in what they espouse. Some are explicitly associated with a religion, others are not. To describe people as antievolutionists is to identify them more by what they don't believe than by what they do. Such a description implies that their positive claims are in fact secondary to their negative ones, that their primary concern is not to promote their account of life but to cast doubt on evolution. The label also suggests a commonality of cause between these groups and implies, sometimes rightly, historical and intellectual linkages between them.

In the context of the American controversies over education, three primary antievolutionary positions merit further discussion. The first of these was *religious creationism*. Creationism is the belief in the separate creation of species (or, in some cases, of family groups of species) and in the distinctly special creation of human beings. This position is most frequently associated with a literal interpretation of the biblical narrative; however, there are a variety of creationisms: those that interpret the "days" of creation in the Bible as six twenty-four-hour periods; those that interpret those "days" as indeterminate eras (which may span millions or billions of years); those that posit long durations or gaps between the individual days of creation; and those that posit a long natural history of the world that predates the recent account of new creation as testified in Genesis.[3]

A second position, *creation science*, is closely related to creationism, but whereas creationism may be regarded as a religious tenet that relies solely on scripture to substantiate its claims, creation science confirms the conclusions of creationism through the observation and interpretation of nature. Supporters of creation science, such as the Institute for Creation Research and Answers in Genesis, claim that it is a form of science—science that is motivated by religion but is not itself religious.

According to the organization most responsible for its advocacy, the third position, *intelligent design*, refers to a claim that "certain features of the universe and of living things are best explained by an intelligent cause, not an undirected process such as natural selection." The phrase *intelligent design* is vague enough to describe a variety of religious beliefs past and present (not all of which are irreconcilable with evolution). But the advocates of this modern-day intelligent design claim as a central aspect of their theory the insufficiency of evolution. Unlike creationists or creation scientists, intelligent design advocates often do not specify the identity of the designer or designers, nor do they explain how "design" is implemented in the material world. To a much greater extent than the other variants of antievolutionism, intelligent design is deliberately noncommittal about its positive claims, and opposition to evolution figures more centrally to its remit.[4]

Each of these antievolutionary positions has been the subject of a federal court case (respectively, *Epperson v. Arkansas* in 1968, *McLean v. Arkansas Board of Education* in 1982, and *Kitzmiller v. Dover Area School District* in 2005). Not surprisingly, each of the later court cases cites the precedents established in the ones that came earlier. These rulings set the stage for understanding the historical and intellectual connections between these three forms of antievolutionism that have attempted to influence the content of American science education.

Indeed, antievolutionism has been primarily concerned with the content of science education for as long as it has been an organized social movement. Religious objections to evolutionary accounts of human origins actually predate Charles Darwin and the 1859 publication of his *Origin of Species*. But these objections did not coalesce into a movement with actual policy goals and the means to enact them until the 1910s and 1920s. What prompted an organized antievolution movement was the development of a new curricular subject, biology, which included evolution along with several other core principles of life sciences. Antievolution in the United States took shape as a school movement that made use of the religious rhetoric against Darwin's theory part of a larger crusade against the expansion of compulsory schooling. This crusade culminated in the famous Scopes trial in Tennessee in 1925.[5]

When antievolutionism is seen as a school movement, its agnotological status is even more evident. Schools are spaces of knowledge-making in which juvenile minds learn facts and theories and, more crucially, methods and skills necessary to discover and evaluate claims to knowledge that they will encounter throughout their lives. School antievolution movements have generally been unsuccessful in having their alternative explanations for life taught or in having evolution removed. Nonetheless, the evolution controversies have prompted responses that have altered the ways the science classroom functions as a place for learning critical judgment and discovery skills.[6]

A century after American high school biology classes first started teaching evolution in a systematic way, organized opposition to evolution is as strong as it has ever been. It has diversified beyond its concerns over the schools and the direct lobbying of school officials and state legislators. From the staging of public conferences, the savvy use of social media, and perhaps most notably, the creation of public spaces like museums, organized antievolutionism is not only a school movement. The organizations that cast doubt on evolution have a variety of goals. As mass communications researcher Edward Caudill pointed out in a recent history of organized American creationism, the history of antievolution must be understood "not just as a science-religion issue, but as a political movement."[7]

In terms of their efforts to influence educational practice, each of these distinct movements has focused more on opposition to evolution than on advocacy of their specific alternatives; consequently, it makes sense to discuss these antievolutionary movements as sharing common aims and strategies of doubt-making rather than to treat them separately. Across these three major antievolutionary movements, there are common techniques for the active construction of ignorance about evolution; in some cases there's even historical evidence for the appropriation of strategies from one of these movements to others. Indeed, anecdotal cases of individuals who have attempted to implement antievolutionary theories in the classroom show much misunderstanding and slippage between these three. In 2006 a schoolteacher in Lebec, California, who attempted to introduce an elective course in "Philosophy of Intelligent Design" submitted a syllabus primarily consisting of materials produced or endorsed by the Institute for Creation Research, a proponent of creation science.[8]

Rather than treating each of these antievolutionary movements separately, it makes more sense to consider them as part of a historically coherent group engaged in several different types of ignorance-making. Looking at antievolutionism from this perspective may also illuminate the ways in which it seems to differ from other cases of active ignorance construction in science. In the tobacco industry's denial of tobacco's carcinogenicity and some politicians' rejection of climate change science, for example, emphasis has been on the production of doubt within a scientific context. Doubt manufactured about evolution's credibility has followed a similar path; however, there are also forms of epistemic debate in the evolution battles that are seen less frequently in other cases of science denial. This is at least partly because of the role religion plays in informing some variants of antievolutionism and because of the legal and political focus on schools.

In addition to the first tier of agnotological practice—the propagation of arguments against the truth of evolution within a scientific context—three other tiers of ignorance-making are frequently invoked in antievolutionism. The second tier of agnotological practices comprises efforts by antievolution movements to question the terms of debate that evolution supporters have proposed, including claims of bias or epistemic prior assumptions about the nature of evidence and its interpretation. Related to these is the claim raised by some antievolutionists, as well as many others who seek to reconcile science and religion, that science is only one of many "ways of knowing," one inherently driven by doubt. The third tier of agnotological practices has to do with arguments about the nature of "science" and what it means for an answer to questions about the diversity and complexity of life to be categorized as scientific. These arguments raise doubts, not about the facts or theories themselves, but about how knowledge claims ought to be classified. The final tier for discussion concerns the moralization of knowledge, that the truth of one's belief can be judged by the personal or social morality that is a consequence of that belief.

Have *All* the Evidence?

Early in their journey, visitors to the Creation Museum in Petersburg, Kentucky, come to a "dig site" in which an animatronic figure brushes away dirt from large ersatz bones. The visual message, which the museum's creators have been emphasizing since its inception, is simple: dinosaurs are not a reason to doubt the Creation story. But the accompanying text also introduces three different agnotological narratives that are further developed as one continues through the museum. The sign accompanying this scene reads: "Dinosaur fossils don't come with tags on them telling us how old they are, where they lived, what they ate, or how they died. We have to figure that out from a few clues we find. But because we never have all the evidence, different scientists can reach very different conclusions depending on their starting assumptions."[9]

There are several claims embedded in this one brief text, claims that set conditions for knowledge that evolution over the course of a long period occurred and raise questions about how one can ever have that kind of knowledge. Answers in Genesis, the organization that built the Creation Museum, endorses a strict young-earth interpretation of the Genesis account of creation. The aim of the museum is to promote creation science, to claim that a proper interpretation of geological, biological, and cosmological evidence supports the claim that the universe is only thousands of years old.

The statement that the "fossils don't come with tags on them telling us how old they are" is a claim that evidence doesn't constrain interpretations to a single opinion; that apparently old fossils can be reconciled with a young earth and one has warrant to legitimately doubt the old-earth interpretations of evidence. This kind of statement is not a construction of doubt that calls into question the existence of evidence, but it does attempt to substantiate doubt about the sufficiency of that evidence to support the evolutionary hypothesis.

In the epilogue to their 2010 book *Merchants of Doubt*, Naomi Oreskes and Erik Conway take note of a new view of science promulgated by organized movements against certain scientific conclusions, a view that science is supposed to provide undeniable certainty. With the scientific community using statistical tools like *p* values and with talk of 95 percent confidence levels and 5-sigma criteria, certainty of the type imagined is unrealistic. Epistemologists have raised a variety of criticisms of the proposition that certainty is a required condition for knowledge, and yet the claim that in the absence of certainty there is a corresponding absence of knowledge, or a necessity for doubt, is a staple of these agnotological arguments. As Oreskes and Conway point out, a certainty requirement is even more unreasonable when practical decision making is expected.[10]

In the case of making decisions about what students ought to learn about biology, the certainty requirement is an untenable one, yet it is present in the rhetoric of several forms of antievolutionism. In the opening text from the Creation Museum, it is implicit in the statement that "we never have all the evidence." The idea that one can ever have *all* the evidence is beyond hyperbole. Yet in Dover, Pennsylvania, where in 2004 the school district endorsed reading a statement encouraging students to explore intelligent design, again the ostensible lack of evidence was stressed. The text read to Dover students stated, in part: "Because Darwin's Theory is a theory, it continues to be tested as new evidence is discovered. The Theory is not a fact. Gaps in the Theory exist for which there is no evidence."[11]

This idea of "gaps" or "missing links" has long been part of the argument against evolution. Many religious thinkers dispute whether a "god of the gaps" is theologically satisfying. The god-of-the-gaps argument is a theological argument rooted in the dualism between knowledge and ignorance. According to this argument, there is an epistemic gap in knowledge of a process or a chain of causality; if it appears that the gap is intractable, then it is reasonable to accept God as an explanatory cause. One major criticism of this theory with respect to evolution has been that later scientific discoveries have closed gaps in ignorance. Even if some gaps remain, the argument in favor of a deity weakens with every positive discovery. Another theological criticism of the god-of-the-gaps argument is the claim that a God whose existence is testified by ignorance is a God about which there can be no further positive knowledge. Without additional sources of knowledge about that God, one cannot make claims of the uniqueness, benevolence, or power of such a deity. For religious creationists, who invoke "gaps" as an argument against evolution and assert the positive conclusions of theology based on scripture, this second objection is of less concern, but scriptural evidence is often seen as outside the realm of "science."[12]

Invocations of the god-of-the-gaps fallacy have most frequently been directed against the arguments of modern-day intelligent design. In his book *Darwin's Black Box*, published in 1996, biochemist Michael Behe proposed the concept of *irreducible complexity*, which provides an argument against the sufficiency of evolution to account for complexity seen in life. In modern biology, the variation that occurs over the course of generations often happens by way of mutations, or singular changes to the genetic material—the DNA. At this level, the DNA double helix functions as a code with discrete letters. A mutation causes a change in one of these letters. Behe argued that, if each step of the accumulation of needed mutations was not favored by natural selection, then a gradual incremental form of evolution without guiding intelligence was untenable.

Behe takes as one example the flagellum, the rotorlike whip that allows some single-cell organisms to propel themselves. Increased movement brings with it increased likelihood of encountering food and a greater chance of continued survival. Despite the energy and material expended on building a flagellum, it is in many environments a beneficial adaptation. However, Behe argues, if the flagellum with its several parts is only partially evolved, then it is not beneficial at all. Unless it can be shown how each step along the path of evolution from a tailless cell to one with a flagellum provides an evolutionary advantage and persists, we can't give a naturalistic explanation of complexity. If the complexity isn't beneficial and gradual, Behe calls it irreducible: "Even though we are told that all biology must be seen through the lens of evolution, no scientist has ever published a model to account for the gradual evolution of this extraordinary molecular machine."[13]

Since the publication of *Darwin's Black Box*, several models of flagellum evolution have been published. Regardless of whether these models are shown to be correct accounts of the historical evolution of flagellum, they demonstrate that in this instance ignorance (in the form of unexplained complexity) is not intractable (in the form of purported irreducibility). More generally, claims of intractable ignorance highlight the way epistemic requirements are elevated to increase skepticism of an evolutionary account. Many critics of Behe state that his expectation that evolution operates solely by gradual changes is both a misleading description of how evolution works and an unreasonable demand on those who argue that naturalistic processes are in principle sufficient. The expectation that evolution is wholly determined by singular changes to genes ignores much of the research on gene expression and regulation and other evolutionary factors that only operate upon whole populations. The selective focus has prompted some to call for moving away from a gene-centered vision of evolution, though not in a non-naturalistic way.[14]

When evolutionary theory predicts that a population of fishes gradually evolved into amphibious tetrapods about 380 million years ago, the evolution skeptic rejects the prediction, citing a lack of fossil evidence that joint-limbed fish ever existed. In 2004 fossils were found of what was later named *Tiktaalik roseae*, a lobe-finned fish with "a shortened skull roof, a modified ear region, a mobile neck, a functional wrist joint, and other features that presage tetrapod conditions." The article announcing the discovery does not suggest that *Tiktaalik* was a direct ancestor of modern tetrapods, but that it is evidence that some transitional species of fish-tetrapods existed at the time predicted by models of vertebrate evolution.[15]

The response from organized antievolutionism has been to question both the interpretation of the evidence and whether this sufficiently fills in the gap between fish and tetrapods. A 2008 post on the Discovery Institute's blog *Evo-*

lution News and Views states that "not all evolutionary paleontologists are convinced that these bones were the precursors to real tetrapod digits." This language suggests that there is doubt among scientists about the significance of *Tiktaalik*. In fact, this response is an example of manufacturing doubt by a strategy of demanding greater standards of certainty. In this case, for a transitional fossil to count as evidence, it must also carry evidence of its direct relation to later tetrapods. For a gap to be considered closed, *Tiktaalik* must fill it, and the now smaller gaps on either side must also be closed. For the intelligent design of the Discovery Institute, which accepts both the old age of life on Earth and the idea of common descent (effected by an interventionist intelligence rather than by natural processes), the gaps in natural evolutionary processes are the locations where an intelligence intervenes. For Answers in Genesis, there is no such thing as missing links, because there are no links between distinctly created families of species: "Whatever else we might say about *Tiktaalik*, it is a fish." Scientists discussing *Tiktaalik* stated that the concept of missing links was misleading because it falsely led the public to "a mistaken emphasis on the single intermediate fossil as the key to understanding evolutionary transitions." Answers in Genesis quoted this statement to claim support from scientists in their assertion that the concept of missing links was a fallacy.[16]

The gaps argument has two parts. It claims that gaps exist in the explanatory account given by evolutionary processes, and it claims that in the absences of naturalistic explanations a supernatural explanation is warranted. The first claim concerns the active construction of ignorance, the assertion of a gap in knowledge either by disputing the validity, quality, or interpretation of evidence or by raising the expectations of the kind of evidence that will suffice to count as an explanation. Both of these techniques are described in other forms of science skepticism. The demand that gaps must be filled in, and that after doing so there are gaps between the gaps, is equivalent to the demand of certainty. The attempt to create the impression that scientific opinion has not reached consensus or that alternate interpretations of evidence (no matter how infrequently held) deserve equal hearing is a way of disputing the validity of evolutionary knowledge.

You of Little Faith, Why Did You Doubt?

Returning to the dig site, visitors not only are warned that "we never have all the evidence" but also are subsequently told that "different scientists can reach very different conclusions depending on their starting assumptions." Throughout the tour, the museum's narrative focuses substantially on the idea of different "starting assumptions" (leaving aside for now discussion of whether the people making these assumptions are all "scientists"). These starting assumptions are at the outset epistemological in nature. Starting assumptions have to do with where knowledge comes from, how it can be known, and what roles God, people, and nature play in the practice of science.

Answers in Genesis presents two possible choices for starting assumptions, distilled into the phrases *God's Word* and *human reason*. The first posits an epistemic hierarchy in which knowledge that comes directly to people from God is infallibly true and secondary forms of knowledge, which do not come directly from God, must be interpreted in light of what is already taken to be certain. Because God is good, all-powerful, and all-knowing, knowledge that comes to us directly from God is also certain. The theology of Answers in Genesis also holds the Bible as the direct revelation of God; the words of God's Word are therefore inherently true. The Bible gives direct accounts of the process of animal and plant creation, as well as a chronology of that creation and of subsequent events. Therefore, there can be no doubt about the age of the earth or of the separate creation of humanity from other animals.

The epistemic world in which God's Word is taken as the starting assumption is not one in which all knowledge is completely certain. There are many details of the process of creation and its continuing function about which scripture does not speak. Creation science can tell us about the hydrology of the flood, about the diversification of life within precreated kinds after the flood, or about fossil interpretation. Creation scientists interpret the placement of fossils within various geological strata in ways that are consistent with given truths known to be certain (e.g., that the universe is only thousands of years old). To do this they rely on the biblical account of the flood as a catastrophic event that significantly changed the face of the earth. They also argue against the prevailing assumption held by scientists that the laws of nature have been fundamentally unchanged for long periods of time. For example, creation scientists publish, in their own venues, articles arguing against constancy of decay rates in radioisotopes, which have long been used to assess the ages of materials. As one such article argues, "Since God is the origin of physical principles, it would be wrong to state that He must act in a certain way. However, Scripture is a reliable record of His actual creation. The models considered here merely point out some unnecessary assumptions involved in interpreting radioactive decay: half lives may not have been constant."[17]

In the God's Word epistemic regime, any hypothesis that contradicts the absolutely certain truths of scripture is logically impossible. Creationists can insist upon this as a way to reject claims inferred from human observation and interpretation. Creation scientists claim that their starting assumption gives them greater confidence that their principles of collecting, observing, and interpreting natural evidence gives accurate knowledge, though that knowledge is not certain in the same way that God's direct word is.

In contrast to the epistemic starting assumption that God's Word is the basis of all certainty, the assumption that human reason is the only way to obtain knowledge is the starting point for others (e.g., evolutionists). In this conception, knowledge is obtained through what we can see and interpret, but such a method can only give us at best uncertain forms of knowledge, always lacking complete evidence and always subject to the fallacies of human reason—our capacity to be deceived and to deceive ourselves.

In the Creation Museum's dichotomy, the epistemic starting assumption of human reason is not just restricted to knowledge about nature; it also includes knowledge about religion and morality. Thus, interpretations by "liberal" religious movements that treat scripture as a text to be judged by human readers (not as inerrant and divinely authored) are inherently fallible claims to uncertain forms of knowledge. Moral relativism, the claim that there can be no certainty about ethical knowledge, is depicted by the Creation Museum as leading to a world "where anarchy or another godless system is hailed as king."[18]

As one journeys through the museum, one's choice of starting assumption moves from an epistemic to a moral and then a religious viewpoint. The choice of human reason, it is suggested, means that even those who still accept some role for the Bible in their worldview subordinate its content to their own human judgments. Consequently, they only read the biblical text to reinforce their own biases, rejecting the parts they find outdated. The choice of human reason as the root basis for evaluating all claims of knowledge is analogous to the decision of Adam and Eve to reject God's admonition against eating from the tree of knowledge and to choose to be deceived into thinking that their own reasoning was fallible. The choice of human reason as epistemically prior to God's Word is no more nor less than the Original Sin itself.

Modern intelligent design makes no explicit claims about the role of human reason in understanding the nature of the designer as a deity, nor does it claim that it is connected in any way to scripture. Nevertheless, one of the key tactics of the intelligent design movement has been to critique the epistemological assumptions behind evolutionary science. In one of the founding texts of the modern movement, *Darwin on Trial* (1991), Philip Johnson argues that "*Scientific* naturalism" starts with "the assumption that science, which studies only the natural, is our only reliable path to knowledge."[19] By 1995 Johnson had refined this argument into a critique of "methodological naturalism." Much like the Creation Museum's framing of interpretation of the world as a choice between two epistemic systems, Johnson contrasted methodological naturalism with something he labeled "theistic realism": "A methodological naturalist defines science as the search for the best naturalistic theories. A theory would not be naturalistic if it left something (such as the existence of genetic information of consciousness) to be explained by supernatural cause. Hence all events in evolution (before the evolution of intelligence) are assumed to be attributable to unintelligent causes."[20]

Thus, the assumption that naturalistic explanations are the only possible scientific explanations presupposes that intelligent design is impossible. This, Johnson argues, is not the result of an empirical scientific study, but rather a presupposition that the supernatural be the basis for an explanation. "The theistic realist," he suggests, "assumes that the universe and all its creatures were brought into existence by God. Theistic realists expect this fact of creation to have empirical, observable consequences that are different from the consequences one would observe if the universe were the product of nonrational causes."[21]

While subsequent language of intelligent design advocacy moves further away from explicit talk of God as the creator in lieu of some unspecified intelligence, this critique of methodological naturalism is repeated in more recent works advocating intelligent design. This critique no longer is represented with something as religious-sounding as "theistic realism." In his 1999 book *Intelligent Design*, William Dembski argues that the dichotomy is not between naturalistic and supernaturalistic causation: "This is the wrong contrast. The proper contrast is between *undirected natural causes* on the one hand and *intelligent causes* on the other." He goes on to say that "design has no prior commitment to supernaturalism."[22]

If ignorance is regarded as the obverse of knowledge, then efforts to change the "rules of science" by proposing new knowledge producing methods are also efforts to redefine the scope and nature of ignorance. Raising doubt about the age of fossils or about the sufficiency of evidence to bridge a gap is a way of legitimating ignorance *within* the existing rules. That kind of doubt-making work done by antievolutionists is quite similar to that in other science-denial fields (such as those discussed by Oreskes and Conway). But this form of agnotology that attempts to change the rules of science—arguing against the naturalistic basis of science or the priority of human reason over revelation—is largely unseen in other areas of merchandized doubt. This difference might be readily attributed to the fact that the opposition to other areas of science has not been so deeply and historically connected to religion.[23]

Many advocates of evolution would accept the claim that science is not the only way to knowledge, that there are different ways of knowing. This multivalent approach to knowledge has been common among self-proclaimed religious believers who accept evolution, such as biologists Kenneth Miller and Francis Collins. In what is likely the most famous formulation of this claim, Stephen Jay Gould posited that science and religion constituted "non-overlapping magisteria," or NOMA. Gould wrote: "The net of science covers the empirical universe: what is it made of (fact) and why does it work this way (theory). The net of religion extends over questions of moral meaning and value. These two magisteria do not overlap, nor do they encompass all inquiry." What Gould is describing is not just a demarcation in existing knowledge but a demarcation in ignorance (seen here as naïvely nonobtained knowledge). In describing the extension of these hypothetical "nets" over questions, rather than just answers, Gould posits that some unknown truths are inherently within the range of possible discovery by science, while others are not. In particular, moral truths are within the range of potential discovery by religious modes of inquiry, and not by science.[24]

Gould's NOMA concept has been criticized both by antievolutionists and by New Atheists, who insist that science is indeed the only way of obtaining knowledge. Yet it has also influenced advocates of evolution-religion reconciliation, like Miller. To use Gould's language of "nets," one interpretation of the position of creation science groups like the Institute for Creation Research and Answers in Genesis is that your science net will only come up full of knowledge if you pay attention to where God tells you to cast it.

"Science Falsely So Called"

In addition to raising doubts about the sufficiency of human reason and methodological naturalism as ways of knowing about the world, in addition to arguing for a different understanding of the nature of ignorance and knowledge and how human actors can attain it, a tertiary level of agnotological strategy on the part of both creation science and intelligent design advocates involves reinterpreting the demarcation between science and nonscience. Antievolutionists construct doubts about the conclusions of scientists by denying the interpretation of evidence or by raising standards of certainty. They take issue with the claim that science is a sufficient tool for obtaining knowledge about the diversification and complexity of life, and they also raise questions about the classification of knowledge-producing methods. Organized antievolutionism argues that its ways of knowing are not only valid but also part of the category of "science." This aspect of antievolutionary agnotology has intensified in response to the legal rulings of the 1980s.

The first legal successes of antievolutionism took place in the 1920s, most famously with the passage of the Tennessee antievolution law that culminated in the Scopes trial in 1925. The Tennessee law remained on the books until the 1960s; it was repealed shortly before a nearly identical law in Arkansas was declared unconstitutional by the Supreme Court.[25] The 1968 ruling in *Epperson v. Arkansas* declared that antievolution laws "cannot be defended as an act of religious neutrality. Arkansas did not seek to excise from the curricula of its schools and universities all discussion of the origin of man. The law's effort was confined to an attempt to blot out a particular theory because of its supposed conflict with the biblical account literally read."[26]

The use of science to confirm or conform to the young-earth creation account had already been attempted, but it was not seen as central to creationism. In 1961 *The Genesis Flood*, by John C. Whitcomb and Henry M. Morris, was first published. The book gave an account of evidence in nature for the recent occurrence of a worldwide flood "based upon the presupposition that the Scriptures are true." As historian Ronald L. Numbers explains, creationists in California attempted to add "creation in scientific terms" to the state's curriculum. "The transmogrification of creationism from religion to science," notes Numbers, "took place in direct response to the events in California, which encouraged creationists to believe that they could squeeze into science classrooms simply by shedding superfluous biblical weight." In the wake of the *Epperson* ruling, creation science became much more relevant to creationism, primarily as a means of teaching creation in schools. Creation science was not using the Bible as the warrant for its claims that the earth was young, that man and animals had been separately created, and that there had been a worldwide flood. Instead it was claiming that by beginning with nature, one could (and indeed ought to) reach conclusions that agreed with the Bible.[27]

In 1981 Arkansas passed a law requiring "balanced treatment" of creation science and "evolution science." The law required that if one of the two topics were taught, then the other must be taught with equal time. Of course, one could "balance" treatment by teaching neither topic, and it's likely that at least some proponents of the law saw it more as a way to reestablish the overturned antievolution provision than as an active mandate to also teach creation science. The law was quickly challenged in court.

The US District Court ruling that invalidated the Arkansas Balanced Treatment Act, *McLean v. Arkansas Board of Education*, is striking because it ruled not only that the law allowing creation science constituted an establishment of religion but also that creation science did not meet the necessary criteria to be counted as a science. The court reasoned in part: "Since creation science is not science, the conclusion is inescapable that the only real effect of Act 590 is the advancement of religion."[28] The *McLean* ruling also set out specific criteria for what constituted science:

1. It is guided by natural law;
2. It has to be explanatory by reference to natural law;
3. It is testable against the empirical world;
4. Its conclusions are tentative, i.e. are not necessarily the final word; and
5. It is falsifiable.[29]

Although the criteria have since been debated by philosophers of science, the five-point framework established a judicial precedent that has shaped evolution controversies ever since. In order to satisfy the legal requirements of the First Amendment prohibition on the establishment of religion, an antievolutionary theory needs to demonstrate that it is in fact a science. The need to do so has led to a new form of agnotological strategy, the manufacture of doubt about the nature of science and about the criteria for determining a theory as scientific.[30]

This representation of antievolutionary work as science has been emphasized since the *McLean* ruling (especially after that judgment was upheld by the US Supreme Court in a separate creation science case, *Edwards v. Aguillard*, in 1987). The need to be considered scientific is the reason why the introductory text from the Creation Museum refers to people working from different starting assumptions as "different scientists."[31]

Contestation over the category of science has been of particular importance to advocates of intelligent design. Dembski's *Intelligent Design* states that while "design theorists take the question *Which is correct, naturalistic evolution or intelligent design?* as a perfectly legitimate question, it is not treated as a legitimate question by the Darwinian establishment." Dembski argues that this is because the establishment's conception of science is too narrow: "Science, according to the Darwinian establishment, by definition excludes everything except the material and the natural. It follows that all talk of purpose, design, and intelligence is barred entry from the start." One solution, pace Gould, would be to argue that the question is legitimate but outside the realm of science, that it is a question that rightly falls under the net of philosophy or religion. However, Dembski, citing Philip Johnson, observes that "science is the only universally valid form of knowledge within our culture." If the question of intelligent design is culturally relevant, it can't be bracketed off as a part of a way of knowing that is ignored by some of the population.[32]

Dembski doesn't directly acknowledge that the necessity to be classed as a science comes not only from demands for cultural respect but also from legal demands (at least insofar as one advocates intelligent design as an alternative to evolution for purposes of the schools). As philosopher Barbara Forrest and biologist Paul R. Gross have revealed, the Discovery Institute, a think tank that promotes intelligent design research, devised an explicit strategy known as the "wedge" to bring about the acceptance of intelligent design *as a science* for the purposes of effecting broader cultural change informed by a religious perspective. In ways quite in line with organized science skepticism seen in other fields, the wedge strategy called on the Discovery Institute to create resources for teachers, to shift public opinion through the creation of popular media, and most importantly, to get articles supportive of intelligent design accepted and published in legitimate scientific journals. The institute both receives substantial donations and makes grants and offers fellowships to scholars supportive of their agenda. This money has allowed the institute's fellows to promote their views, advertise their books, organize conferences, and create events for "opinion makers."[33]

This agnotological tactic is not always present in ignorance about other forms of science. By the late nineteenth century, organized norm-enforcing scientific communities played a role in marginalizing theories labeled "pseudoscience," or quackery. Historians Daniel P. Thurs and Ronald L. Numbers have pointed out that throughout the nineteenth century the biblically allusive phrase *science falsely so called* was commonly used (and was often used by creationists to describe evolution). By the mid-twentieth century the deliberate enforcement of scientific discipline and the presentation of science itself as a community practice with strong norms that guarded against dangerous pseudoscience became more central to the ways in which science earned much of its cultural authority. Writing in the wake of *McLean*, philosopher of science Larry Lauden, who had strongly criticized the legal assertion of criteria for science in that trial, though he was not supportive of creation science, declared that "terms like 'pseudo-science' and 'unscientific' . . . are just hollow phrases that do only emotive work for us." To dismiss the emotive affect of the pseudoscientific label is to ignore the performative nature of demarcation. If, as the *McLean* verdict claims, "science is what scientists do," then the very fact of scientists declaring a theory or movement "unscientific" makes its status unscientific.[34]

It's precisely this performative process that has compelled organized antievolution movements to proclaim their scientific identity. Just as other areas of scientific agnotology have seen ways of manufacturing doubt about scientific claims through the production of claims "complicating the causal links," doubt about the scientific status of a theory or movement can also be manufactured. Doubt about a theory or movement's scientific status can be produced by repeated assertions of creation science's and intelligent design's scientific status; by the use of technical language and facts to make the substance of claims less easily scrutinized (relying on the agnotological assumption that science itself is an elite practice inaccessible to the unschooled); by identifying the supporters and researchers in the fields of antievolution as *scientists* (in some cases, by using their scientific expertise in unrelated or tangential fields as an imprimatur for their statements on evolution); and by characterizing evolution science or a Darwinian establishment as dogmatic, close-minded, and ripe for eclipse in a triumphalist historical narrative.[35]

It's in these latter two modes of arguing for the scientific status of intelligent design and creation science that the final mode of agnotological practice in antievolutionism—judging a form of knowledge by focusing on its consequence on the knower, particularly on moral grounds—is best evidenced.

You Will Know Them by Their Fruits

"If it could be shown that man, instead of being made in the image of God, is a development of beasts we would have to accept it, regardless of its effect, for truth is truth and must prevail. But when there is no proof we have a right to consider the effect of the acceptance of an unsupported hypothesis." Thus William Jennings Bryan argued on the pages of the *New York Times* in 1922. Though Bryan had already been lecturing on "the menace of Darwinism," the *New York Times* piece helped make antievolutionism more visible to scientists.[36]

Bryan's antievolutionism was much more complicated than his opponents' caricatures of him would lead one to believe. It was even more complex than many biographies and historical accounts of the Scopes trial assert. As many people have noted, Bryan was no devotee to a young earth. But this fact did not put him at odds with many of his fellow antievolutionists. At times, Bryan invoked chemistry as an argument against evolution, stating that if there was no inherent natural tendency toward greater complexity at the chemical level, then it couldn't occur at the biological one. He also criticized the sufficiency of evidence behind evolutionary claims. Bryan's greatest concern, however, was that Darwinian evolution, as he understood it, presupposed that everything had to occur through gradual and natural processes. Bryan was concerned that evolution, much like the intelligent design critique of methodological naturalism, ruled out the possibility of supernatural interventions. Unlike proponents of modern intelligent design, he wasn't talking about a designer implementing the flagellum; rather, he was concerned with miracles like the revelation of scripture, the virgin birth, and most important, the Resurrection. Without the possibility of miracles, the Bible could no longer be taken as true and there could be no certain knowledge about morality. Without the Resurrection and its redemptive promise, there would be no religious basis for just behavior. For Bryan, *theistic evolution* was an oxymoron.[37]

Bryan thought that evolution was wrong on scientific grounds, but his advocacy for removing it from the classrooms didn't come simply from a concern that school content be accurate. In 1916 psychologist James Leuba published a study that Bryan cited in his *New York Times* article. According to Bryan, many people became agnostics or atheists as they became scientists, and university students were becoming less religious as a result of learning about ideas like Darwin's.[38]

The idea that learning about evolution can shape moral behavior has served as a proxy argument against the mandate that it be taught. In subsequent iterations of antievolutionism, this moralization of epistemology has taken a variety of other forms. The tactic largely shifted away from students themselves and toward depicting perpetrators of evil or immoral acts as being bereft of moral sense because of Darwinism, or in some cases, acting in evil ways because they were in fact *using* Darwinism.

The use of the label "Darwinism" to describe evolutionary theory is related to the strategy of moral judgment. Some historical accounts by antievolutionists point to Darwin's alleged atheism, or claims that he was racist, as consequences of his embrace of evolution. In recent years, this has led to a surge in historical scholarship intended to rescue Darwin from these pervasive mythologies. Both Answers in Genesis and the Discovery Institute have published books and articles emphasizing a connection between evolution and the practice of eugenics, the attempt to alter the makeup of a population through reproduction control, such as sterilization of those deemed "unfit." The Creation Museum offers exhibits claiming that the rejection of God's Word and the embrace of human reason enabled racist practices like slavery and Jim Crow laws.[39]

Richard Weikart, a senior fellow at the Discovery Institute, recently authored *From Darwin to Hitler*, which argues (as implied by the book's title) that the grotesque acts of the Third Reich were the application of Darwinian evolution. Weikart's study implicates Darwin in the atrocities that came decades after his death and implies that to affirm Darwin's theory is to validate such horrors.[40]

For the Creation Museum, the connection between believing in evolution and immorality is even more fundamental. Pollution, for instance, is viewed as the natural consequence of prioritizing human reason over God's Word. That very act is identified with the fall from grace and the banishment from Eden. To embrace evolution is to embrace the decision made by Adam and Eve to defy God. "Graffiti Alley," in the Creation Museum, depicts a variety of social ills (e.g., homosexuality, abortion, and school shootings) as direct consequences of rejecting God's Word. By the end of one's trip through the Creation Museum, the epistemic choice between starting assumptions becomes a deeply moral one. It is the gradual shift in focus from the epistemic differences to the moral ones that is part of the museum's extraordinary effort to control its narrative and the visitor's experience.[41]

The Tree of Knowledge and the Roots of Ignorance

Agnotology in the sphere of antievolution occurs on several levels, from the promotion of facts seeming to contradict or refute scientific claims, to arguing for different standards for producing and ratifying knowledge, to calling into question the application of the status of "science" to particular knowledge claims. There can be no question that the effort to build ignorance about evolution has been successful in the United States and to a lesser extent in much of the rest of the world. According to a 2012 Gallup poll, about 46 percent of Americans surveyed stated that they most agreed with the statement that "God created humans pretty much in their present form at one time within the last 10,000 years or so." The two alternative choices were, "human beings have developed over millions of years from less advanced forms of life, but God guided this process" (32%) and "human beings have developed over millions of years from less advanced forms of life, but God had no part in this process" (15%). These numbers have shifted only slightly since Gallup began asking the question around the time of the *McLean* trial in the early 1980s.[42]

Such surveys tell us little about the specific ways that antievolution has succeeded, and the responses depend heavily on the way choices are phrased. The responses do not show which of the several agnotological strategies are most effective for particular audiences. In a Pew poll from 2013, only 33 percent agreed with the statement "humans existed in present form since the beginning," compared with the 60 percent who agreed with the statement "humans have evolved over time." It's unlikely that Gallup or Pew deliberately engages in push polling, but imprecisely phrased questions tend to skew responses. If someone agrees that humans developed over millions of years and that God guided that process, describing that position with the conjunction *but* implies that this is a somewhat tenuous position to hold.[43]

What is somewhat clearer from the data, and from anecdotal cases, is that there's quite a bit of slippage in how the public understands the specifics of antievolutionism. Organized antievolution has done a better job of creating doubt about evolution than it has of advocating for the specific details of its alternatives. Yet over the course of several decades, despite new advances in evolutionary science and despite new developments by evolution's opponents, the overall public knowledge seems to have changed very little.

In 1973 evolutionary biologist Leigh Van Valen proposed an evolutionary mechanism that he dubbed the "Red Queen Hypothesis," alluding to the chess game in *Through the Looking-Glass*, in which participants must constantly run in order to stay still. Evolution, even in periods without apparent change or without extinctions, is always a moving target. This description is equally fitting for the ongoing battles over evolution. Van Valen wrote: "From this overlook we see dynamic equilibria on an immense scale, determining much of the course of evolution by their self-perpetuating fluctuations." Likewise, we see a dynamic equilibrium of knowledge and ignorance, both being generated in new varieties and in increasing sophistication, yet the overall balance of belief has remained stable. As evolutionary biology has adapted to changes in its prevailing theories over 150 years, and as antievolutionism has seen a new species emerge from the surviving remnants of each legal catastrophe, an end to the battle between knowledge and ignorance seems distant. One might not be reassured by the Red Queen's prognosis that "no species can ever win, and new adversaries grinningly replace the losers," yet this ongoing battle is itself a form of evolution.[44]

NOTES

1. Robert N. Proctor, "Agnotology: A Missing Term to Describe the Cultural Production of Ignorance (and Its Study)," in *Agnotology: The Making and Unmaking of Ignorance*, ed. Robert N. Proctor and Londa Schiebinger (Stanford, CA: Stanford University Press, 2008), 4.

2. In a study of Jehovah's Witnesses missionaries operating in Chontal-speaking regions of Mexico, anthropologist Jena Barchas-Lichtenstein describes a sermon in which missionaries provide arguments against evolution in part of a discussion of creationism, despite the unlikelihood of its hearers having been made aware of evolutionary theory in any prior context. Jena Barchas-Lichtenstein, "'When the Dead are Resurrected, How are We Going to Speak to Them?': Jehovah's Witnesses and the Use of Indigenous Languages in the Globalizing Textual Community" (PhD diss., UCLA, 2013); and e-mail correspondence, 23 May 2014.

3. For a comprehensive account of the varieties of creationism, see Ronald L. Numbers, *The Creationists: From Scientific Creationism to Intelligent Design* (Cambridge, MA: Harvard University Press, 2006).

4. "Explaining the Science of Intelligent Design," accessed 29 May 2014, http://www.intelligentdesign.org.

5. Adam R. Shapiro, "Civic Biology and the Origin of the School Antievolution Movement," *Journal of the History of Biology* 41 (September 2008): 409–33.

6. Adam R. Shapiro, *Trying Biology: The Scopes Trial, Textbooks, and the Antievolution Movement in American Schools* (Chicago: University of Chicago Press, 2013), 157–65.

7. Edward Caudill, *Intelligently Designed: How Creationists Built the Campaign against Evolution* (Urbana: University of Illinois Press, 2013), 1.

8. *Kenneth Hurst et al. v. Steve Newman et al.* Original complaint archived at http://ncse.com/files/pub/legal/hurst/20060110_original%20complaint.pdf (accessed 29 May 2014). Original syllabus of the class archived at https://web.archive.org/web/20060218164252/http://mountainenterprise.com/IntelDesignSyl/IntelDesignSyllabus051209.htm (accessed 29 May 2014).

9. The text of this sign is visible on the "Dig Site" page of the Creation Museum's "virtual tour," accessed 29 May 2014, http://creationmuseum.org/whats-here/photo-preview/.

10. Naomi Oreskes and Erik Conway, *Merchants of Doubt: How a Handful of Scientists Obscured the Truth on Issues from Tobacco Smoke to Global Warming* (New York: Bloomsbury, 2010), 267, 156–57; Baron Reed, "Certainty," *Stanford Encyclopedia of Philosophy*, Winter 2011, http://plato.stanford.edu/entries/certainty/; Oreskes and Conway, *Merchants of Doubt*, 271.

11. *Tammy Kitzmiller et al. v. Dover Area School District et al.*, 400 F.Supp.2d 707 (M.D. Pa. 2005).

12. Thomas Dixon, *Science and Religion: A Very Short Introduction* (Oxford: Oxford University Press, 2008), 44–46.

13. Michael Behe, *Darwin's Black Box: The Biochemical Challenge to Evolution*, 2nd ed. (New York: Free Press, 2006), 70.

14. Nicholas J. Matzke, "The Edge of Creationism," *Trends in Ecology and Evolution* 22 (2007): 566; David Dobbs, "Die Selfish Gene, Die," *Aeon*, 3 December 2013, rev. 13 December 2013, http://aeon.co/magazine/nature-and-cosmos/why-its-time-to-lay-the-selfish-gene-to-rest/.

15. Edward B. Daeschler, Neil H. Shubin, and Farish A. Jenkins Jr., "A Devonian Tetrapod-Like Fish and the Evolution of the Tetrapod Body Plan," *Nature* 440 (6 April 2006): 757.

16. Casey Luskin, "The Rise and Fall of Tiktaalik? Darwinists Admit 'Quality' of Evolutionary Icon is 'Poor' in Retroactive Confession of Ignorance (Updated)," *Evolution News and Views* (blog), accessed 29 May 2014, http://www.evolutionnews.org/2008/09/the_rise_and_fall_of_tiktaalik011701.html; Per Erik Ahlberg and Jennifer A. Clack, "A Firm Step from Water to Land," *Nature* 440 (6 April 2006): 747; Dr. David Menton, "*Tiktaalik* and the Fishy Story of Walking Fish, Part 2," 23 May 2007, https://answersingenesis.org/extinct-animals/tiktaalik-and-the-fishy-story-of-walking-fish-part-2/.

17. Eugene F. Chaffin, "Accelerated Decay: Theoretical Models," in *Proceedings of the Fifth International Conference on Creationism*, ed. R. L. Ivey (Pittsburgh: Creation Science Fellowship, 2003), reprinted at http://static.icr.org/i/pdf/technical/Accelerated-Decay-Theoretical-Models.pdf, 8.

18. Michael Belknap, "Graffiti Alley and the Decay of the West," 1 July 2013, https://answersingenesis.org/morality/graffiti-alley-and-the-decay-of-the-west/.

19. Philip E. Johnson, *Darwin on Trial*, 3rd ed. (Downer's Grove, IL: InterVarsity, 2010), 145.

20. Philip E. Johnson, *Reason in the Balance: The Case against Naturalism in Science, Law and Education* (Downer's Grove, IL: InterVarsity, 1998), quoted in *Creationism's Trojan Horse: The Wedge of Intelligent Design*, by Barbara Forrest and Paul R. Gross (Oxford: Oxford University Press, 2004), 19.

21. Ibid.

22. Stephen C. Meyer, *Darwin's Doubt: The Explosive Origin of Animal Life and the Case for Intelligent Design* (New York: HarperOne, 2013), 383–403; William A. Dembski, *Intelligent Design: The Bridge between Science and Theology* (Downers Grove, IL: InterVarsity, 2000), 259.

23. Meyer, *Darwin's Doubt*, 382.

24. Kenneth R. Miller, *Finding Darwin's God: A Scientist's Search for Common Ground between God and Evolution* (New York: Cliff Street Books, 1999); Francis Collins, *The Language of God: A Scientist Presents Evidence for Belief* (New York: Free Press, 2006); S. J. Gould, "Non-overlapping Magisteria," *Natural History* 106 (1997): 16–22.

25. Edward J. Larson, *Trial and Error: The American Controversy over Creation and Evolution* (New York: Oxford University Press, 1985).

26. *Epperson v. Arkansas*, 393 US 97 (1968).

27. Numbers, *Creationists*, 225, 271.

28. *McLean v. Arkansas Board of Education*, 529 F.Supp. 1255, 1258–64 (ED Ark. 1982).

29. Ibid.

30. See esp. Michael Ruse, "Creation-Science is Not Science"; Larry Lauden, "Commentary: Science at the Bar—Causes for Concern"; and Ruse, "Response to the Commentary: *Pro Judice*," all in *Philosophy of Science: The Central Issues*, ed. Martin Curd and J. A. Cover (New York: Norton, 1998), 38–61.

31. *Edwards v. Aguillard*, 482 US 578 (1987).

32. Dembski, *Intelligent Design*, 117, 118.

33. Forrest and Gross, *Creationism's Trojan Horse*, 15–34, 154.

34. Daniel P. Thurs and Ronald L. Numbers, "Science, Pseudoscience, and Science Falsely So-Called," in *Philosophy of Pseudoscience*, ed. Massimo Pigliucci and Maarten Boudry (Chicago: University of Chicago Press, 2013), 125; Michael Gordin, *The Pseudo-Science Wars: Immanuel Velikovsky and the Birth of the Modern Fringe* (Chicago: University of Chicago Press, 2013), 10; Larry Laudan, "The Demise of the Demarcation Problem," in *Physics, Philosophy and Psychoanalysis: Essays in Honor of Adolf Grunbaum*, ed. Robert S. Cohen (Dordrecht, Netherlands: D. Reidel, 1983), 125.

35. Naomi Oreskes and Erik M. Conway, "Challenging Knowledge: How Climate Science Became a Victim of the Cold War," in Proctor, *Agnotology*, 67; see also Oreskes and Conway, *Merchants of Doubt*, 270–71, for discussion of historical mythmaking in other instances of scientific agnotology.

36. William Jennings Bryan, "God and Evolution," *New York Times*, 26 February 1922, 1; Bryan, *The Menace of Darwinism* (New York: Fleming H. Revell, 1922).

37. For examples of differing depictions of Bryan's antievolutionism, see Michael Kazin, *A Godly Hero: The Life of William Jennings Bryan* (New York: Knopf, 2006); Edward J. Larson, *Summer for the Gods: The Scopes Trial and America's Continuing Debate over Science and Religion* (New York: Basic Books, 1997); and Caudill, *Intelligently Designed.*

38. Bryan, "God and Evolution"; James H. Leuba, *The Belief in God and Immortality: A Psychological, Anthropological and Statistical Study* (Boston: Sherman French, 1916).

39. Adrian J. Desmond and James R. Moore, *Darwin's Sacred Cause: Race, Slavery and the Quest for Human Origins* (Chicago: University of Chicago Press, 2011).

40. Richard Weikart, *From Darwin to Hitler: Evolutionary Ethics, Eugenics, and Racism in Germany* (New York: Palgrave Macmillan, 2004).

41. Belknap, "Graffiti Alley and the Decay of the West."

42. "Evolution, Creationism, Intelligent Design," accessed 29 May 2014, http://www.gallup.com/poll/21814/evolution-creationism-intelligent-design.aspx.

43. "Public's Views on Human Evolution," 30 December 2013, http://www.pewforum.org/2013/12/30/publics-views-on-human-evolution.

44. Leigh Van Valen, "A New Evolutionary Law," *Evolutionary Theory* 1 (July 1973): 21.

5

Environment

KEVIN C. ELLIOTT

The environmental health sciences have been the locus for particularly intensive efforts to promote various forms of ignorance throughout the twentieth and twenty-first centuries. In fact, much of the groundbreaking historical research on the deliberate construction of ignorance has come from scholars like Robert Proctor and Naomi Oreskes, who have studied industry efforts to mislead the public about tobacco and climate change. Proctor's studies of the tobacco industry appear to have played an important role in his development of the study of culturally induced ignorance or doubt, which he termed *agnotology*. This chapter builds on the extensive body of previous scholarship in this area by elucidating the historical development of seven common ignorance-making strategies that it is important for educators to recognize.[1]

The strategies employed by the tobacco industry throughout the twentieth century make for an especially valuable case study, because the industry took a number of ignorance-making strategies developed by various groups and took them to new levels of sophistication. Seven types of strategies are especially noticeable in this and other cases in the environmental health sciences:[2]

1. Selectively funding research projects and individuals
2. Withholding research findings
3. Generating doubt about existing studies and evidence
4. Silencing, marginalizing, and/or harassing opposing figures, including scientists
5. Selectively emphasizing and framing research questions and results
6. Strategically martialing traditional scientific concepts and norms

7. Developing public relations (PR) and lobbying campaigns to promote preferred messages

Both before and after the tobacco industry made such extensive use of these strategies, a number of other interest groups also employed them. Institutions involved in research on asbestos, lead, vinyl chloride, benzene, chromium, atrazine, acid rain, and dioxin employed various combinations of these strategies. Recent efforts to discredit the evidence for anthropogenic climate change have brought many of these strategies to even greater social prominence. We will see that a number of individuals and institutions appear repeatedly across these case studies. By elucidating these strategies and the institutions that support them, we can reflect on approaches for responding to them.

Analyzing the ignorance-making strategies employed in the environmental context can also yield at least two insights for those seeking to understand and improve public education. First, this analysis highlights the efforts of private interests to cultivate their own institutions and practices for educating the public (and in many cases *miseducating* the public) about environmental issues. Second, given these large-scale efforts at swaying public opinion to serve private interests, we need to rethink what is involved in educating the public in basic scientific literacy for the twenty-first century. Scholars in science studies have recently emphasized that it is at least as important for citizens to develop *critical science literacy*—an understanding of how science works and the social forces that shape it—as it is for them to learn about basic scientific facts and concepts. Analyzing ignorance-making strategies in the environmental context makes the need for critical science literacy abundantly clear, and it provides some of the historical understanding that is necessary for attaining this literacy.[3]

Critical science literacy, including an understanding of the sorts of ignorance-making strategies discussed here, has become especially important as citizens and governments seek to address major environmental issues like climate change. Geographer Daniel Bedford recently argued that the concept of agnotology can be helpful for teaching students about climate change, because they can learn to identify misinformation that is fomented in public opinion pieces, novels, movies, and other media presentations. Bedford's paper was challenged by critics who accused him of failing to teach about the complex nature of the earth's climate system and promoting uncritical deference to consensus science. While Bedford's critics are correct that scientific consensuses can be wrong, they fail to take account of the ways in which private interest groups attempt to "manufacture uncertainty" and attack consensus science in misleading ways. As Bedford himself emphasized, some aspects of climate science are well established, while others are much less clear. If citizens are to incorporate scientific information effectively in their decision making, they need to understand the strategies used by interest groups to mislead them about the levels of support for different scientific claims.[4]

Tobacco

As early as the 1930s, German scientists provided evidence that cigarette smoking causes lung cancer, and the Nazis ran campaigns against smoking. But this research gained limited traction in the United States until the 1950s. December 15, 1953, is often held up as a fateful day; it was the day when the presidents of several major cigarette companies met with public relations expert John Hill in New York City to plot a campaign to protect their products against a growing flood of research indicating that they were harmful. During the 1950s a combination of animal studies, epidemiology, in vitro studies, and chemical analyses of tobacco smoke resulted in a consensus among public health professionals that cigarette smoking was associated with lung cancer.[5]

In response to this growing body of evidence indicating that cigarettes caused cancer, Robert Proctor highlights a three-pronged ignorance-making approach used by the tobacco industry: "Its goal at many points was to *generate* ignorance—or sometimes false knowledge—concerning tobacco's impact on health. The industry was trebly active in this sphere, feigning its *own* ignorance of hazards, while simultaneously affirming the *absence of definite proof* in the scientific community, while also doing all it could to *manufacture ignorance on the part of the smoking public*."[6] This combination of approaches can be further broken down into seven more detailed strategies for constructing public ignorance about tobacco's dangers.

The first major strategy employed by the tobacco industry was *to strategically fund research projects and individuals that could help challenge claims* about the hazards of cigarettes. During his meetings with tobacco officials, John Hill emphasized that they needed to engage in a long-term effort to fund research that would serve their PR goals. The industry responded by forming a Tobacco Industry Research Committee (TIRC, later renamed the Council for Tobacco Research or CTR), which sponsored $282 million worth of research between the mid-1950s and the late 1990s. To put this in perspective, in 1981 the American Cancer Society and the American Lung Association gave $300,000 to research, whereas the TIRC/CTR provided $6.3 million. Unfortunately (from the public health perspective), Proctor calls the TIRC/CTR projects "red herring research" or "decoy research" or "distraction research," because they were carefully designed to avoid collecting any evidence that smoking caused cancer. The studies explored basic questions about the mechanisms of disease and investigated possible ways in which heredity, infections, nutrition, hormones, and stress might contribute to cancer.[7]

This strategy of funding selective research projects had multiple benefits for the industry. First, it allowed the tobacco companies to generate goodwill among government officials and the public at large by affirming that they were investigating the problem. Second, it equipped them with a body of information that they could use to challenge claims about tobacco's harmfulness and complicate subsequent debates. Third, it enabled the companies to generate a group of experts who could serve, consciously or unconsciously, as helpful witnesses for them in PR campaigns and court cases. The industry aimed "to seed the universities with research money and grow an unassailable crop of advocates."[8]

A second strategy employed by the tobacco industry to promote ignorance about their product's harmfulness was *to withhold research findings*. By the early 1960s, tobacco industry scientists had concluded both that smoking caused cancer and that nicotine was addictive, but the industry failed to pass this information on to the public. The addictiveness of nicotine was not recognized by the broader scientific community until the 1980s, and the tobacco industry refused to acknowledge it until the 1990s. In addition to withholding these general conclusions, the industry suppressed a great deal of other information and research. For example, R. J. Reynolds closed down its "Mouse House" research center in 1970 because it was starting to obtain information that could connect smoking with emphysema. The firing of the twenty-six scientists at this facility became known as the "Mouse House massacre." Similarly, Proctor reports that Philip Morris shut down its Nicotine Program in 1984 because of the worry that "the research could expose the company to the charge of having demonstrated addiction."[9]

The tobacco industry's third key strategy was *to generate doubt about existing studies and evidence*. Internal industry documents show that key officials adopted this approach intentionally. In 1971 the vice president of the tobacco industry's main lobbying and PR organization wrote in a memo that their goal was to generate "doubt about the health charge without actually denying it." Other officials claimed that "doubt is our product" and that their primary strategy objective was "to maintain doubt on the scientific front." This approach came straight from Hill, who insisted that "scientific doubts must remain." In order to generate this doubt, the industry did not typically rely on the public grants provided by the TIRC/CTR. Instead, the CTR provided funding for "Special Projects," which "allowed the industry to ask questions they didn't want anyone to know they were asking." These projects often involved research designed to discredit the evidence associating adverse health effects with smoking.[10]

A fourth ignorance-promoting strategy employed by the tobacco industry was *to silence, marginalize, or harass scientists and other figures* who presented unfavorable research results. This strategy helped the industry to achieve at least two goals: (1) it diminished the credibility and slowed down the work of offending scientists; and (2) it discouraged other scientists from actively attacking Big Tobacco. Stanton Glantz is a good example of a prominent scientist who suffered industry attacks. A professor at the University of California, San Francisco, he was a crucial figure in the identification of secondhand smoke as a cause of heart disease. After he published a review in the journal *Circulation* in 1991 highlighting the health effects of secondhand smoke, the industry paid several scientists between $3,000 and $4,000 each to write letters challenging Glantz's work. A year later, biostatistician Nathan Mantel published an article criticizing the *Circulation* paper and invoiced Philip Morris $135,065 for his work on secondhand smoke in 1991. Several years later, after Glantz received a grant from the National Cancer Institute (NCI), Philip Morris developed and implemented a plan to question his credibility and integrity. This plan involved running advertisements, publicly challenging the NCI's decision to fund Glantz's project, reaching out to allies on Capitol Hill, and writing to Glantz's superiors at the University of California. Industry-related groups even initiated an unsuccessful lawsuit against the university in an effort to derail and challenge his work.[11]

A fifth strategy employed by Big Tobacco was *to selectively emphasize and frame research questions and results*. Naomi Oreskes and Erik Conway describe an excellent example of this strategy. They note that after Hill met with the tobacco executives in 1953, his PR firm, Hill and Knowlton, developed a list of fifteen questions about the hazards of tobacco that the industry could present to journalists. For example, why did cancer rates vary between cities even when their smoking rates were similar, and why were men experiencing the greatest rise in lung cancer rates even though cigarette use was increasing most rapidly among women? According to Oreskes and Conway, "None of the questions was illegitimate, but they were all disingenuous, because the answers were known." But this strategy of posing disingenuous questions via the mass media worked very well: "The industry had realized that you could create the impression of controversy simply by asking questions, even if you actually knew the answers and they didn't help your case."[12]

The tobacco industry's sixth major strategy was *to strategically martial traditional scientific concepts and norms* in ways that advanced their efforts to generate ignorance. For example, the industry was able to exploit the traditional scientific virtues of demanding high standards of evidence and critically evaluating new conclusions. The industry cleverly challenged evidence from human epidemiology studies (because of the potential for methodological flaws), while insisting that laboratory studies in animals did not come close enough to replicating the human condition. Therefore, their stubborn refusal to accept any evidence for tobacco's health impacts could be spun as an insistence on maintaining an open mind and demanding the best possible scientific evidence. The industry was also able to exploit the traditional norms of "objectivity," "balance," and "fairness," which have played an important role in both science and the media. Tobacco apologists frequently persuaded journalists to present their preferred messages (even when they did not reflect the views of mainstream scientists) by arguing that this was necessary to maintain balance and fairness.[13]

The seventh strategy of the tobacco industry was *to develop PR and lobbying campaigns to promote preferred messages.* In many ways, the other six strategies discussed here were all components of the industry's overarching efforts to generate powerful PR messages to disseminate to the public, to scientists, to physicians, and to government officials. For example, the efforts to generate favorable research, to cast doubt on worrisome findings, and to formulate strategic questions were all parts of Hill and Knowlton's masterful PR strategy. Guided by Hill and Knowlton, tobacco industry officials disseminated their preferred messages through their own pamphlets and booklets, through scientific journals, through advertisements, through front groups and think tanks, and through their inside connections with the leaders of major news organizations.[14]

Jon Christensen argues that the industry's overall PR messages were delivered in four major phases. First, the industry sought to "fight science with science." This phase involved arguing that industry was seriously studying the issues and that factors other than tobacco were important for understanding lung cancer. The second phase was a "doubt is our product . . . truth is our message" phase, in which the industry primarily sought to sow doubt about the mounting evidence for tobacco's effects and to keep some level of controversy alive. In the third phase, "undermining science," the focus shifted from casting doubt on specific findings to raising questions about entire fields of science, such as epidemiology and risk analysis. The final phase, "it's not news," involved a major change. The industry ultimately had to acknowledge (after decades of obfuscation) that the health effects of tobacco were well known, and it shifted to billing itself as a "responsible manufacturer of a risky product." Altogether, the *Public Relations Journal* regarded the lobbying strategy put together on behalf of the tobacco industry one of the "most formidable public relations/lobbying machines in history."[15]

Ignorance across the Environmental Health Sciences

The way the ignorance-generating strategies employed by the tobacco industry have been employed across numerous other areas of the environmental health sciences is striking. Even before the tobacco industry, the asbestos and lead industries withheld information about the adverse health effects of their products (strategy 2). According to David Michaels, the risks of asbestos were already becoming known in the first decades of the twentieth century, but asbestos exposure now causes one hundred thousand deaths per year around the world because companies like W. R. Grace and the Johns-Manville Corporation kept on manufacturing asbestos-containing products while denying the risks. Similarly, even though the adverse effects of lead were well known, General Motors and DuPont launched tetraethyl lead in gasoline in the 1920s. From the very beginning, they were forced to engage in a coordinated campaign to promote ignorance about the potential human and environmental health effects of their products.[16]

These sorts of activities continued across many other industries in subsequent decades. In accordance with the second strategy (withholding research findings), the manufacturers of vinyl chloride (the precursor to PVC, a widely used plastic) engaged in a particularly brazen effort to suppress information when their product began to come under scrutiny in the early 1970s. In 1973 the National Institute of Occupational Safety and Health (NIOSH) made a formal request for information about the health effects of vinyl chloride.

Shortly before industry officials met with NIOSH, they were apprised of alarming evidence from European studies indicating that experimental animals were developing tumors when exposed to vinyl chloride at levels of half the exposure limit set by the Occupational Safety and Health Administration. The industry officials withheld this evidence and reported enthusiastically afterwards that their meeting had lessened the chances of "precipitous action by NIOSH on vinyl chloride."[17]

Strategy 3 (casting doubt on previous studies) also became increasingly prominent throughout the later decades of the twentieth century as "product defense" companies began to specialize in reanalyzing previous data that seemed worrisome to industry. Two examples come from studies of benzene, an organic compound found in petrochemicals like gasoline, and chromium, a chemical element used to form stainless steel and other corrosion-resistant materials. When NIOSH performed a major epidemiological study in the 1970s to quantify the relationships between benzene exposure and leukemia, the industry brought in the esteemed risk-assessment expert Dennis Paustenbach and his company, ChemRisk, to reanalyze the data. ChemRisk changed the exposure estimates in the study to worst-case scenarios, making it possible to assert that benzene exposure might lead to leukemia only at "very high concentrations." When it turned out that the new dose-response curve still demonstrated some increased cancer risk at low exposure levels, the oil industry again hired Paustenbach (now working for a different company, Exponent) to alter the study's assumptions. More recently, David Michaels documented how the product-defense company ENVIRON manipulated studies of chromium, splitting the data into two parts and altering exposure categories to eliminate what would have been very worrisome results for the industry.[18]

There are also numerous cases of interest groups working to silence, harass, or marginalize opposing scientists (strategy 4). Irving Selikoff, the leading researcher on the health effects of asbestos, was vilified by the industry as "a media zealot who exaggerated the risks of asbestos on the back of bogus medical qualifications and flawed science." Herbert Needleman, the University of Pittsburgh professor who performed pioneering research on the neurological effects of lead, had to fight through multiple inquiries and accusations of research misconduct by industry-connected scientists. More recently, internal documents released through a lawsuit indicate that UC Berkeley scientist Tyrone Hayes was targeted by Syngenta, manufacturer of the controversial herbicide atrazine. Following Hayes's findings that atrazine might contribute to the feminization of frogs, Syngenta commissioned a psychological profile on him and explored other strategies that could be used against him, including investigating his wife. The same documents reveal that Syngenta hired a detective agency to investigate the scientists on the Environmental Protection Agency (EPA) Science Advisory Panel (SAP) studying atrazine and examined the personal life of a judge who might be assigned to an atrazine case. The atrazine case also illustrates strategy 1 (strategically funding research projects and individuals), insofar as Syngenta secretly paid a number of academics to question its risks and highlight its economic value.[19]

The fifth strategy (selectively emphasizing and framing research results and questions) is prominent in the Reagan administration's response to the problem of acid rain. In the face of increasing evidence that acid rain was a significant environmental problem, the White House's Office of Science and Technology Policy (OSTP) formed a panel to examine the issue. According to Oreskes and Conway, the procedure for producing, editing, and disseminating the report was riddled with irregularities designed to weaken its message. For example, it included a special appendix not approved by the committee that was written by Fred Singer, a physicist who was specially picked for the panel by the OSTP, apparently with the goal of promoting the antiregulatory policies of the Reagan administration. The executive summary was also changed at the request of the OSTP after the panel had signed off on a previous version. The altered summary allegedly "changed the tenor of the report," softening its tone and shifting the policy recommendation to control sulfur dioxide emissions from the beginning to the end in order to lessen its impact. Finally, the report was withheld from the public (strategy 2) until after a crucial congressional vote. Congressman Norman D'Amours claimed that the White House wanted to avoid giving legislators "the ammunition we needed to push acid rain controls through Congress."[20]

The sixth strategy, employed in numerous subsequent environmental health controversies as well as in the tobacco case, was to martial traditional scientific concepts and norms for ignorance-generating purposes. A particularly vivid example is the development of The Advancement of Sound Science Coalition (TASSC), a front organization that was originally created by one of Philip Morris's PR firms but that ultimately lobbied against a wide range of environmental regulations. The demand for "sound science" before enacting regulations was rhetorically brilliant, because TASSC could claim that it was merely calling for high-quality scientific evidence—even though it was often either demanding unreasonably high standards of evidence or raising disingenuous criticisms of high-quality evidence. One of TASSC's typical strategies was to label any scientific work that threatened industry's interests as "junk science" performed by "junk scientists." (In reality, six of the individuals labeled "junk scientists" by TASSC won the highest award bestowed by the American College of Epidemiology.) This rhetoric of demanding sound science has proved immensely valuable in the legal arena as well. According to Carl Cranor, regulated industries have persuaded judges to demand unacceptably high standards of evidence in lawsuits against their products by insisting that these forms of evidence are necessary for good scientific practice.[21]

As in the tobacco case, a seventh strategy used by interest groups to promote public ignorance about a range of environmental health threats was to develop PR and lobbying campaigns to promote their preferred messages. This approach is illustrated vividly in the case of the dioxin family of chemicals, which are by-products of the chlorine industry that have been implicated in a range of health effects, including cancer. The chlorine industry faced a significant PR problem because dioxins are "perhaps the most potent [carcinogens] ever tested" and because the chemical group in general developed a reputation as "one of the most toxic substances known to humans." In part to counteract this problem, the Chemical Manufacturers Association created the Chlorine Chemistry Council to engage in PR and lobbying. In 1994, a year after its founding, the council obtained $12 million from the association, as well as $120 million from member companies, and it had hired two PR firms to augment its own staff. One of these firms, Mongoven, Biscoe, & Duchin (MBD), specializes in gathering intelligence on environmentalists; according to Sharon Beder, MBD "keeps the Chemistry Council up to date on 'anti-chlorine activists' and their strategies, and advises on ways to counter them." The PR strategies proposed by MBD included hiring another PR firm to "reach out to editorial boards" to promote the industry's perspective on dioxins, enlisting scientists to ask "pointed questions" at conferences, working with pediatric groups to counteract concerns about health effects in children, and finding prominent physicians to advance the chlorine industry's case with medical groups.[22]

The dioxin case illustrates the sophistication with which powerful interest groups wielded combinations of ignorance-generating strategies by the final decades of the twentieth century. In accordance with strategy 2 (withholding data), the US Centers for Disease Control (CDC) abandoned a study of the health effects of Agent Orange under the pretense that the available data were inadequate. A subsequent congressional hearing concluded that the data were not a genuine problem and that the study was "flawed and perhaps designed to fail," apparently because of concerns to protect the Department of Defense and the chlorine industry. In accordance with strategy 3 (casting doubt on previous studies), in 1990 the industry hired scientists to reanalyze an earlier study of dioxin; they looked through the slides of tissue taken from rats exposed to dioxin and contended that there was evidence for 50 percent fewer tumors than originally counted. The chlorine industry used this reanalysis to push the EPA toward weakening its dioxin regulatory standards. Indeed, the ties between the chemical industry and the EPA were so close that EPA chief Anne Gorsuch Burford resigned from her post in 1983 in part because of the agency's questionable handling of dioxin-contaminated sites, and her successor, John Hernandez, resigned because of evidence that he had allowed the industry to alter an EPA report on dioxin.[23]

But the chlorine industry's strategies do not stop there. In accordance with strategy 5 (selectively presenting and framing results), the industry organized its own conference on dioxin in 1990 and hired a communications company to publicize the results. The press release reported a consensus conclusion that dioxin was "much less toxic to humans than originally believed," which outraged some of the scientists at the conference who had not come to that conclusion. Perhaps most strikingly, the major chemical companies Monsanto and BASF funded their own research on dioxin and published it in prominent journals like the *Journal of the American Medical Association* and *Scientific American* (strategy 1). These studies were later found to incorporate fraudulent manipulation of the data in order to minimize findings of harm. When a prominent environmental journalist attempted to expose the fraud, he was silenced with the threat of a costly lawsuit (strategy 4).[24]

Climate Change

After reviewing the development of these seven ignorance-generating strategies in numerous areas of environmental health science, it is enlightening to see how they are now being employed in the case of climate change. Because it is one of the greatest challenges facing contemporary society, there is much to be gained by putting contemporary debates over climate change in the context of previous efforts to generate public ignorance. Oreskes and Conway have made it much easier to provide this contextualization thanks to their recent book *Merchants of Doubt*, which tells the story of the George C. Marshall Institute and its role in challenging the evidence for climate change. Founded by right-wing physicists Robert Jastrow, Frederick Seitz, and William Nierenberg in order to defend president Reagan's strategic defense initiative (SDI) in the mid-1980s, the institute later focused much of its attention on challenging environmental problems.

In June 1988 the prominent climate modeler James Hansen brought climate change to the nation's attention by testifying to congressional hearings both that the earth had warmed discernibly over the past decade and that it was very unlikely that the warming was the result of natural causes. These hearings, together with the formation of the Intergovernmental Panel on Climate Change (IPCC) and other political pressures, created significant momentum in the George H. W. Bush administration to address climate change. According to Oreskes and Conway, the Marshall Institute played an important role in halting that momentum. Bill Nierenberg worked with others to produce a white paper on climate change and provided a briefing to members of the White House staff. Apparently the paper had a chilling effect on action to address climate change. A member of the Office of Cabinet Affairs claimed that "everyone has read it. Everyone takes it seriously." Stephen Schneider, a prominent climate scientist at Stanford University, lamented that White House chief of staff, John Sununu, held the report up "like a cross to a vampire."[25]

This incident illustrates the profound influence of strategy 7 (the use of PR campaigns and think tanks) on the climate change debate. Robert Jastrow claimed that "it is generally considered in the scientific community that the Marshall report was responsible for the [George H. W. Bush] Administration's opposition to carbon taxes and restrictions on fossil fuel consumption" and that the Marshall Institute "is still the controlling influence in the White House." But as in the tobacco case, this overarching strategy depended on the other strategies discussed here. For example, the Marshall report relied heavily on selectively framing information (strategy 5). It argued that the twentieth-century warming trend identified by Hansen didn't track carbon dioxide emissions and that solar activity was responsible for the warming. But this was a disingenuous argument, because the analysis by Hansen's team showed that solar activity was probably responsible for some warming but that it did not explain the overall warming as successfully as a combined model that incorporated carbon dioxide emissions and volcanic activity in addition to solar activity. Moreover, Schneider pointed out that if the climate were as sensitive to solar radiation as the Marshall report argued, then it would be extremely sensitive to changes in greenhouse gases as well.[26]

Unfortunately, the climate change case also vividly illustrates strategy 4 (harassing and marginalizing opposing scientists). In 1994 climate scientist Ben Santer was asked to be the convening lead author for the eighth chapter of the second IPCC report on climate change. At the time, Santer was an early-career climate scientist performing pioneering work to show that the observed warming of the earth was caused primarily by greenhouse gas emissions rather than by other causes. He later won a MacArthur "Genius Grant" and numerous distinctions from his employer, the Department of Energy. But when the IPCC report came out in 1996, scientists from the Marshall Institute falsely accused Santer of making inappropriate last-minute changes to his chapter after it had been approved. (His changes were actually made according to IPCC protocol, in collaboration with his coauthors, in response to last-minute peer-review comments.) His critics wrote to members of Congress, pushed to get Santer fired from the Department of Energy, and published a highly publicized op-ed piece in the *Wall Street Journal* accusing him of fraud. According to Santer, he was also accosted at a Capitol Hill briefing by an oil industry lobbyist who "got up in [his] face, turned beet red and [started] screaming" about the changes made to the chapter. The American Meteorological Society was sufficiently concerned about the false charges made against him that they published "An Open Letter to Ben Santer" in their journal defending him against the *Wall Street Journal* accusations.[27]

The fossil fuel industry also fought against the IPCC report, by funding its preferred researchers (strategy 1). For example, Patrick Michaels received money from the fossil fuel industry to help publish a quarterly newsletter, the *World Climate Review*, which challenged mainstream climate science and was circulated for free to members of the Society for Environmental Journalism. He also worked as a consultant for a coal-mining industry group to promote the notion that continued burning of fossil fuels would benefit agricultural production. In 1995 he was called by the Republican majority as a "star witness" for congressional hearings on the IPCC report. In accordance with strategy 3, Michaels insisted that his own analyses showed the IPCC models to be problematic but that the IPCC had ignored his concerns. Unfortunately, he was either ignorant or employing strategy 5 (selectively presenting information), because IPCC modelers showed that his analysis contained an "elementary flaw."[28]

The story told by Oreskes and Conway is merely the tip of the iceberg, however. They focus on the Marshall Institute and the early years of the climate change countermovement (CCCM). The sociologists Aaron McCright and Riley Dunlap trace the same ignorance-generating strategies through the first decade of the twenty-first century, showing how conservative think tanks, activists, and government officials funded contrarian scientists (strategy 1), suppressed research (strategy 2), misrepresented and manipulated scientific findings in an effort to promote doubt (strategy 3), and intimidated or threatened opposing scientists (strategy 4). Robert Brulle recently analyzed the key institutions and funding sources that underpin these activities. Using IRS data, he analyzed the annual income of 91 important CCCM organizations funded by 140 different foundations. He found that the total average annual income for these 91 organizations was more than $900 million. The CCCM organizations receiving the most foundation support were large conservative think tanks such as the American Enterprise Institute, the Heritage Foundation, the Hoover Institution, the Manhattan Institute, and the Cato Institute. The largest donors to these institutions included the Scaife, Bradley, Koch, Howard, Pope, Searle, and Templeton Foundations. To illustrate the power of these groups in environmental health debates, Oreskes and Conway point to a study indicating that 92 percent of the fifty-six "environmentally skeptical" books published in the 1990s were linked to right-wing foundations, along with all of the thirteen skeptical books published in the 1980s.[29]

Significantly, Brulle found that two "donor directed" foundations, Donors Trust and Donors Capital, have become major sources of funding for the CCCM. These foundations have decreased the transparency of the funding process so that organizations and individuals can donate to them and designate how the gifts should be used, but without making the gifts public. In effect, these donor-directed foundations create a "black box," such that individuals and institutions can fund preferred organizations without acknowledging who they are supporting. For example, it appears that the Koch Affiliated Foundations and ExxonMobil may have begun routing their donations through Donors Trust and Donors Capital, since their gifts to CCCM organizations decreased significantly at the same time that donations to these entities increased. The network of institutions working to manufacture uncertainty about climate change is exceedingly well funded, including many of the most powerful conservative think tanks and foundations, but with a trend toward decreased transparency about the institutions involved.[30]

Common Institutions and Individuals

There are at least two benefits to developing this historical taxonomy of ignorance-generating strategies. One is that it highlights the need to increase citizens' critical science literacy so that they are less likely to be misled by these strategies. The other is that this chapter clarifies a new form of public education (or *miseducation*) that merits additional scrutiny. That is, private interest groups have cultivated a network of individuals, think tanks, journals, and PR organizations that they use to generate ignorance about environmental health threats. To better understand this form of public education, it is valuable to identify the individuals and institutions that are common to the multiple cases discussed here. For example, it is not surprising that similar strategies were employed in most of these environmental health controversies, because the same PR firm contributed to most of them. Hill and Knowlton worked not only for the tobacco industry but also for those dealing with asbestos, lead, vinyl chloride, the ozone hole, and a host of other environmental threats.[31]

Other PR firms, such as Burson-Marsteller and MBD, have also worked across numerous environmental health issues. It is no accident that MBD helped the chlorine industry to "spy" on those opposing dioxin, because MBD and another PR firm, the Foundation for Public Affairs, engage in this practice for a variety of groups (including Monsanto, Dupont, Philip Morris, and Shell Oil) against a wide array of environmental activists. Similarly, TASSC defended not only tobacco smoke but also pesticides, saccharin, fast food, and other potential health threats. The product-defense companies that have spun off from the PR industry also appear over and over again in many different instances of environmental miseducation. As David Michaels puts it, "Having cut their teeth manufacturing uncertainty for Big Tobacco, scientists at ChemRisk, the Weinberg Group, Exponent, Inc., and other consulting firms now battle the regulatory agencies on behalf of the manufacturers of benzene, beryllium, chromium, MTBE, . . . perchlorates, phthalates, and virtually every other toxic chemical in the news today."[32]

The product-defense companies have generated an additional set of institutions to meet their needs, namely, specialized journals to publish their results. These journals were created because these companies needed peer-reviewed venues that would publish their work (so as to gain the legitimacy afforded by peer review) but would choose reviewers who were unlikely to challenge questionable scientific analyses performed in order to obtain results favorable to the study sponsors. According to David Michaels, these venues include journals like *Indoor and Built Environment* (created by the tobacco industry), *Regulatory Toxicology and Pharmacology* (published by a society sponsored by pharmaceutical, tobacco, and pesticide companies), and the *Journal of Occupational and Environmental Medicine*. He laments that his scientific colleagues shrug and say, "It's all part of the game. . . . We know what these papers are worth," but members of Congress, journalists, judges, and citizens do not know enough of their background to dismiss these sources of information.[33]

Michaels also shows how scientists who are affiliated with these product-defense companies can end up manufacturing uncertainty in multiple cases. For example, he notes that he became interested in the work of Daniel Roth and Paul Levy because they reanalyzed data sets in an effort to dismiss health concerns about beryllium. He investigated their other work and found that Levy had also been hired by R. J. Reynolds to reanalyze data on lung cancer and secondhand smoke and Roth had worked with Philip Morris to dismiss links between smoking and lung cancer. Roth had also reanalyzed data on the health effects of airborne pollutants from coal-burning power plants and on risks of vehicle rollovers. Together, Roth and Levy had even challenged evidence concerning the relationship between alcohol consumption and breast cancer. Michaels points out that similar scientists and strategies turn up in one doubt-manufacturing case after another.[34]

In their book on climate-change denial, Oreskes and Conway provide perhaps an even more fascinating story of the scientists who have specialized in manufacturing uncertainty. The leaders of the Marshall Institute—Robert Jastrow, Fred Seitz, and Bill Nierenberg—were physicists who wanted to form an organization to counter the left-wing claims of the Union of Concerned Scientists about national-defense issues. But both before and after the founding of the institute, these scientists worked on a range of other issues as well. For example, after he retired from his post as president of Rockefeller University, Fred Seitz was hired by R. J. Reynolds to run a major grants program. Much of the resulting work was high-quality basic research, but as we have already seen, these sorts of programs were designed by the tobacco companies to produce "an extensive body of scientifically, well-grounded data useful in defending the industry against attacks."[35] They were also designed to generate helpful experts. In the 1970s academics Bruce Owen and Ronald Braeutigam advised industry executives that this was an important strategy to pursue:

> Regulatory policy is increasingly made with the participation of experts, especially academics. A regulated firm or industry should be prepared whenever possible to co-opt these experts. This is most effectively done by identifying the leading experts in each relevant field and hiring them as consultants or advisors or giving them research grants and the like. This activity requires a modicum of finesse; it must not be too blatant, for the experts themselves must not recognize that they have lost their objectivity and freedom of action.[36]

In addition to running this grants program for Big Tobacco, Seitz also served as an advisor to TASSC and worked with the Marshall Institute on climate change skepticism. He was the author of the letter to the *Wall Street Journal* accusing Ben Santer of fraud.

Similarly, Bill Nierenberg worked on other environmental health issues before joining the Marshall Institute. He was the leader of the acid rain review panel discussed above as an example of how the Reagan administration selectively framed information (strategy 5). As early as 1980, he was also involved in reports on climate change and led the crucial Marshall Institute briefing on climate change for the George H. W. Bush White House in 1989. Fred Singer, the physicist who worked with Nierenberg on the Reagan administration's acid rain panel, was also affiliated with the Marshall Institute and played a major role in climate change denial, including attacks on Ben Santer. But Oreskes and Conway also document how he worked on campaigns to challenge acid rain and the ozone hole, and he worked with Philip Morris to develop TASSC and the "sound science vs. junk science" movement.[37]

It is perhaps not surprising that a group of right-wing physicists would get involved in defending national-defense initiatives like the SDI. After all, many left-wing physicists came to regret their discipline's involvement in developing nuclear weapons during the Cold War, and the Marshall Institute founders were partly reacting against this left-wing movement. But it is somewhat surprising that these physicists ended up seeking to generate uncertainty about such a wide variety of environmental health issues—ones that they arguably did not have appropriate expertise to address—including tobacco, acid rain, the ozone hole, and climate change. Oreskes and Conway have hypothesized that the link for these physicists between national-defense issues and environmental issues is that they were passionate Cold Warriors who feared big government and attacks on individual liberties. Insofar as most of the environmental health issues discussed in this chapter have generated regulations that limit particular forms of liberty, Oreskes and Conway think that these physicists came to regard environmentalism as a threat comparable to the threat of communism.[38]

It is also possible that these physicists were influenced by more subtle aspects of their disciplinary background. The sixth strategy identified here (strategically martialing traditional scientific norms and concepts) can be employed not only in a cynical and manipulative way but also in a more genuine (but nevertheless misguided) way. Perhaps the physicists discussed here were influenced by their experience with the high levels of certainty and predictability available to them in physics. It is possible that they did not fully appreciate the ways in which other areas of science—especially the environmental sciences—often depend on evidence of a more complex and fragmentary character, requiring the integration of multiple lines of evidence that are less compelling when taken individually. It is possible that the disciplinary background of these scientists, combined with their ideological suspicion of environmentalism, contributed to their skeptical perspective. This skepticism was then amplified into a powerful social force because of the significant amount of money and power available to the industries that did not want to address environmental threats.[39]

Responding to the Manufacture of Ignorance

The historical taxonomy of ignorance-generating strategies in this chapter can also be valuable as a source of ideas for responding to these strategies and the new institutions for public education that private interests have developed. As members of the public and policymakers become more familiar with these strategies, they may want to encourage two general approaches for addressing them. First, policies can be enacted to make it more difficult to implement the strategies. Second, policies can be changed to create different incentive structures so that powerful interest groups will have less motivation to implement the strategies.[40]

One way to make it more difficult to employ the strategies is to increase transparency throughout the process of generating science on environmental health issues. For example, if those performing new research (strategy 1), reanalyzing old research (strategy 3), accusing others of misconduct (strategy 4), disseminating selective information (strategy 5), and performing PR activities (strategy 7) were more consistently required to disclose their sources of funding, it would facilitate greater scrutiny of these activities. Refereed journals can play an important role in creating policies to promote this sort of transparency. An important additional step is to require that the data from privately funded science be made publicly available; at present, publicly funded data are available for private interests to reanalyze (strategy 3), but the data from privately funded science are not similarly accessible. Finally, more steps could be taken to prevent the withholding of adverse information about product safety (strategy 2); these steps could include legally requiring that the information be disclosed, setting significant penalties for failure to disclose, and limiting the use of excuses related to "confidential business information" (CBI).[41]

Another way to make it more difficult to employ these ignorance-making strategies is to take crucial policy-relevant research out of the hands of those who have the greatest financial stakes in the outcome. For example, there could be requirements that crucial studies to be used for regulatory purposes must be designed, contracted, and interpreted by an agency such as the EPA or even the Organization for Economic Cooperation and Development (OECD). Manufacturers of regulated products could still be required to pay for the studies, but they would not be able to control the design and interpretation of the research. While it would presumably be politically challenging to implement such an approach, a more modest step would be to make sure that there is ample public funding on policy-relevant topics in environmental health so that crucial research studies do not remain primarily in the hands of those who may be motivated to generate ignorance. In some cases, it may also be possible to design collaborative studies in which industry, government agencies, and environmental groups work together to design and interpret studies. Deliberative bodies or panels created by relatively neutral entities such as the National Academy of Sciences (NAS) can also play a valuable role in evaluating the quality of previous studies and the overall weight of evidence on debated topics.[42]

In addition to rendering the ignorance-making strategies more difficult to implement, there may be ways to decrease the incentives for employing them. One approach is to alter regulatory policies so that they depend less on detailed scientific information that lends itself to the strategies discussed throughout this chapter. The Massachusetts Toxic Use Reduction Act (TURA) of 1989 is often held up as a model in this regard. In an effort to circumvent endless debates about the precise toxicity of various chemicals, the TURA created a list of substances about which there was cause for concern. Companies that used large quantities of the substances were then required to disclose the amounts used and to investigate alternatives to using them. The benefit of this approach was that it took some of the emphasis off fighting over the toxicity of these products and placed more emphasis on looking for potentially safer alternatives. A lesson to be drawn from this and other cases is that regulations requiring companies to publicly disclose potentially harmful products that they use and emit can be helpful and easier to enact than regulations that ban the products completely. While disclosure requirements do not completely eliminate incentives to manufacture public uncertainty about the harmfulness of particular products (as can be seen in the case of tobacco), industry groups may sometimes find it more convenient to look for less controversial ways of doing business than to fight negative public opinion.[43]

Another approach for lessening the incentives to manufacture uncertainty (besides altering regulatory policies) is to reframe public debates about environmental issues. For example, science policy experts such as Roger Pielke and Daniel Sarewitz have recently argued that public policy discussions about environmental problems should focus less on nitty-gritty details about scientific evidence and more on the values that should underlie environmental policy decisions. Sarewitz points out that large-scale environmental problems like climate change will almost always be complex enough that interest groups will be able to find some scientific evidence that supports their positions. (Even when this evidence is not immediately forthcoming, it can typically be manufactured.) Therefore, rather than engaging in endless debates about the precise likelihood that specific problems will arise, it may be more fruitful to ask questions about our values: for example, In order to address this potential problem, could we take preventive actions that are not overly costly to society? or Are there ways of reframing this potential problem so that we can turn it into an opportunity to make our society better? Some proponents of the precautionary principle suggest that one of its goals is to promote this sort of shift from science-based questions to value-focused questions.[44]

Toward Critical Science Literacy

This chapter provides a historical taxonomy of seven prominent strategies for generating ignorance about environmental health issues.

These strategies were employed with tremendous skill by the tobacco industry, they were employed in a range of other cases both before and after tobacco, and they are now being used to dispute the evidence for anthropogenic climate change.

Elucidating these strategies yields at least two lessons for those seeking to understand and improve public education. First, private interests seek to cultivate their own institutions for educating the public (and in some cases *miseducating* the public) about environmental issues. Scientists, think tanks, scholarly societies, and conferences are funded by those with deep pockets in an effort to generate desired scientific findings and to attack opposing findings. Second, in response to these large-scale efforts at swaying public opinion, it is important to rethink what is involved in providing the public with basic scientific literacy in the twenty-first century. As we face global environmental challenges like climate change, we can no longer afford to ignore institutions and processes that actively work toward the obfuscation of well-supported scientific conclusions. Citizens in the twenty-first century need to be educated in critical science literacy. A free and democratic society now requires a citizenry capable of discerning the difference between science conducted in the public interest and studies conducted for private gain.

NOTES

1. Robert N. Proctor, *Golden Holocaust: Origins of the Cigarette Catastrophe and the Case for Abolition* (Berkeley: University of California Press, 2012); Naomi Oreskes and Erik Conway, *Merchants of Doubt: How a Handful of Scientists Obscured the Truth on Issues from Tobacco Smoke to Global Warming* (New York: Bloomsbury, 2010); Robert Proctor, "Agnotology: A Missing Term to Describe the Cultural Production of Ignorance (and Its Study)," in *Agnotology: The Making and Unmaking of Ignorance*, ed. Robert Proctor and Londa Schiebinger (Stanford, CA: Stanford University Press, 2008), 11.

2. This taxonomy of strategies is based loosely on a set of approaches for influencing science that is described in Tom McGarity and Wendy Wagner, *Bending Science: How Special Interests Corrupt Public Health Research* (Cambridge, MA: Harvard University Press, 2008). While this particular taxonomy is unique, it is similar to other lists of strategies in, for example, Oreskes and Conway, *Merchants of Doubt*; Proctor, *Golden Holocaust*, 290–92; and Aaron McCright and Riley Dunlap, "Anti-reflexivity: The American Conservative Movement's Success in Undermining Climate Science and Policy," *Theory, Culture & Society* 27 (2010): 100–133.

3. On the concept of critical science literacy, see Susanna Priest, "Critical Science Literacy: What Citizens and Journalists Need to Know to Make Sense of Science," *Bulletin of Science, Technology & Society* 33 (2013): 138–45.

4. Daniel Bedford, "Agnotology as a Teaching Tool: Learning Climate Science by Studying Misinformation," *Journal of Geography* 109 (2010): 159–65. For criticism of his article, see David Legates, Willie Soon, and William Briggs, "Learning and Teaching Climate Science: The Perils of Consensus Knowledge Using Agnotology," *Science & Education* 22 (2013): 2007–17. For Bedford's response, see Daniel Bedford and John Cook, "Agnotology, Scientific Consensus, and the Teaching and Learning of Climate Change: A Response to Legates, Soon and Briggs," ibid., 2019–30. On the importance of clarify-

ing the levels of support for different scientific claims, see Daniel McKaughan and Kevin Elliott, "Backtracking and the Ethics of Framing: Lessons from Voles and Vasopressin," *Accountability in Research* 20 (2013): 206–26.

5. Oreskes and Conway, *Merchants of Doubt*, 15; Jon Christensen, "Smoking Out Objectivity: Journalistic Gears in the Agnogenesis Machine," in Proctor and Schiebinger, *Agnotology*, 266; Proctor, *Golden Holocaust*, 232–33.

6. Proctor, "Agnotology," 13–14, emphasis in the original.

7. Oreskes and Conway, *Merchants of Doubt*, 15–16, 25; Proctor, *Golden Holocaust*, 257–61; Proctor, "Agnotology," 14–15.

8. Proctor, "Agnotology," 14; Oreskes and Conway, *Merchants of Doubt*, 13–14; D. Zegart, *Civil Warriors* (New York: Delacorte, 2000), 290.

9. Oreskes and Conway, *Merchants of Doubt*, 21; Proctor, *Golden Holocaust*, 263–64.

10. Proctor, "Agnotology," 17; Oreskes and Conway, *Merchants of Doubt*, 16; Proctor, *Golden Holocaust*, 276.

11. Anne Landman and Stanton Glantz, "Tobacco Industry Efforts to Undermine Policy-Relevant Research," *American Journal of Public Health* 99 (2009): 47–53.

12. Oreskes and Conway, *Merchants of Doubt*, 18.

13. Proctor, "Agnotology," 11–12; Oreskes and Conway, *Merchants of Doubt*, 19–20; Christensen, "Smoking Out Objectivity," 275–76.

14. Oreskes and Conway, *Merchants of Doubt*, 16.

15. Christensen, "Smoking Out Objectivity," 268–69; Kristin Shrader-Frechette, *Taking Action, Saving Lives: Our Duties to Protect Environmental and Public Health* (New York: Oxford University Press, 2007), 67.

16. Michaels, *Doubt Is Their Product: How Industry's Assault on Science Threatens Your Health (New York: Oxford University Press, 2008)*, 12–18, 23–24, 39–42; Gerald Markowitz and David Rosner, *Deceit and Denial: The Deadly Politics of Industrial Pollution* (Berkeley: University of California Press, 2002), 94, 108–38.

17. Michaels, *Doubt Is Their Product*, 35–36; Markowitz and Rosner, *Deceit and Denial*, 186–90. For a litany of other cases in which information about potentially toxic substances was suppressed, see Shrader-Frechette, *Taking Action, Saving Lives*, 53.

18. Michaels, *Doubt Is Their Product*, 46, 75, 107–8.

19. Jock McCulloch and Geoffrey Tweedale, "Shooting the Messenger: The Vilification of Irving J. Selikoff," *International Journal of Health Services* 37 (2007): 619; Markowitz and Rosner, *Deceit and Denial*, 136; Clare Howard, "Special Report: Syngenta's Campaign to Protect Atrazine, Discredit Critics," *Environmental Health News*, 17 June 2013, http://www.environmentalhealthnews.org/ehs/news/2013/atrazine.

20. Oreskes and Conway, *Merchants of Doubt*, 77–100.

21. Shrader-Frechette, *Taking Action, Saving Lives*, 81; Michaels, *Doubt Is Their Product*, 58; Oreskes and Conway, *Merchants of Doubt*, 150–52; Carl Cranor, *Toxic Torts: Science, Law, and the Possibility of Justice* (Cambridge: Cambridge University Press, 2008).

22. Shrader-Frechette, *Taking Action, Saving Lives*, 52; Sharon Beder, *Global Spin: The Corporate Assault on Environmentalism*, rev. ed. (White River Junction, VT: Chelsea Green, 2002), 141, 151, 156.

23. Shrader-Frechette, *Taking Action, Saving Lives*, 52; Beder, *Global Spin*, 147.

24. Beder, *Global Spin*, 143, 148. For other cases in which scientists, journalists, or activists were sued or harassed, see Shrader-Frechette, *Taking Action, Saving Lives*, 55–56.

25. Oreskes and Conway, *Merchants of Doubt*, 184.

26. Ibid., 187–90.

27. Ibid., 1–3, 197–209.

28. Ibid., 203.

29. McCright and Dunlap, "Anti-Reflexivity," 111–20; Robert Brulle, "Institutionalizing Delay: Foundation Funding and the Creation of U.S. Climate Change Counter-Movement Organizations," *Climatic Change* 122 (2014): 681–94; Oreskes and Conway, *Merchants of Doubt*, 236.

30. Brulle, "Institutionalizing Delay," 687–94.

31. Michaels, *Doubt Is Their Product*, 45.

32. Beder, *Global Spin*, 135–36; Shrader-Frechette, *Taking Action, Saving Lives*, 56, 81; Michaels, *Doubt Is Their Product*, 46.

33. Michaels, *Doubt Is Their Product*, 53–55. The journal *Regulatory Toxicology and Pharmacology* is published by the International Society for Regulatory Toxicology and Pharmacology, which has been criticized by organizations like the Center for Science in the Public Interest for its proindustry orientation and industry sponsorship. See, e.g., "Re: Regulatory Toxicology and Pharmacology," 19 November 2002, https://www.cspinet.org/new/pdf/final_letter_academic_press_rtp.pdf.

34. Michaels, *Doubt Is Their Product*, 50–52.

35. Oreskes and Conway, *Merchants of Doubt*, 12–13.

36. Bruce Owen and Ronald Braeutigam, *The Regulation Game* (New York: HarperCollins, 1978), 7.

37. Oreskes and Conway, *Merchants of Doubt*, 85–94, 126–30, 143.

38. Ibid., 248–55.

39. See, e.g., Cranor, *Toxic Torts*.

40. These two approaches are also discussed in Kevin Elliott, "Financial Conflicts of Interest and Criteria for Research Credibility," *Erkenntnis*, 79 (2014): 917–937.

41. Shrader-Frechette, *Taking Action, Saving Lives*, 83. Calls for greater transparency about sources of research funding can be found in McGarity and Wagner, *Bending Science*, 236; and Michaels, *Doubt Is Their Product*, 245. For transparency in research-misconduct allegations, see Kristin Shrader-Frechette, "Research Integrity and Conflicts of Interest: The Case of Unethical Research-Misconduct Charges Filed by Edward Calabrese," *Accountability in Research* 19 (2012): 220 42. Evidence that disclosure of conflicts without further steps may be fairly ineffectual can be found in Kevin Elliott, *Is a Little Pollution Good for You? Incorporating Societal Values in Environmental Research* (New York: Oxford University Press, 2011), 94–95; and Elliott, "Scientific Judgment and the Limits of Conflict-of-Interest Policies," *Accountability in Research* 15 (2008): 1–29. For calls for making privately funded data available, see McGarity and Wagner, *Bending Science*, 241–46; and Michaels, *Doubt Is Their Product*, 251. For a general ethical argument for making environmental research data available to the public, see Pat Soranno, Kendra Cheruvelil, Kevin Elliott, and Georgina Montgomery, "It's Good to Share: Why Environmental Scientists' Ethics Are Out of Date," *BioScience* 65 (2015): 69–73. On requiring disclosure of adverse information, see McGarity and Wagner, *Bending Science*, 246–51; and Michaels, *Doubt Is Their Product*, 249–50.

42. For more information about having a government agency perform regulatory studies, see Elliott, "Financial Conflicts of Interest"; and David Volz and Kevin Elliott, "Mitigating Conflicts of Interest in Chemical Safety Testing," *Environmental Science and Technology* 46 (2012): 7937–38. For calls for public funding of policy-relevant science, see, e.g., Elliott, *Is a Little Pollution Good for You?*, 102–13. For discussion of collaborative approaches, see Elliott, "Financial Conflicts of Interest."

43. See Elliott, "Financial Conflicts of Interest"; and Michaels, *Doubt Is Their Product*, 257–58.

44. Roger Pielke Jr., *The Climate Fix: What Scientists and Politicians Won't Tell You about Global Warming* (New York: Basic Books, 2010); Daniel Sarewitz, "How Science Makes Environmental Controversies Worse," *Environmental Science & Policy* 7 (2004): 385–403; Mary O'Brien, "Alternatives Assessment: Part of Operationalizing and Institutionalizing the Precautionary Principle," in *Protecting Public Health and the Environment: Implementing the Precautionary Principle*, ed. Carolyn Raffensperger and Joel Tickner (Washington, DC: Island Press, 1999), 207–19; Joel Tickner, "A Map toward Precautionary Decision Making," in ibid., 162–86.

PART II / Mythologizing Ignorance

6

Class

DANIEL PERLSTEIN

In the 1930s the public schools of Arthurdale, West Virginia, were among the most famous in the United States.[1] The political and economic crisis of the Great Depression had revitalized progressive education in America, and Arthurdale, a new community for refugees from West Virginia's coal fields, offered an ideal setting in which to realize progressivism's pedagogical and democratic ideals. There, educators sought to demonstrate that a progressive curriculum of active learning that built on students' interest and experience could serve even those students whose imaginations and capacity for self-direction had been distorted by poverty and alienation. John Dewey thought the schools possessed "extraordinary significance for education," Eleanor Roosevelt championed them, and countless educational pilgrims made their way to Arthurdale.[2]

Although the area around what would become Arthurdale sat on one of America's richest coal veins, in the 1920s the coal boom created by World War I had turned to bust. Unemployed miners and their families survived in paper-walled shacks located near fetid, sewage-filled creeks. Hunger, disease, violence, and stunted growth were children's common experience.[3] "Hopelessness and despair," observed Fisk University economist Homer Morris, settled on the mine camps "as a dense smoke-laden fog." Their inhabitants, in the eyes of Eleanor Roosevelt, "were like people walking around dead. They were alive, but they were dead as far as any real living was concerned."[4]

Soon after entering the White House, Franklin Delano Roosevelt initiated the creation of Arthurdale and other federal "subsistence homesteads." New Deal reformers anticipated that at Arthurdale unemployed miners would be hired to clear the land and build their own homes and community buildings and then support themselves by farming small plots and working part time in a factory. Therefore, government action would not only bring relief to dispossessed Americans but also restore their sense of dignity and self-direction through what Roosevelt termed a new "rural industrial" way of American life.[5]

Although school policy was marginal to the New Deal, Arthurdale was a notable exception. Dispossessed refugees from the mines, Eleanor Roosevelt told a resettlement conference, "must be taught to live." To accomplish this, the project would offer "a social experiment in community life, which centers around its school."[6]

In a nation where public housing, relief, and schooling for the poor have often been spartan and punitive, Arthurdale was exceptionally generous and humane. Grounded in the premise that the poor had no less a right than the rich to lives of beauty, dignity, freedom, and grace, Arthurdale's effort to adapt progressive education to the poor constituted a landmark in the history of American education.

Progressive education has not been without critics. They have argued that its neglect of disciplinary knowledge fostered anti-intellectualism and lowered academic standards, that its avoidance of direct instruction constituted a bias toward more privileged students, that it was rooted in a racist theory of history, and that its claims of self-direction articulated new forms of discipline that reinforced corporate power.[7]

This chapter adds an agnotological critique to such criticisms. It argues that a mythic vision of pioneering shaped progressive education, distorting its core enactments of self-direction and activity and serving more to discourage than to foster thinking about the central conflicts of social and economic life. While Arthurdale served working-class people, it did so by encouraging them to forget their experience of working-class life and to ignore class relations and modern economic life.

Progressivism, Homesteading, and the New Deal

Progressive education emerged in the late nineteenth century as part of a broader reform movement. A complex amalgamation of social welfare, social solidarity, and social control, progressivism constituted an accommodation of industrial capitalism as well as a critique of it. Although the New Deal dramatically increased the administrative capacity of the American state, its relationship to progressivism represented continuity more than change.[8]

Just as New Dealers echoed progressive ideals, progressive educators in turn appealed to frontier-myth images of nineteenth-century, self-directed pioneers, who had lit out to the frontier to remake their lives and communities. In the twentieth century, countless American urbanites and intellectuals continued to envision new rural communities as a means of restoring individual autonomy and interpersonal connection in the face of the alienation and economic divisions of modern urban life.

Modern calls to move "back to the land" combined nostalgic agrarian yearnings and humanitarian notions of relief with new forms of social engineering. Leading resettlement advocate and irrigation engineer Elwood Mead, as historian Paul Conkin argues, personified the dream "of combining social reclamation with land reclamation, social engineering with irrigation engineering." The idea was not to have individual homesteaders stampeding into newly opened territories. Rather, modern resettlements were to be carefully designed in advance through the bureaucratized selection of settlers, expert planning of community facilities, and scientific direction of farming. Back-to-the-land advocates saw no conflict in simultaneously seeking individual autonomy and a vastly expanded state. They called for "government paternalism" that would catalyze "a movement comparable to pioneer days" on the "new frontier."[9] In the 1930s, American intellectuals and reformers extolled rural life as a shield against economic and social disaster. Arthurdale reflected their vision of a new, bureaucratized manifestation of America's pioneer spirit.[10]

A New Deal for Progressive Education

Progressive education reflected many of the same ideals as the resettlement movement. Progressives typically argued that education should build on and extend students' experience and intrinsic motivation or interest. It should do so through active "learning by doing" that connected classrooms to real life tasks, rather than rewarding the acquisition of discipline-based knowledge.[11]

For John Dewey and other progressive educators, the entirety of children's experience formed the basis of learning. If that learning was judged by students' ability to impact their experience and the environment in which it occurred, then education and social action became a mutually constituting, iterative process: learning shaped what one did and this "doing" in turn shaped what one learned.

Like back-to-the-landers, progressive educators looked to America's rural past to remake its present. Pioneer life made manifest both the ways humans have met their fundamental needs and the self-direction and social solidarity progressives wished to revitalize in contemporary life. Dewey argued that the educational value of pioneer living far surpassed that of the American school. The homestead was, he wrote,

> the center in which were carried on, or about which were clustered, all the typical forms of industrial occupation. The clothing worn was for the most part not only made in the house, but the members of the household were usually familiar with the shearing of the sheep, the carding and spinning of the wool, and the plying of the loom. . . . The entire industrial process stood revealed, from the production on the farm of the raw materials, till the finished article was actually put to use. Not only this, but practically every member of the household had his own share in the work. . . . In all this there was continual training of observation, of ingenuity, constructive imagination, of logical thought, and . . . training in habits of order and of industry, and in the idea of responsibility, of obligation to do something, to produce something, in the world.[12]

By reenacting the "typical relations of humanity" with the simplified and observable methods of an earlier age, Dewey argued, children could begin to develop a sense of the social processes through which their own needs were met. As they studied mathematics through carpentry, biology through growing plants, and chemistry through dying fabrics, students explored the evolving ways humans collectively answered their basic needs for "food, shelter, habitation, clothing, and industrial occupations."[13]

Progressive educators' focus on pioneer America reflected their understanding of social and economic life. They argued that the commodification of industrial goods and the specialization of labor required for large-scale manufacturing rendered people, in Dewey's words, ignorant of "the technical, intellectual and social relationships" by which their most basic needs were fulfilled. Like traditional classrooms, in which the extrinsic reward of a grade supplied the main motivation to master classroom tasks, modern industry deprived workers of a sense of intrinsic purpose in their work. Rendered ignorant of the meaning of their work, people in related occupations developed group ambitions and class-consciousness. Social conflicts overwhelmed the self-sufficiency and cohesiveness that had characterized rural communities. Feelings of class identity and division, according to this view, reflected limited consciousness and therefore might be overcome by education, which simultaneously fostered a harmonious society and promoted individual growth.[14]

In the 1934 inaugural issue of the flagship journal of progressive social reconstructionists, aptly named *The Social Frontier*, Charles Beard quoted Noah

Webster to celebrate pioneering's democratic power. When "New England pioneers 'came to a new country . . . [whose] whole soil was unreclaimed from barbarism,'" Beard argued, they heralded the Jeffersonian ideal that America should be "an agricultural country, that the land should be parceled out among independent farmers, each looking to the soil at his feet, the labor of his hands, and the sun in heaven for his livelihood."[15]

Progressive educators advocated lessons about the "here and now" rather than vague abstractions and generalities. "The most direct way of vitalizing [schools'] work," John Dewey and Evelyn Dewey argued, was "through closer relations with local interests and occupations . . . to meet the special needs and definite features of local life." And yet, they saw pioneer life as the crystallization of contemporary conditions rather than as a retreat from real life. This celebration of pioneering epitomized progressive educators' belief that focus on the activities and techniques through which needs were met, rather than on the broader social and economic relations entailed by systems of production, would foster the creation of a more harmonious society. Amid the political and economic crises and conflicts of the 1930s, many educators and activists became convinced that learning that fostered and relied upon the capacity to transform one's immediate environment was especially suited to the interests of working-class students and could help them build a new, cooperative social order.[16]

In response to George Counts's famed 1932 call for educators to expose the bankruptcy of the capitalist system through lessons critiquing the social and economic order, educator and Dewey protégée Elsie Clapp countered that people's commitment to "social reconstruction" was "not produced by imparting information about different conditions or by gathering statistical data about what exists." Rather, "the process of living and learning and doing" was itself "revolutionizing." Arthurdale was the expression of this hope that new communities centered on progressive schools could revitalize America's pioneer spirit.[17]

Progressive Education at Arthurdale

Reformers chose Elsie Clapp to serve as the head of Arthurdale School and director of community affairs. Born to a privileged family in 1879, Clapp was a sophomore at Vassar College when Dewey's *School and Society* left her spellbound by the vision it described. After studying under Dewey and then serving as his teaching assistant at Columbia University, Clapp taught at a number of elite private progressive schools. She was also active in campaigns for women's and workers' rights. In 1929 she became a rural high school principal, and she gained a national reputation as an authority on progressive education in rural communities.[18]

People left stranded in the coal camps, Clapp observed, "are unlikely to have come off unscathed." In order to foster individual health and community revitalization, the school program had to include more than academic work or even the provision of assorted services. Instead, the school served as the center of community life. It sponsored community clubs, dances, and health clinics. Mothers learned about diet and nutrition as they prepared meals in the school kitchen. The high school shop teacher answered questions about the care and use of tools; the biology teacher guided men unfamiliar with farming; the math teacher helped manage the cooperative store and market. The joint endeavor to meet people's needs through canning lessons or improved hygiene led to specific improvements, but in Clapp's eyes community schooling was above all the act of sharing the life and interests of community; it unified educational experience with participation in civic and social life. "Acting on the conviction that education is equivalent to living and learning," she reported to the United States Resettlement Administration, teachers "are themselves members of the community, living in it, participating in every phase of its life and activity. They function as community educators as well as school teachers, dealing with adults directly as well as through their children, re-conditioning their thinking by helping them realize their problems and assisting in their work upon them." Schooling was transformed from a public agency delivering a service into a cooperative enterprise.[19]

For Arthurdale's children, school began in the fall of 1934. They studied village life even as the Arthurdale village was being constructed just beyond the schoolhouse doors. On frequent field trips, they made observations of their fathers and other homesteaders planting crops, threshing buckwheat, and building houses and barns. And as their parents answered questions about what they were doing, the adults became teachers. While a portion of the school day was devoted to arithmetic, reading, and writing lessons, formal study built on and provided tools for students' explorations of their environment, rather than field trips serving as enrichment activity at the margins of the classroom. Students' records, Clapp concluded, made "clear how much and how quickly the children learned from activities going on around them and from what their families were making and doing."[20]

Children's exploration of their new surroundings not only served as the basis for academic achievement; it helped heal the scars of life in the mining camps and served as an introduction to cooperation in community life. When they first arrived at the resettlement, Clapp observed, the children experienced "recurrent periods of grieving and fear and lassitude that had no apparent immediate cause." But after they lived in Arthurdale for a while, things changed. "Negative thoughts and attitudes," Clapp maintained, "do not flourish here. This is virgin soil, and tomorrow is another day."[21]

At Arthurdale, Clapp focused the curriculum on the same pioneer activities whose description had enthralled her as a college student reading Dewey's *School and Society*. Studies of pioneer life centered on an old log cabin on the Arthurdale site. There, children learned to dye wool with pokeberries, walnuts, and acorns, to dip candles, and to cook pioneer style. They built fiddles and other instruments as they studied Appalachian history, music, and culture. With the help of high school students and adults from the Mountaineer Craftsman's Cooperative, they made split-log benches, tables, a cradle, and a churn. Pioneer diaries served as texts for reading instruction and inspired the creation of dramas about pioneer life; through their "flax notebooks," students developed their writing. Repairing the cabin and re-creating the pioneer experience, Clapp argued, children learned carpentry and arithmetic in the process of "solv[ing] problems that dealt with real things."[22]

Despite their commitment to building education on students' experiences, educators' celebration of pioneer life constituted an agnotological curriculum that did more to distract students from reflecting on their actual lives than to clarify current conditions. Like the back-to-the-landers, progressive educators found pioneer living more authentic than contemporary society. Hinting at the way claims about economic life intersected with vaguer cultural anxieties, Clapp contrasted the ephemeral nature of "the things 'cash money' could buy" with "the enduring means of existence" manifest in "the homely implements and utensils for planting crops, making barns and houses and furniture, and cooking and canning" and re-created in Arthurdale schools.[23]

"Pioneer study," as fourth-grade teacher Elisabeth Sheffield argued, not only illuminated the past; it

> interprets present-day living to these children. They are living in a time long after the industrial revolution. They press a button for light, buy shoes or a dress—or, at least, goods—at a store. They take it all for granted. But after they

have carded, spun, and woven wool, they start to understand more about materials; after they have tanned a hide, they begin to know what goes into the making of a pair of shoes; after they have cooked over an open fire, a stove has more value for them."[24]

Pioneer life was meant to foreshadow students' own experience. When the children "studied about the experience of the settlers who came into West Virginia," Clapp believed, "they were in effect learning about their own families' pioneering in taking up a homestead and starting life again at Arthurdale." Even faced with the systemic dislocations of the Depression, Arthurdale's educators looked to rural, preindustrial frontier life as the embodiment of America's cooperative, democratic spirit. Educators imagined that reanimating that spirit would enable Arthurdale residents "to cope successfully with their immediate problems and develop their abilities in the direction of their own affairs."[25]

Community, Progressive Education, and the Experience of Inequality

The ignoring of class antagonism through invocations of the frontier shaped not only what was studied and done in Arthurdale but also what was not studied and not done. A curriculum based on activity, interest, local conditions, and social life might have led Arthurdale educators to focus on the structure of the coal industry or the impact of the Depression on students' recent misery and the continuing misery of their neighbors.

Arthurdale's creators were familiar with the political economy of coal, but in their commitment to revitalizing refugees from the coal fields they both overstated the degree to which homesteaders had been incapacitated by life in the mining camps and obscured the causes of their dispossession. Arthurdale drew many of its residents from nearby Scott's Run. As Eleanor Roosevelt acknowledged, it had earned the name "Bloody Run" for the militancy of its miners and their strikes. In the summer of 1932 a thousand men and women had marched from Scott's Run to Morgantown, where they had faced off against state police and deputies armed with tear gas and automatic rifles. Among the First Lady's justifications of federal intervention in the area was that without some indication of government interest and concern, conditions in the mining camps were "nearly the kind that breed revolution."[26]

Focused on the activities through which pioneers met their daily needs, Arthurdale students did not study the common coal company practice of paying workers in scrip rather than cash, limiting them to company stores that locked them into permanent debt. Those who considered shopping at nearby stores with lower prices were often threatened with termination. "The coal company became the ruling body of the people I'll call 'slaves,'" former Scott's Run inhabitant Sidney Lee argued. It "controlled the entire life of the miner and his family. . . . How could any stable form of human life be built upon such an insecure foundation?" When owners closed the mines, miners' families were left "stranded" in the camps not so much by the psychic scars of camp life as by the intended workings of the camp economy, which left them destitute.[27]

Arthurdale's curriculum also left students ignorant of the state's role in enforcing corporate domination. They might have considered what activities might have been appropriate in the case of two teenage boys from a mine camp who broke into a store and stole food because they had none at home. (A judge lectured them on the virtues of honesty and the wickedness of the young before sentencing them to two years in prison. He had sentenced camp residents in twenty cases for similar "crimes.")[28]

Arthurdale educators had a ready model of class-conscious progressive education. In November 1932, at a time when Clapp served on the journal's advisory board, *Progressive Education* described West Virginia summer play schools of the socialist children's organization Pioneer Youth of America, established with the cooperation of the West Virginia Mine Workers Union and the League for Industrial Democracy. Project educators reasoned that children could not help but reflect on hunger, eviction, police power in defense of class privilege, and the need for a new social order. Topics addressed through child-centered activities and exploration of the built environment included wage cutting, strikes, evictions, unemployment, and social change. At Arthurdale, on the other hand, lessons on West Virginia's extractive industries turned students' attention to a 1790 iron furnace; the word *coal* was scarcely mentioned.[29]

Like progressive reformers more broadly, Arthurdale's creators, in their commitment to fostering social harmony, assumed the very harmony they sought and so excluded working-class activists from the project. Government officials screened Arthurdale applicants for church attendance, sobriety, "moral character . . . intelligence, perseverance and foresight." They even demanded references from former bosses in order to assess financial prudence and reliability at work.[30]

Community, Culture, and Agnotology

Just as Arthurdale introduced the study of modern industrial life through the exploration of archaic, small-scale technologies of production, it encouraged cultural activity and appreciation by focusing on preindustrial West Virginian folk traditions. Instead of focusing on industrial working-class life or modern popular media, Clapp promoted a sentimental and sanitized version of local culture. She argued that "knowledge must be built slowly by extension of their experience and understanding." And yet, Arthurdale teachers discounted the actual experience of students and families. Life in the coal camps, Clapp reasoned,

> had obscured to [Arthurdale residents] their common heritage; and being in the shadow of urban Morgantown, they had been disturbed by the radio, the movies, and by bourgeois cultural standards. If the homesteaders' sojourn in the mines had been much longer, there might have been at Arthurdale no basis for culture; but since they had grown up on farms, and had in Arthurdale come back to the land, their cultural roots were hardy and were beginning to find again a non-sulfuric and fostering soil.[31]

Arthurdale music and drama director, Fletcher Collins, deemed the influence of the coal mines on residents' culture "infinitesimal compared with that of historic tradition . . . passed down to them . . . by generations of their ancestors." Noting residents' English and Scotch-Irish origins, Collins traced their tradition to the Celtic and Germanic elements of western European civilization. Embodied in the old English ballads they sang, the square dances they held, and the fiddles they made and played, "historical tradition is certainly for them not an escape from the present; it is rather a background for their lives, an intuitive recognition of its importance to them as an underlying bed rock, as an orientation, as a form of security." "Once articulated," Collins concluded, this tradition "can be communicated, and homogeneity on a regional or national scale can follow."[32]

Arthurdale educators' celebration of pioneering and of pioneers' imagined culture encouraged students to ignore class conflicts and relations in multiple ways. The first white settler in what would become Arthurdale came to establish a plantation. John Fairfax, whom George Washington advised to buy the tract, surrounded his own house with cabins for his slaves, one of which survived. Instead of introducing the study of slavery, Old Watt's cabin served as "background to the pioneer study program," as Elisabeth Sheffield suggested, "satisfying a certain craving [students] had, just coming from the desolate mine camps, for possessing something of their own. . . . Their heritage is recognized. . . . [They can be] conscious and proud of [their] past."[33]

Celebrations of pioneer life, Elsie Clapp echoed, enabled adults to "recall . . . the life they had known on farms long ago in their childhood." The slave cabin that served as a focal point of their children's pioneer study "assured them that the life they had left when they went into the mines was theirs still." Beneath the surface of coal camp despondency "in this mine territory in the Appalachian region," Clapp claimed, "we have discovered the foundation for an indigenous culture."[34]

Arthurdale's history of slave labor was not so much forgotten as purged in what historian Adam Hochschild calls an "active deed" of "deliberate forgetting." Clapp imagined—and taught her students—that Fairfax had "built" Old Watt's cabin, though it seems unlikely that the master would do the manual labor for the slave. One student wrote that the area that became Arthurdale "was all woods, except that there was a little cleared space where the cabin was." "The earliest houses around Arthurdale," another reflected, "were undoubtedly log cabins." A student described his grandfather building a cabin much like Old Watt's. The cabin helped students and teachers remake slavery in the image of the pioneer. In so doing, it helped working-class whites ignore the racial privilege that moderated their class oppression.[35]

The invocation of common cultural bonds—fiddle music, square dances, and the like—was a way for working-class whites to deploy, but not notice, their whiteness. Just as Arthurdale excluded applicants who had the wrong attitude toward work, it excluded those with the wrong origins. A quarter of the applicants to become homesteaders were black, many of them native West Virginians and farmers, but only native-born whites were selected for Arthurdale. Almost all were Scotch-Irish or Pennsylvania German in "upbringing and racial background," predisposing them, Clapp argued, to generosity and good humor. Black activists launched an unsuccessful national campaign to integrate Arthurdale. Given the importance progressives attached to education as a means of revealing and strengthening social bonds across superficial differences, Deweyan reformers might have followed the lead of black activists and sought to mitigate white hostility rather than to accommodate it. Instead, the invocation of pioneering and white ethnic culture elevated racial difference above working-class solidarity.[36]

Community Schooling in National Context

Arthurdale served as the model for community-focused progressive schools in marginalized communities across the United States. In Arthurdale itself, however, progressive education proved short-lived. Community schooling depended on the willingness of teachers to work long hours in uncertain conditions. Repudiating commercially produced materials, progressive theorists envisioned teachers constructing lessons from local resources, an ideal of teaching as preindustrial craft work that mirrored back-to-the-land celebrations of pioneering. High teacher turnover undermined progressive practices. Moreover, even when educational activities enabled students to gain familiarity with productive processes and the science that underlay them, students' ability to apply that understanding in the real world depended upon economic forces that the homestead could not control.[37]

Ironically, the very commitment to fostering residents' capacity to articulate their desires contributed to the eclipse of community schooling. As the initial glow of resettlement wore thin, malaise set in. Although cooperative practices were nurtured in Arthurdale, policy was made in Washington. Like the Roosevelts and other Americans, residents believed in the myth of pioneer yeomanry. Homesteaders, to Clapp's dismay, "furthered personal ends" as much as the common good. They wanted the homes the government funded, but they also wanted the government out of their lives. Although progressive educators claimed to build on students' experience, they ignored parents' experience of school. Arthurdale's approach, Eleanor Roosevelt's confidante Marion Dickerman would recall, "was a bit too progressive for the homesteaders. They began to feel that their children were not being taught the fundamentals—the three r's—as they should be." While progressive educators envisioned children making meaningful lives in the resettlement community, parents envisioned schooling that would open up the world to their children. As residents' hopes grew, their notion of connecting learning to life included their children having access to college, a stance requiring traditional, academic discipline–based courses and not the idiosyncratic community-focused curriculum.[38]

In the end, though, Arthurdale's curriculum did not determine the project's fate. Although Arthurdale confirmed for John Dewey the value of education based on encounters with matters "local, present, and close by," not the "pale and shadowy" abstractions that often passed for social study, the history of Arthurdale itself could not be adequately explored locally and up close.[39]

As socialist activist and Eleanor Roosevelt biographer Joseph Lash has put it, Arthurdale's creators refused "to subordinate human values to cost consciousness." The project's expansive educational program cost substantially more than typical public schools. Reformers considered such luxuries as landscaping with flowering shrubs and indoor toilets fitting even for the poor. Residents were paid a living wage for their work building the community. Right-wing critics seized on early instances of project mismanagement to combat all New Deal programs. The project was the butt of reporters' jokes for years. A continent away from West Virginia, the *Los Angeles Examiner* denounced Arthurdale as part of its campaign to discredit socialist Upton Sinclair's gubernatorial campaign in California.[40]

Arthurdale's schools, in the eyes of financier and philanthropist Bernard Baruch, had demonstrated "what can be done especially in the way of salvaging and redeeming old and young people." Still, Baruch stressed that the premise that homesteaders would regain self-sufficiency and dignity rested on economic conditions beyond their control. No private company was willing to move its factory to Arthurdale, and Congress blocked the construction of a post office furniture factory there lest the government create competition for private companies. As the project's political and economic costs grew, Arthurdale lost the support of Franklin Roosevelt and other key policymakers. In 1936 Clapp and many of the progressive teachers left, the progressive community-school model was abandoned, and gradually progressive practices faded.[41]

Unlearning by Doing

Arthurdale and its schools allowed displaced miners and their families to see that they could remake their individual and collective lives, and Arthurdale, through the agency of its schools, gave hope to the dispirited. Community schooling brought the social nature of activity into the structure and curriculum of the school and enriched the lives of all Arthurdale's residents. Through the efforts of Clapp and her fellow teachers, homesteaders achieved a level of consciousness about communal life that gave real substance to invocations of local democracy. Schooling in Arthurdale fostered individual growth and revitalized collective intellectual and social life among people whose lives had been devastated by the 1930s crisis of capitalist production. It demonstrated that self-directed and self-actualizing activity rooted in progressive theory and practice could inform the education of impoverished, marginalized youth.

And yet, that same pedagogy created curricular silences about class in the evolution of Arthurdale and the broader society in which it was situated. The focus on activity at the center of pioneer study distracted attention from the relations of production. Industrial workers and slaves were not alienated, because they lacked knowledge about the place of their work in the production process. Exploration of the unfolding ways humans met their needs could not illuminate the exploitation they experienced. Rather, the celebration of pioneer economic and cultural life and the attempt to re-create it in Depression-era America served to discourage students' engagement with their most pressing political and economic questions.

Whatever blind spots progressive community-school teachers possessed, they created a far more democratic model of school than was, or is, the norm. The model relied on a romanticized, mythical vision of community bonds that effaced class conflicts in American life. Its genuinely humanizing and democratizing potential was mirrored by the silences by which it reinforced relations of domination.

At Arthurdale, the reinvention of the past served efforts to imagine a less contentious present. This active unlearning required sustained effort. As Adam Hochschild suggests, such deliberate forgetting "is not something passive; it is an active deed. . . . It is not a moment of erasure, but a turning of things upside down, a strange reversal of the victimizer mentally converting himself to victim." And yet, Arthurdale's white working-class residents were both the objects and the subjects of this process, both victims and beneficiaries. Just as elite reformers constructed a pioneer narrative that ignored the formative influence of class relations and oppression in the lives of miners, white miners used that narrative to simultaneously ignore and deploy their racial privilege. As New Deal notions of pioneer autonomy overwhelmed New Deal visions of an expansive welfare state, the federal government sold the project's homes to residents at greatly discounted prices, creating a pleasant white town. In short, "selective forgetting," as literary theorist Roland Barthes suggests, is a form of "distractive" history in which "presence and absence" intermingle. In this process, Norman Klein concludes, "fictions are built into facts" and "distraction dominates a story."[42]

Invocations of self-direction, learning by doing, and education connected to the practical tasks of real life possess an appeal that significantly influenced twentieth-century educational theory and practice. They continue to foster our most democratic pedagogies. The history of Arthurdale points to the need to imagine democratic communities not in an idealized past but in a transformed future. Such a project cannot discourage students from thinking about the complexities of class; it must encourage them.

NOTES

1. Portions of this chapter are adapted from Daniel Perlstein, "Community and Democracy in American Schools: Arthurdale and the Fate of Progressive Education," *Teachers College Record* 97 (1996): 625–50 (copyright 1996 by Teachers College Record, www.tcrecord.org, and used by permission); and Daniel Perlstein and Sam Stack Jr., "Building a New Deal Community: Progressive Education at Arthurdale," in *"Schools of Tomorrow," Schools of Today: What Happened to Progressive Education?*, ed. Susan Semel and Alan Sadovnik (New York: Peter Lang, 2005), 213–36.

2. John Dewey, foreword to *Community Schools in Action*, by Elsie Clapp (New York: Viking, 1939), vii; Blanche Wiesen Cook, *Eleanor Roosevelt*, vol. 2, *The Defining Years, 1933–1938* (New York: Viking, 1999), 139–52; Clapp, *Community Schools in Action*, 190.

3. C. J. Maloney, *Back to the Land: Arthurdale, FDR's New Deal, and the Costs of Economic Planning* (Hoboken, NJ: John Wiley & Sons, 2011), 40–56; Elsie Clapp, *The Use of Resources in Education, Part Two, Educational Use of Resources in West Virginia, at Arthurdale* (New York: Harper & Brothers, 1952); Perlstein, "Community and Democracy in American Schools," 626.

4. Homer Morris, *The Plight of the Bituminous Coal Miner* (Philadelphia: University of Pennsylvania Press, 1934), 110; Eleanor Roosevelt, *The White House Press Conferences of Eleanor Roosevelt* (New York: Garland, 1983), 115.

5. Elsie Clapp, *Arthurdale—a School* (Washington, DC: United States Resettlement Administration, 1936), 1; Franklin Roosevelt, "Actualities of Agricultural Planning," in *America Faces the Future*, ed. Charles Beard (Boston: Houghton Mifflin, 1932), 348; Clapp, *Use of Resources*, 6. For a critique of the resettlement program, see Diane Ghirardo, *Building New Communities: New Deal America and Fascist Italy* (Princeton, NJ: Princeton University Press, 1989).

6. Clapp, *Community Schools in Action*, 115–16; Clapp, *Use of Resources*, 23; E. Roosevelt, *White House Press Conferences*, 115; Joseph Lash, *Eleanor and Franklin* (New York: New American Library, 1973), 541. On the New Deal's limited attention to education, see Harvey Kantor and Robert Lowe, "Class, Race, and the Emergence of Federal Education Policy: From the New Deal to the Great Society," *Educational Researcher* 24 (April 1995): 6–7.

7. On lowered academic standards, see David Angus and Jeffrey Mirel, *The Failed Promise of the American High School, 1890–1995* (New York: Teachers College Press, 1999). On bias toward privileged students, see Basil Bernstein, "Class and Pedagogies: Visible and Invisible," in *Class, Codes and Control*, vol. 3 (London: Routledge & K. Paul, 1975), 116–56; and Lisa Delpit, "Skills and Other Dilemmas of a Progressive Black Educator," *American Educator* 20 (Fall 1996): 10–11. On progressive education's theory of history, see Thomas Fallace, "Repeating the Race Experience: John Dewey and the History Curriculum at the University of Chicago Laboratory School," *Curriculum Inquiry* 39 (2009): 396; and Frank Margonis, "John Dewey's Racialized Visions of the Student and Classroom," *Educational Theory* 59 (2009): 17–39. On self-discipline, see Thomas Popkowitz, "Dewey, Vygotsky, and the Social Administration of the

Individual: Constructivist Pedagogy as Systems of Ideas in Historical Spaces," *American Educational Research Journal* 35 (1998): 535–70. On corporate power, see Clarence Karier, "John Dewey and the New Liberalism: Some Reflections and Responses," *History of Education Quarterly* 15 (1975): 438; and Karier, "Liberalism and the Quest for Orderly Change," ibid. 12 (1972): 60.

8. Michael McGerr, *A Fierce Discontent: The Rise and Fall of the Progressive Movement in America, 1870–1920* (New York: Oxford University Press, 2005); Daniel Rodgers, *Atlantic Crossings: Social Politics in a Progressive Age* (Cambridge, MA: Harvard University Press, 1998); Shelton Stromquist, *Re-Inventing "The People": The Progressive Movement, the Class Problem, and the Origins of Modern Liberalism* (Urbana: University of Illinois Press, 2006); Patrick Reagan, *Designing a New America: The Origins of New Deal Planning, 1890–1943* (Amherst: University of Massachusetts Press, 1999).

9. Paul Conkin, "The Vision of Elwood Mead," *Agricultural History* 34 (1960): 90; Maloney, *Back to the Land*, 61; M. Wilson, "The Place of Subsistence Homesteads in Our National Economy," *Journal of Farm Economics* 16 (1934): 81.

10. Dona Brown, *Back to the Land: The Enduring Dream of Self-Sufficiency in Modern America* (Madison: University of Wisconsin Press, 2011).

11. John Dewey and Evelyn Dewey, *Schools of To-morrow* (New York: Dutton, 1915), 250, 292.

12. John Dewey, *The School and Society and The Child and the Curriculum* (Chicago: University of Chicago Press, 1900), 22–24.

13. Katherine Camp Mayhew and Anna Camp Edwards, *The Dewey School: The Laboratory School of the University of Chicago, 1896–1903 (New York: D. Appleton-Century Company, 1936)*, 29–30.

14. John Dewey, *Democracy and Education: An Introduction to the Philosophy of Education* (New York: Macmillan, 1916), 98; Evelyn Dewey, *New Schools for Old: The Regeneration of the Porter School* (New York: Dutton, 1919).

15. Charles Beard, "Property and Democracy," *Social Frontier* 1 (October 1934): 14.

16. Lucy Sprague Mitchell, *Here and Now Story Book* (New York: Dutton, 1921); J. Dewey, foreword, viii; Dewey and Dewey, *Schools of To-morrow*, 205–6.

17. George Counts, "Dare Progressive Education Be Progressive?," *Progressive Education* 9 (1932): 261; Elsie Clapp, "Learning and Indoctrination," ibid., 270; Clapp, *Use of Resources*, 14–16; Clapp, *Community Schools in Action*, 67, 88.

18. Sam Stack Jr., *Elsie Ripley Clapp (1879–1965): Her Life and the Community School* (New York: Peter Lang, 2004), 92–100; Clapp, *Community Schools in Action*, 67.

19. Perlstein, "Community and Democracy in American Schools," 629–30; Clapp, *Arthurdale—a School*, 9.

20. Charles Pynchon, "School As Social Center," *New York Times*, 5 May 1935, sec. 9, 23; Clapp, *Use of Resources*, 23, 30–35; Stack, *Elsie Ripley Clapp*, 193; Perlstein and Stack, "Building a New Deal Community," 218.

21. Clapp, *Use of Resources*, 34–35, 97; Pynchon, "School As Social Center."

22. Clapp, *Use of Resources*, 44; Clapp, *Community Schools in Action*, 145; Stack, *Elsie Ripley Clapp*, 195.

23. Clapp, *Use of Resources*, 41; Clapp, *Community Schools in Action*, 251.

24. Clapp, *Community Schools in Action*, 143.

25. Clapp, *Use of Resources*, 41; Clapp, *Arthurdale—a School*, 25, 28.

26. Eleanor Roosevelt, *This I Remember* (New York: Harper & Brothers, 1949), 126; Mauritz Hallgren, *Seeds of Revolt: A Study of American Life and the Temper of the American People during the Depression* (New York: Knopf, 1933), 58–59; William Brooks, "Arthur-

dale—a New Chance," *Atlantic Monthly*, February 1935, 199; Eleanor Roosevelt, *If You Ask Me* (New York: Appleton-Century, 1946), 27; Clarence Pickett, *For More Than Bread: An Autobiographical Account of Twenty-Two Years' Work with the American Friends Service Committee* (Boston: Little, Brown, 1953), 47, 62–63, 66; Lash, Eleanor and *Franklin*, 528.

27. Maloney, *Back to the Land*, 14; Morris, *Plight of the Bituminous Coal Miner*, 100; Sidney Lee, *"and the trees cried"* (Morgantown, WV: privately printed, 1991), 45–46.

28. Morris, *Plight of the Bituminous Coal Miner*, 114.

29. Agnes Sailer, "Miners' Children at Play," *Progressive Education* 9 (1932): 507–8; Clapp, *Community Schools in Action*, 160.

30. Tamara Hareven, *Eleanor Roosevelt: An American Conscience* (New York: Da Capo, 1975), 95; Roosevelt, "Homestead—W. Virginia, 1935" (manuscript intended for Liberty Magazine), 3, Eleanor Roosevelt Papers, Franklin Delano Roosevelt Presidential Library, Hyde Park, NY, Box 3032; Cook, *Eleanor Roosevelt*, 138–39.

31. Clapp, *Arthurdale—a School*, 20; Clapp, *Use of Resources*, 41.

32. Clapp, *Arthurdale—a School*, 25–26.

33. Clapp, *Community Schools in Action*, 143–45, 252.

34. Clapp, *Use of Resources*, 41; Clapp, *Arthurdale—a School*, 25.

35. Adam Hochschild, *King Leopold's Ghost: A Story of Greed, Terror, and Heroism in Colonial Africa* (New York: Houghton Mifflin, 1999), 295; Clapp, *Community Schools in Action*, 251, 253.

36. Perlstein, "Community and Democracy in American Schools," 637, 639. Eleanor Roosevelt was one of the few whites to actively support the black activists' Arthurdale campaign. Cook, *Eleanor Roosevelt*, 139–43. For a fuller account of Arthurdale's racial politics, see Perlstein, "Community and Democracy in American Schools," 637–42. For an account of interracial conflict and solidarity in the coal fields, see Joe William Trotter Jr., *Coal, Class, and Color: Blacks in Southern West Virginia, 1915–1932* (Urbana: University of Illinois Press, 1990), 9, 111, 263.

37. *Report of the Survey of Arthurdale School* (Morgantown: College of Education, West Virginia University, 6 May 1940), 31; Martin Berman, *Arthurdale, Nambe, and the Developing Community School Model: A Comparative Study* (PhD diss., University of New Mexico, 1979), v; John Collier, "Subsistence Homesteads under Public Works," *Indians at Work*, November 1933, 33; Clapp, *Community Schools in Action*, 347.

38. Kenneth Davis, *FDR*, vol. 2, *The New Deal Years, 1933–1937* (New York: Random House, 1995), 354; Pickett, *For More Than Bread*, 58–59; Perlstein, "Community and Democracy in American Schools," 644.

39. J. Dewey, foreword, viii.

40. Lash, *Franklin and Eleanor*, 532; E. Roosevelt, *This I Remember*, 130; Paul Conkin, *Tomorrow a World: The New Deal Community Program* (Ithaca, NY: Cornell University Press, 1959), 250; Davis, *FDR*, 2:122; Cook, *Eleanor Roosevelt*, 144–46; E. Roosevelt, *White House Press Conferences*, 265; Clapp, *Community Schools in Action*, 347.

41. Perlstein, "Community and Democracy in American Schools," 644; Maloney, *Back to the Land*, 167.

42. Hochschild, *King Leopold's Ghost*, 295; Norman Klein, *The History of Forgetting: Los Angeles and the Erasure of Memory* (London: Verso, 1997), 16; Maloney, *Back to the Land*, 193–94.

7

Identity

EILEEN H. TAMURA

Since the 1960s, the popular press has promoted myths about the Asian American success story. These myths typically portray Asian Americans as models of behavior, academic achievement, and success. *Newsweek* captured one of the many waves of this popular sentiment in 1971 when it famously depicted Japanese Americans as "outwhiting the whites," even though the publication noted that juvenile delinquency among this population had risen a bit and that some lived in poverty.[1] Academics have also contributed to the mythmaking. In the 1980s, psychologist Philip E. Vernon went so far as to state that Asians had larger heads than Caucasians and Africans, and he concluded that head size correlated with intelligence. The sociologist Peter Rose, accepting the stereotype, wrote in 1997 that Asian Americans had moved from being "pariahs to paragons."[2]

These and other ostensibly scholarly statements have provided authority to the thesis, further perpetuating what I call the model minority identity meme. While this meme—based on disinformation and selective memory—has moved some Asian Americans beyond past hostilities and discrimination against them, it has shrouded significant underlying prejudices and ignored the needs of struggling Asian American students. The model minority identity meme, which emerged in the 1940s and continues to this day, conveys a half-truth perpetuated by those who combine claims supported by evidence with claims recognized as false or misleading. Moreover, the meme has been problematic when Asian Americans have internalized it, using it to the detriment of themselves and other Asian Americans. As such, the evolution of Asian American identity has intersected with a long external as well as internal history of socially constructed ignorance through a process of mythmaking about the model minority.

Origins of the Model Minority Identity Concept

Some have argued that hints of the model minority identity, before the concept had a name, began to appear during and after World War II. Before the war, white Americans often viewed Asian Americans, the most numerous of whom were ethnic Chinese and Japanese, as inassimilable yellow peril. The Chinese were targets of brutal mob attacks—in 1871 in Los Angeles, in 1877 in San Francisco, and in 1885 in Rock Springs, Wyoming. The Chinese Exclusion Act of 1882, renewed in 1892 and again in 1902, barred them from entering the United States, and the Supreme Court cases of Takao Ozawa and Bhagat Singh Thind denied Asian Americans the right to naturalize because they were deemed to be nonwhite. When it came to property rights, a staple of the American market tradition, Asian immigrants were barred from becoming naturalized citizens and forbidden from owning or leasing land. At the same time, those born in the United States and who were therefore US citizens were excluded from mainstream society in areas such as housing and employment. At the outbreak of World War II, 120,000 Nikkei—ethnic Japanese—were incarcerated as enemy aliens in War Relocation Authority camps, even though 64 percent of them were US citizens.[3]

During the war, however, camp administrators attempted to Americanize their Nikkei subjects. The first major thrust, not wholly successful, was the resettlement program, an effort to scatter young adults throughout the country to expose them to white middle-class mores. The second major thrust, much more successful than the first, was military service, an initiative by the government to transform Japanese Americans from outcasts to model citizens. Working in tandem with these external forces were many Nikkei themselves, parents and their children, who sought acceptance and assimilation and pressured the nonconforming among them to behave in ways that demonstrated their respectability to white Americans.[4]

Like Japanese America, Chinese America felt the impact of World War II, but in contrasting ways. China's status as an ally of the United States brought positive reactions toward ethnic Chinese. They successfully advocated for full integration into American society and won the removal of laws excluding Chinese from entering the United States. The improvement in status for Chinese America was accompanied by a movement within, toward Americanism, symbolized by the shift from the immigrant press written in Chinese to the Chinese American press written in English. As with Japanese Americans, a new generation of Chinese Americans pushed for full acceptance within mainstream America.[5]

In the 1950s and 1960s, mainstream newspapers became an important source from which people learned about the emergent model minority identity meme. As the civil rights struggle gained momentum, for example, the *Chicago Daily News* published a controversial letter to the editor in 1957 urging African Americans to emulate Japanese Americans. The letter stated that Japanese Americans had improved their socioeconomic status through "spiritual and moral values alone," not from "pressure groups[s]." This prompted a response from Abe Hagiwara, of the Japanese American Citizens League (JACL), who reminded readers that the JACL had itself served as a pressure group working to end racial discrimination. Other critics of the *Chicago Daily News* letter pointed to African Americans' two centuries of enslavement, which neither the Nikkei nor any other ethnic group had experienced.[6]

Six years later another article, penned by the Japanese American journalist Howard Imazeki, appeared in the newspaper *Hokubei Nichibei* and was reprinted in the *San Francisco Examiner.* While recognizing their long history of slavery and their important struggle for equal rights, Imazeki nonetheless chastised African Americans for lacking the effort to "better themselves." The article created a firestorm of opposing views. While some Japanese Americans lauded Imazeki for "hitting the nail on the head," others denounced him for his "self-righteousness" and for failing to understand that African Americans experienced more racism than Japanese Americans did. Still others called on all ethnic minorities to join African Americans in the movement for civil rights.[7]

These and other such public statements that compared Japanese Americans with African Americans continued through the sixties, the civil rights movement, serving as political fodder against demands made by African Americans and their supporters. At the same time, the controversy rehabilitated the image of Asian Americans toward what would develop as the model minority identity meme, crystalized by the demographer William Petersen's well-received essay in the *New York Times* in 1966, "Success Story, Japanese-American Style." The article's instant and widespread appeal led Petersen to expand his ideas into a book, *Japanese Americans: Oppression and Success.* In both publications, Petersen wrote glowingly of Japanese Americans' accomplishments. A careful read of his article and book, however, reveals his political agenda. As he noted, "This book is less a study of Japanese Americans per se than an analysis of the puzzle that they present to theorists of ethnic relations." In other words, Petersen used his discussion of the Nikkei to compare them with other ethnic minorities, referring frequently to blacks, in particular, in his book.[8]

Despite his praise for Japanese Americans, Petersen also advanced the idea that they should be characterized as foreign regardless of their US citizenship. Like other commentators, Petersen pointed to so-called traditional Japanese cultural traits such as duty, perseverance, accommodation, self-sacrifice, devotion to family and education, and political moderation. He wrote that Japanese Americans "could climb over the highest barriers our racists were able to fashion in part because of their meaningful links to an alien culture." In contrast, African Americans were "thoroughly American." In other words, Japanese Americans were able to achieve what they did precisely because of their foreignness. They were "alien citizens," US citizens and yet appearing foreign in the eyes of European Americans. This European American view of Asian Americans as "perpetual foreigners" has continued up to the present.[9]

Petersen's article and subsequent book persuaded many to see the apparent success of the Nikkei as a refutation of the argument that structural problems in American society held ethnic minorities back. The refutation extended to claims about institutional racism and that such structural problems did not cause urban poverty and violence. As Petersen and his circle saw it, deficiencies within ethnic minorities themselves—such as attitudes, behaviors, and beliefs—caused these problems. Referring to African Americans, Petersen wrote, "For with the substantial improvements gained during the past two decades, Negroes shifted their reference group: once pleased to have risen above their fathers' status, now many blacks are aggressively dissatisfied that they have not yet achieved full equality with whites." He disagreed with such aspirations: "One consequence of massive civil-rights programs has been to exacerbate racial conflict, to encourage the rise of black violence and white backlash." Petersen's critics countered his comments on history and identity by asserting that he had failed to mention white violence toward blacks and other ethnic minorities both before and during the civil rights era. Critics also feared that making this false comparison between Asian Americans and African Americans would only encourage racial animosity between the two groups.[10]

During the same year that saw the publication of Petersen's article, 1966, another one appeared, this time in *U.S. News & World Report*, and the purported subject was Chinese Americans. Once again, Asian Americans were used as a tool to disparage African Americans. "At a time when it is being proposed that hundreds of billions be spent to uplift Negroes and other minorities," stated the article, "the nation's 300,000 Chinese Americans are moving ahead on their own, with no help from anyone else." They have become "a model of self-respect and achievement in today's America."[11]

The lesson from such news articles became increasingly clear: Asian Americans of both Chinese and Japanese ancestry occupied the same role. Moreover, feelings of unease toward the rise of activism and protests—stimulated by the push for civil rights and the call for black power—led white Americans to posit that both Japanese Americans and Chinese Americans were models for the supposed lower-class culture of African Americans. The message communicated, in short, was that black America should learn from Asian America about how to succeed in mainstream America.[12]

The Hollowness of the Identity Meme

While Asian Americans have endured racism and discrimination, it is a fallacy to compare their experiences to those of African Americans. This is not to lessen the seriousness of racial attacks on Asian Americans but to point out the uniqueness of African American history: two centuries of slavery that included violent treatment, the breakup of families, the systematic rape of women, the annihilation of culture, and the denial of literacy. For the hundred years that followed slavery, Jim Crow dominated the South. In states where a small proportion of African Americans worked as professionals or business people, they had to endure segregation and discrimination in education, housing, and employment. To be sure, this long, sordid history of racism against African Americans does not negate the violent racism against Asian Americans. Rather, it serves as a reminder that conflating the two based on minority status fuels mythmaking and the social construction of ignorance.[13]

Even though Asian Americans may have benefited from the fallacy, many who embodied "success stories" rejected the implications such Horatio Alger–like stories taught. Daniel K. Inouye, an Asian American who would later become the most senior US senator, expressed the view shared by others critical of cross-ethnic comparisons. As keynote speaker at the 1968 Democratic National Convention, during a time of heightened racial tensions, Inouye spoke unequivocally of the "more than 100 years of systemic racist deprivation" African Americans experienced after slavery. He continued,

> As an American whose ancestors came from Japan, I have been accustomed to a question . . . 'Why can't the Negro be like you?' First, although my skin is colored, it is not black. . . . The color of my skin does not ignite prejudice that has smoldered for generations. Second, although my grandfather came to this country in poverty, he came without shackles; he came as a free man enjoying certain constitutional rights under the American flag. Third, my grandfather's family was not shattered as individual members of it were sold as chattel. . . . And fourth, although others of my ancestry were [incarcerated] behind barbed wire during World War II, neither my parents nor I were forced by covenants and circumstances to live in ghettos.[14]

Inouye's commentary was just one of many in the highly charged decade of the sixties. Energized by Black Power, civil rights movements, and anti-Vietnam protests, Asian American activists, most of them college students, called for a Yellow Power Asian American movement. These activists debunked the notion that Asian Americans were politically moderate. Amy Uyematsu, a leading Yellow Power proponent, wrote in a 1969 call to arms that Asian Americans were too quick to accept white middle-class attitudes toward blacks, which saw them needing "to 'adapt' to the system" and emulate other nonwhites, namely, Asian Americans. Advocating a "yellow political power," Uyematsu warned her readers of injustices and the dangers of ignoring continual problems of urban poverty and its attendant health issues. At California State University, Long Beach, the Asian American Student Alliance questioned "whether we want to accept or be accepted by a society whose values are corrupted by greed," a society that "perpetuates racism, war, and oppression." Its statement read, "We question our 'success' when we see young Asian brothers and sisters getting strung out on drugs, getting busted, fighting each other, and even committing suicide." At the University of California, Santa Cruz, the Third World Political Alliance declared that they were "allowed to succeed economically, academically, and socially—but only to a certain extent. . . . Asian Americans have not been completely accepted regardless of how long they have been in this country."[15]

These Asian Americans felt a kinship with other subordinated nonwhites—African Americans, Latinos, and Native Americans. And like the other nonwhite groups, they protested the absence of academic study of their own history. As a result of these protests, beginning in the late 1960s, programs in ethnic studies, black studies, Chicano studies, and Asian American studies appeared in colleges and universities across the country.[16]

Despite these critiques, the model minority identity meme continued to serve as a lesson for the general public, a bludgeon against claims of institutional racism, and a tool in the propagation of myths and ignorance-making. In the decades following the 1960s and continuing into the twenty-first century, this identity meme has expanded to include all Asian Americans.

Diversity within Asian America

From the 1970s on, the model minority concept and the umbrella term *Asian American* homogenized a diverse people that comprised at least twenty-four different ethnic groups. Among them were groups not usually thought of, for example, Bangladeshis, Malaysians, Pakistanis, and Sri Lankans. In addition to ethnic diversity, the class background of Asian immigrants varied considerably. As a result of the Immigration and Nationality Act of 1965, two types of migrants entered the United States. Highly educated Asians constituted one group. The other was made up of unskilled or semiskilled workers with little schooling, such as Vietnamese, Hmong, Cambodians, and Laotians who arrived after the Vietnam War. Because of this "dual-chain migration," Asian Americans who entered the country represented divergent socioeconomic and educational backgrounds. Since the 1990 Immigration Act, the US has given preferential status to highly educated and highly skilled immigrants. This employment-preferences category has favored immigrants from China, Taiwan, and Korea.[17]

For decades the homogenization process obscured the educational needs of Asian Americans. Racial stereotypes, such as the model minority "success story," lumped diverse individuals together. Disaggregated achievement data about post-1965 Asian American children and youths who differed in ancestral lands, cultural values, socioeconomic levels, and parental levels of schooling, however, showed that students' performance varied widely. In particular, students from low-income immigrant families, English-language learners, and Cambodian, Hmong, Laotian, and other Southeast Asian students struggled in school. The model minority identity meme had destructive consequences for these students, who lacked needed academic and social-service support, including effective language and literacy instruction, appropriate curricula, and culturally responsive teaching. Some studies of the late-twentieth- and early twenty-first-century have shown that these students experienced disproportionately high pressure and expectations to do well in all areas, contributing to "passive and negative attitudes towards learning." Moreover, as a result of the identity meme, many schools failed to monitor dropout rates of their Asian American students and ignored the high percentage that left school without graduating. In Seattle in the 1986–87 school year, for instance, 12 percent of Vietnamese and 18 percent of other Southeast Asian students dropped out of high school. Such statistics attracted less attention than the dropout rates of other minorities. Because it was assumed that all Asian Americans were doing well, individuals who needed assistance were not recognized.[18]

Aggregated data on Asian American schooling also contributed to the persistence of homogenization and the model minority identity meme. Data from the 1980s and 1990s, for example, indicated that adult Asian Americans had more schooling than adult non-Asians, including white Americans. However, these data were misleading. Specifically, in 1990, while US-born Chinese, Japanese, Korean, and Asian Indians attained higher levels of schooling than did whites, US-born Vietnamese and other Asian Americans fell well below whites. A few states had relatively high concentrations of Asian Americans attending universities, but their overall increase nationwide at colleges and universities paralleled the increase of Latinos and African Americans. On the other hand, the data in 1990 for minors showed that both Asian Americans and immigrant Asians generally surpassed other ethnic groups, including whites, in level of schooling. As a result, since 1990 Asians, both US-born and immigrants, have been increasingly visible on university campuses. In short, the aggregated data have long distorted the picture of Asian American academic achievement and have obscured the reality that many of these college students came from families possessing professional and technical skills.[19]

The visibility of Asian Americans in higher education placed them at the center of the affirmative action controversy. In the 1980s, when activists protested what they saw as limits to Asian American admissions by elite universities, critics of affirmative action depicted Asian Americans as victims of an unfair policy that allowed universities to admit less qualified blacks and Latinos. In this way, critics succeeded in recasting and co-opting the debate over affirmative action by making the policy part of the controversy over Asian American university admissions.[20]

In addition to diverting attention away from historic racial injustices, critics of affirmative action used the visibility of Asian Americans in higher education to rekindle the model minority myth. Throughout the late twentieth century and, and continuing to the present, they continued to do so even though the Asian American presence in institutions of higher education actually undermined the myth, for during this period Asian Americans earned less than their white counterparts who had the same level of schooling. Their lower income level led to a pattern of overcompensating for the discrepancy. The American social environment, in which Asian Americans had to attain higher levels of schooling in order to achieve the same level of income as whites, became a strong driving force for their continuation in school. Asian Americans' lower level of income despite having the same level of schooling showed that they had not been fully accepted within mainstream society.[21]

The disparity in earning power between white Americans and Asian Americans was apparent in extensive and comprehensive demographic studies conducted in the 1990s. Similar findings came from the 1995 U.S. Glass Ceiling Commission, which found that Asian Americans earned less than whites even if they had the same qualifications. Other studies came to the same conclusion. Some data showed Asian Americans having the highest median household income, even higher than whites'. However, the data were on household and not individual income, and Asian Americans typically had more workers per household than whites. Additionally, Asian Americans tended to live in urban areas, where the cost of living was above the national average.[22]

Looking more closely at subgroups within the Asian American category complicates the model minority identity meme. According to the 1980 and 1990 censuses, Japanese Americans and Chinese Americans had about the same earning power as white Americans, while Southeast Asian Americans were at the bottom of the economic pyramid. Moreover, wide class differences have existed within specific ethnic groups. Within the Chinese and Korean communities, for example, there was a socioeconomic divide between highly educated professionals and low-skilled workers. Immigrants with a professional background had the resources and social and cultural capital to adapt more easily to the US economic and educational systems.[23]

Buying into the Meme: World War II Incarceration of the Nikkei

Those who used the model minority identity meme—in newspapers and books, in struggles over civil rights and student power, and in the intentional and unintentional obscuring of student needs and statistical data—largely advanced the myth from outside the Asian American community. The meme, however, was also propagated from within by those who had embraced the identity from its beginning in the mid-twentieth century. As a result, two aspects of the model minority identity meme converged with cross-purposes. One was the effort to live up to the idea of being models of behavior in order to be accepted by mainstream society. The other was their status as perpetual foreigners, viewed as such by the larger society who targeted them as disloyal solely because of their ethnoracial background.

During World War II, Nikkei who did not exhibit behaviors "acceptable" to the status quo, such as hard work, perseverance, accommodation, and political moderation, were shunned by their own ethnic community. And the history that the Nikkei passed on to succeeding generations was silent on issues of dissent, protest, resistance, and draft refusal. In fact, the Nikkei were even silent on the incarceration itself, when 120,000 of them living on the West Coast were forced from their homes and livelihoods and taken to the concentration camps.[24]

The embrace of the model minority stereotype is a prime example of the ways in which people have constructed their collective memory to retain certain aspects of the past while erasing others. The model minority identity meme affirmed what many Nikkei believed about themselves and served to strengthen their pride in their ethnic identity. The popular press and news media highlighting the notion of the model minority contributed to the affirmation of this meme. At the same time, it prompted and perpetuated feelings of shame for having been objects of suspicion during World War II, removed from their homes and incarcerated in government camps. For decades after the war, Nikkei said little or nothing about their experiences in exile. Suspicions directed against them implied that as perpetual foreigners they did not meet mainstream expectations of Americanism. This sense of shame also meant that Nikkei who actively resisted incarceration, those who protested loudly, those who renounced their US citizenship, and those who refused the draft while exiled in government camps became outcasts within their own ethnic group. Bearing the label "disloyal," most remained silent about their past, and for decades their actions remained invisible. In stark contrast, Japanese American war veterans were honored for their valor. Here was an instance of selective memory constructing history: glorification for some, erasure for others.[25]

This mind-set caused fissures within the ethnic community and perpetuated misinformation about members' experiences during World War II. A significant outcome was the construction of ignorance among their children. Most Sansei, third-generation Japanese Americans, grew up unaware of their family and ethnic history. Silence from their parents added to the lack of information in US history textbooks on the subject.[26]

More than five decades passed before many Nikkei were comfortable about revealing what had previously been hidden. There were several reasons for this change of heart. For one thing, many youths and young adults born after the war—Sansei and the fourth-generation Yonsei—joined other Asian Americans in rejecting the model minority identity meme and demanding to know more about their histories. For another, researchers began to unearth documents implicating the US government in its mass incarceration of innocent people. For yet another, books and articles appeared on resistance in the camps. And finally, a two-decade redress movement culminated in 1988, when Congress passed a law that provided for an official apology and reparations.

The complexities of three especially sensitive issues eventually came to light as Nisei (second-generation Japanese Americans) spoke of experiences that for decades had not been freely discussed. One was resistance to the draft. Many Nisei refused to be conscripted in protest against the mass incarceration of Japanese Americans; others refused for personal and family reasons. While spending time in federal prisons and for subsequent decades, these draft resisters were stigmatized by other Nikkei who believed that to refuse the draft was unpatriotic and un-American. The experiences of these men who said no, which have only recently come to light, have illuminated the various circumstances that led individuals to dissent and have given their children and future generations a better understanding of the past.[27]

The second issue involved about five thousand Nisei soldiers who were members of the armed forces before the attack on Pearl Harbor. Many of them were eventually sent to the battlefields as replacements for casualties of the now-famous 442nd Regimental Combat Team, organized in 1943, after a year of Japanese Americans being labeled "enemy aliens" by the US War Department. Others refused combat duty because of racist treatment in the army and because their families remained in US concentration camps. These men were assigned to a "Special Organization" that performed menial hard labor and later received less than honorable discharges. Others who refused combat duty were court-martialed, given dishonorable discharges, and sentenced to years of hard labor. These military resisters, who some have called "soldiers of conscience," were ignored and misunderstood for more than sixty years. Ignorance about them facilitated the perpetuation of the model minority identity meme.[28]

The third issue was Nisei renunciation of US citizenship. Protests and dissent occurred in all the camps, and especially so at Tule Lake in northern California. There, strikes, martial law, arrests, imprisonment in the stockade, pressure groups, and violence made it the most turmoil-ridden of the camps. In this context, more than fifty-seven hundred Nisei renounced their citizenship. Like the draft resisters and the soldiers who refused combat duty, renunciants remained for decades invisible to the children born after the war.[29]

Nisei young adults, men and women, renounced their US citizenship for a variety of reasons. Many believed that it was the only way to support and remain with their immigrant parents, who had decided that Japan was the only place where they would be accepted. Others were disillusioned about their future lives in America. Still others had children and parents to care for and had no job prospects. They believed that if they renounced their US citizenship, they would not be forced out abruptly from Tule Lake, the remaining government camp, without any means of livelihood. These and other nonpolitical reasons led to their decisions. While most of them eventually regained their citizenship through a protracted legal process, they remained silent about this episode in their past.[30]

As such, silence contributed to the production of ignorance across generations of Japanese Americans and allowed for the rise and flourishing of the model minority identity meme. This silence continued for more than five decades. At the turn of the twenty-first century, however, former renunciants began to feel more comfortable speaking in public about their past. The successful passage of the redress law, together with the letter of apology from President George H. W. Bush and the reparations, helped to alleviate the shame, confusion, and bitterness that had burdened them for decades. Pilgrimages to the sites of the former concentration camps, in particular to Tule Lake, where most Nisei renounced their citizenship, have been a means of both educating the public about and bringing closure to this history.[31]

Buying into the Meme: Exclusion within Asian American

As the previous section illustrates, the model minority myth not only fostered a perception among Nisei that they indeed fulfilled this stereotype but also perpetuated ignorance among Nikkei about their own history. In Hawai‘i in the 1970s and 1980s this mind-set contributed to controversy over Nisei rejection of reforms advocated by Filipino Americans. This rejection was an instance in which the identity meme led to issues of exclusion within Asian America.

The reform controversy began in the 1970s, when the Oahu Filipino Community Council (OFCC) in Hawai'i, a federation of Filipino organizations, released a report calling for, among other things, an increase in the hiring of teachers of underrepresented ethnic groups. Its focus was primarily on the need to hire more Filipino teachers, but it also called for more teachers of Samoan, Korean, and Native Hawaiian ethnicities. In addition, the report recommended bilingual and bicultural programs in ethnic studies.[32]

The OFCC's report was a response to violence in the public schools resulting from changes in the student body. These changes came about as a result of the Immigration and Nationality Act of 1965, which led to a large yearly influx of new immigrants to the state, of whom about 50 percent were from the Philippines. Not surprisingly, the shifting state demographics brought changes to the public school population. In 1978 there were 15,345 students with limited English proficiency, a substantial increase from the few thousand such students prior to passage of the immigration law. Of these students, 34.5 percent were from the Philippines, the largest group among immigrant students with limited English proficiency. Yet, while the public school student population was changing, the ethnic background of teachers remained static, dominated by Nikkei. In 1978 they constituted 60 percent of all teachers, while about 20 percent of students were Nikkei. At the same time, while 20 percent of students were of Filipino descent, fewer than 3 percent of teachers were ethnically Filipino. Of the newly hired teachers that year, 75 percent were of Japanese or Caucasian ancestry, and only 5 percent were of Filipino ancestry.[33]

The shifting demographics helped to create an unstable environment in the schools, where tensions emerged between local and immigrant students. Complicating matters, fighting erupted between those Filipinos born in Hawai'i, those who were American citizens, and those recently arrived from the Philippines. These clashes became so violent that students were killed in the brawls. Immigrant students felt that they were treated poorly and mocked for their clothes and accents. Locally born Filipino Americans viewed their immigrant counterparts as foreign—loud, inappropriately dressed, and different from what they believed Americans should be like. Not totally accepted themselves by mainstream society, the Hawai'i -born ethnic Filipinos showed how being labeled perpetual foreigners affected them. By holding immigrant Filipinos in contempt, the locally born Filipinos meant to make it clear that they were not themselves foreigners.[34]

The increasing numbers of Filipino students in the schools, as well as the frequency and severity of student fights, led the OFCC to conclude that the public schools needed to hire more ethnic Filipino teachers, who would be able to communicate and relate effectively with Filipino immigrant and local students. With this in mind, the OFCC requested that the Hawai'i State Department of Education (HSDOE) take action on behalf of this and other underrepresented ethnic groups. The HSDOE, however, refused to respond. As a result, the OFCC sent a complaint to the US Department of Health, Education, and Welfare and the US Commission on Civil Rights with a list of civil rights violations that included "unequal treatment in Hawai'i public schools." The OFCC also asked the HSDOE to "hire 1,000 more Filipino teachers by 1978."[35]

The HSDOE moved slowly. By 1983 it had made only minimal changes. It would be more than a decade before it responded favorably to OFCC recommendations. Among those that the HSDOE finally adopted were special recruitment efforts to attract Filipino applicants for teaching positions and providing credit for education and work experience received outside the United States.[36]

The OFCC persisted with its position because of federal actions taken in the 1960s: Title VII of the 1964 Civil Rights Act prohibited discrimination based on race or national origins; President Lyndon Johnson's 1965 Executive Order 11246 prohibited discrimination based on national origin; and President Richard Nixon's 1969 Philadelphia Plan laid out goals and timetables for hiring.[37]

Ironically, Asian Americans, specifically Nisei, children of immigrants, who had themselves been excluded from desired professions during the 1920s and 1930s and who had by the 1970s and 1980s become not only teachers but also public school administrators, proved to be major barriers in the push to hire more ethnic Filipinos. It was not that the Nisei had forgotten what it was like to be children of immigrants and subjects of discrimination. In fact, in explaining his objection to "more proportional ethnic hiring," the Hawai'i state personnel director, a Nisei, referred to past discrimination against the Nikkei. "No doubt about it," he said, "a great number (of civil servants) are [ethnic] Japanese. During the years of World War II, it was difficult to get federal and private employment. The (local) Civil Service had the lowest paying jobs—in fact, the only jobs you could get. They got these jobs, and they worked their way up. That ought to be taken into consideration." Implicit in this response was the idea that the Nisei had not complained but instead had worked hard and eventually risen to positions of influence, and this was the model that ethnic Filipinos should emulate.[38]

The Nisei's comments indicate that he either ignored or lacked an understanding of institutional racism, including structural barriers that blocked the hiring of college-educated immigrant teachers. In this case, the state of Hawai'i during the 1970s required public school teachers to be US citizens. As a result, the only public school jobs available to Filipino immigrants were temporary, part-time work with students in bilingual and Students with Limited English Proficiency (SLEP) classrooms. These jobs came without fringe benefits. In 1986 Congress passed the Immigration Reform and Control Act. Title I of the act forbade employers to discriminate based on citizenship. Nevertheless, some school principals continued to ask immigrant teacher applicants if they were US citizens.[39]

Coupled with the issue of hiring noncitizens was the concern of school principals over accent. As a marker of foreignness, accent may be a neutral or a negative indicator of possible noncitizenship. Upon hearing public school principals criticizing teacher applicants with Filipino accents, the Filipina activist Vicky Bunye noted in 1986, "You walk into any school [in Hawai'i], everybody's got an accent. If it's a French accent, it's OK. If it's Ilocano [Filipino] or Samoan, it's not good enough." Her response did not move principals to change their decisions. One Nisei high school principal defended his position with this explanation: "It's difficult to hire a person whom I do not understand and who does not understand me." Another principal—of mixed Caucasian, Native Hawaiian, and Japanese ethnicity—noted, "When we are trying to get [SLEP] to . . . hear good standard English, I would be remiss as a principal if I didn't assure that."[40]

These principals made no mention of teachers whose first language was Hawai'i Creole, many of whom were teaching in Hawai'i's public schools. Hawai'i Creole is a dominant island language that evolved during the second and third decades of the twentieth century among children of immigrant plantation workers. While English is the lexifier of the creole, the language has a non-English grammatical structure and a unique accent and intonation. That is, Hawai'i Creole has a distinctive sound system that differs from that of standard English. Teachers who grew up speaking Hawai'i Creole, therefore, likely did not model what the principal called "good standard English."[41]

This episode highlights some of the ways in which the model minority identity meme, used to characterize Asian Americans, could also work against them. In this instance the meme created fissures within Asian America itself. As youths and adults, the Nisei had faced anti-Japanese sentiments and job discrimination at the same time that they sought acceptance from mainstream society. Teaching was one area of employment open to them.

Five decades later, having become teachers and then school administrators, and having attained positions of influence within island society, they turned away from another immigrant group. This rejection is a case of a group's reconstructed memory about its own history. They joined the dominant belief that Filipinos who had been teachers in the Philippines were unfit to teach in public schools in the United States. Underlying the refusal to hire immigrant Filipinos was an image of an America that excluded those whose accent and cultural behaviors labeled them as foreign. The Nisei who embraced the model minority meme argued that Filipino teachers should not be hired, and by implication their message was that no special support was needed for Filipino students struggling in school. Filipinos challenged this perspective, calling it a violation of the 1964 Civil Rights Act and affirmative action regulations. This confrontation illuminates an instance of fracture within Asian America. Power changed the social imagination of the Nisei who had moved from the margins of society to its center. The result was a failed promise of Asian American solidarity and a turning away from new possibilities of diversity in knowledge and ideas.[42]

Legacy of an Identity

Since the mid-twentieth century, Asian Americans have undergone a dramatic transformation in the public imagination from "inassimilable yellow peril" to model minority. This change has been advanced, diffused, and debated through a variety of cultural processes, including the popular press, academic writings, political controversy, student movements, social research, and intergenerational exchanges and traditions. The identity meme has had specific outcomes too. It has influenced societywide debates over civil rights and affirmative action policies; it has also encouraged a disregard of educational needs specific to the Asian American community. At bottom, this socially constructed model minority identity meme has distorted and obscured a more complex understanding of realities experienced and understood by Asian Americans. It has constructed a form of ignorance about this community's past that has relied on simple comparisons with other minorities, problematic associations among those with Asian ancestry, and research that has had homogenizing rather than complexifying implications.

Some might argue that the myth of the model minority is an alternate knowledge construct rather than a form of ignorance. But the primary agnotological insight offered in this chapter is that the social construction of ignorance lens provides a useful analytical tool for unpacking issues of identity and memory in history of education. Using this critical lens to examine the Asian American model minority identity meme casts light on the adverse ways in which the meme inserted itself into the education of youths, illustrating how ignorance, created and sustained, provided injurious consequences. Despite well-argued critiques, the meme was promoted as a lesson to those outside and inside Asian America. Deniers of institutional racism succeeded in deflecting constructive discussion and action to rectify harsh inequities in American society. Asian Americans who bought into the meme supported the spread of misinformation about their history at the same time that they rejected those Asian Americans who exhibited behaviors deemed unacceptable. The result was the perpetuation of ignorance about ethnic history, identity, and memory.

NOTES

1. Charles Michener, "Success Story: Outwhiting the Whites," *Newsweek*, 21 June 1971, 23–25.

2. Vernon discussed in Frank H. Wu, *Yellow: Race in America Beyond Black and White* (New York: Basic Books, 2002), 44; Peter I. Rose, *Tempest-Tost: Race, Immigration, and the Dilemmas of Diversity* (New York: Oxford University Press, 1997), 3–5. According to Michael E. Martinez, *Future Bright: A Transforming Vision of Human Intelligence* (New York: Oxford University Press, 2013), brain size is only one among a multitude of factors influencing intelligence, and Albert Einstein's brain was of average size.

3. Ellen D. Wu, *The Color of Success: Asian Americans and the Origins of the Model Minority* (Princeton, NJ: Princeton University Press, 2014), 11–13; F. Wu, *Yellow*, 59–61; Eileen H. Tamura, *Americanization, Acculturation, and Ethnic Identity: The Nisei Generation in Hawaii* (Urbana: University of Illinois Press, 1994), 78–79.

4. E. Wu, *Color of Success*, 13, 16–42.

5. Ibid., 43–71.

6. Ibid., 165. It should be noted that during and after World War II the JACL was intensely disliked by many Nikkei, who accused the organization of betraying fellow

Nikkei. Prior to and during the war, JACL members, in their zeal to promote the idea of Americanism, spied on fellow Nikkei and reported on them to federal authorities.

7. Ibid., 165–68.

8. William Petersen, "Success Story, Japanese-American Style," *New York Times Magazine*, 6 January 1966, reprinted in *Minority Responses: Comparative Views of Reactions to Subordination*, ed. Minako Kurokawa (New York: Random House, 1970), 169–78; Petersen, *Japanese Americans: Oppression and Success* (New York: Random House, 1971), 4–5.

9. Petersen, "Success Story, Japanese-American Style," 177–78; Petersen, *Japanese Americans*, 7. The phrase *alien citizens* is from Mae Ngai, *Impossible Subjects: Illegal Aliens and the Making of Modern America* (Princeton, NJ: Princeton University Press, 2004), 8. The earliest use I have seen of the term *perpetual foreigners* is in F. Wu, *Yellow*, 79.

10. Petersen, *Japanese Americans*, 6–7, 218, 221–36; David Palumbo-Liu, *Asian/American: Historical Crossings of a Racial Frontier* (Stanford, CA: Stanford University Press, 1999), 172, 179–81. Palumbo-Liu presents an extensive analysis of Petersen's laudatory comments about Japanese American achievements. On one hand, Japanese Americans had succeeded despite and because of oppression, and other minorities should be able to do the same. On the other hand, because of the uniqueness of what Petersen called the Japanese "subnation"—with its authoritarian control, lack of individualism, and cultural distinctiveness from the rest of the nation—no other American ethnic group could mimic Japanese America. This characterization of subnation was another example of Petersen's view of Japanese Americans as alien citizens.

11. "Success Story of One Minority in the U.S.," *U.S. News & World Report*, 26 December 1966, 73.

12. E. Wu, *Color of Success*, 170, 172, 206–9. Wu notes that "some commentators, perturbed by blacks' increasingly uncompromising demands for ending racism and poverty, spelled out divergences" between Japanese and Chinese Americans, on the one hand, and African Americans, on the other, "to discredit the latter's insurgency."

13. F. Wu, *Yellow*, 64–65.

14. "Transcript of the Keynote Address by Senator Inouye Decrying Violent Protests," *New York Times*, 27 August 1968, http://www.nytimes.com/packages/pdf/politics/inouye_articles/a.transcript.1968.pdf.

15. William Wei, *The Asian American Movement* (Philadelphia: Temple University Press, 1993), 1–43; Mitchell T. Maki, Harry H. L. Kitano, and S. Megan Berthold, *Achieving the Impossible Dream: How Japanese Americans Obtained Redress* (Urbana: University of Illinois Press, 1999), 57–59; Alice Yang Murry, *Historical Memories of the Japanese American Internment and the Struggle for Redress* (Stanford, CA: Stanford University Press, 2008), 200–206; quotations from E. Wu, *Color of Success*, 246–47.

16. Wei, *Asian American Movement*, 1–43; Daryl Joji Maeda, *Rethinking the Asian American Movement* (New York: Routledge, 2012).

17. The importance of English-language skills, which have varied considerably among Asian immigrants, has also presented one more considerable hurdle for adults and youths unfamiliar with English. *Asian Americans and Pacific Islanders, Facts, Not Fiction: Setting the Record Straight* (New York: National Commission on Asian American and Pacific Islander Research in Education (CARE) and College Board, 2008), 16–21; John Liu and Lucie Cheng, "Pacific Rim Development and the Duality of Post-1965 Asia Immigration to the United States," in *The New Asian Immigration in Los Angeles*

and Global Restructuring, ed. Paul Ong, Edna Bonacich, and Lucie Cheng (Philadelphia: Temple University Press, 1994), 74–99; John M. Liu, Paul M. Ong, and Carolyn Rosenstein, "Dual Chain Migration: Post-1965 Filipino Immigration to the United States," *International Migration Review* 25 (Autumn 1991): 487–513.

18. Valerie Ooka Pang, Peggy P. Han, and Jennifer M. Pang, "Asian American and Pacific Islander Students: Equity and the Achievement Gap," *Educational Researcher* 40 (November 2011): 378–89; Guofang Li, "Other People's Success: Impact of the 'Model Minority' Myth on Underachieving Asian Students in North America," *KEDI Journal of Educational Policy* 2 (2005): 69–86. See also Curtis Chang, "Streets of Gold: The Myth of the 'Model Minority,'" *Harvard Political Review* 14 (1987): 6–9.

19. Lucie Cheng and Philip Q. Yang, "The 'Model Minority' Deconstructed," in *Contemporary Asian America: A Multidisciplinary Reader*, ed. Min Zhou and James V. Gatewood (New York: New York University Press, 2000), 464–67; *Asian Americans and Pacific Islanders, Facts, Not Fiction*, 5.

20. Dana Y. Takagi, *The Retreat from Race: Asian-American Admissions and Racial Politics* (New Brunswick, NJ: Rutgers University Press, 1992), 117–18, 139; Jonathan Kozol, *Savage Inequalities: Children in America's Schools* (New York: Crown, 1991); Kozol, *The Shame of the Nation: The Restoration of Apartheid Schooling in America* (New York: Crown, 2005). Opponents of affirmative action were able to use Asian Americans as a distraction from racial inequalities, a tactic that has continued in schooling to the present. In writing about the 1980s, Jonathan Kozol in *Savage Inequalities* compared urban and suburban schools, revealing stark contrasts in the condition of school buildings, learning environments, and the supply of books and other educational resources. Racial inequities stood out, with poor, essentially ethnic minorities attending inadequate urban schools, and middle-class whites attending well-equipped suburban schools. Fourteen years later, his *Shame of the Nation* showed how conditions in inner-city schools have continued to deteriorate at the same time that resegregation of ethnic minority students has increased steadily, so that by the turn of the twenty-first century blacks and Latinos constituted as much as 98 percent of many urban schools.

21. F. Wu, *Yellow*, 50–51; Deborah Woo, *Glass Ceilings and Asian Americans: The New Face of Workplace Barriers* (Walnut Creek, CA: AltaMira, 2000); Cheng and Yang, "'Model Minority' Deconstructed," 467.

22. Herbert R. Barringer, *Asians and Pacific Islanders in the United States* (New York: Russell Sage Foundation, 1993), 265; F. Wu, *Yellow*, 50, 55; Cheng and Yang, "'Model Minority' Deconstructed," 469–71. It should be noted that a pair of researchers, Arthur Sakamoto and Satomi Furuichi, found no statistically significant support for this conclusion.

23. F. Wu, *Yellow*, 54; Stacey J. Lee, *Unraveling the "Model Minority" Stereotype: Listening to Asian American Youth*, rev. ed. (New York: Teachers College Press, 2009), 15.

24. The literature on the incarceration of Japanese Americans during World War II is massive. A few important publications suffice here: Roger Daniels, *Concentration Camps: North America; Japanese in the United States and Canada during World War II* (1971; reprint, Malabar, FL: Krieger, 1993); Greg Robinson, *A Tragedy of Democracy: Japanese Confinement in North America* (New York: Columbia University Press, 2009); US Commission on Wartime Relocation and Internment of Civilians, *Personal Justice Denied* (Seattle: University of Washington Press, 1997); Peter Irons, *Justice at War: The Story of the Japanese American Internment Cases* (1983; reprint, Berkeley: University of California Press, 1993); Eric Muller, *American Inquisition: The Hunt for Japanese American Disloyalty in World War II* (Chapel Hill: University of North Carolina Press, 2007); Richard

Drinnon, *Keeper of Concentration Camps: Dillon S. Myer and American Racism* (Berkeley: University of California Press, 1987); and Michi Nishiura Weglyn, *Years of Infamy: The Untold Story of America's Concentration Camps*, updated ed. (Seattle: University of Washington Press, 1996).

25. On the healing process of pilgrimages to former camps, see Sachiko Takita, "The Tule Lake Pilgrimage and Japanese American Internment: Collective Memory, Solidarity, and Division in an Ethnic Community" (PhD diss., University of California Los Angeles, 2007), 264.

26. Satsuki Ina, *From a Silk Cocoon*, a documentary film directed by Satsuki Ina and Stephen Holsapple (San Francisco: Hesono O Productions, 2005), DVD.

27. Eric L. Muller, *Free to Die for Their Country: The Story of the Japanese American Draft Resisters in World War II* (Chicago: University of Chicago Press, 2001); Mike Mackey, ed., *A Matter of Conscience: Essays on the World War II Heart Mountain Draft Resistance Movement* (Powell, WY: Western History, 2002); Cherstin Lyon, *Prisons and Patriots: Japanese American Wartime Citizenship, Civil Disobedience, and Historical Memory* (Philadelphia: Temple University Press, 2011).

28. Shirley Castelnuovo, *Soldiers of Conscience: Japanese American Military Resisters in World War II* (Westport, CT: Praeger, 2008); Linda Tamura, *Nisei Soldiers Break Their Silence: Coming Home to Hood River* (Seattle: University of Washington Press, 2012).

29. Donald E. Collins, *Native American Aliens: Disloyalty and the Renunciation of Citizenship by Japanese Americans during World War II* (Westport, CT: Greenwood, 1985); Barbara Takei, "Legalizing Detention: Segregated Japanese Americans and the Justice Department's Renunciation Program," in "A Question of Loyalty: Internment at Tule Lake," special issue, *Journal of the Shaw Historical Library* 19 (2005): 75–105. Most of the renunciants regained their citizenship as a result of the determined efforts by the attorney Wayne M. Collins, who spent twenty-three years helping renunciants restore their citizenship.

30. Eileen H. Tamura, *In Defense of Justice: Joseph Kurihara and the Japanese American Struggle for Equality* (Urbana: University of Illinois Press, 2013), 122–26.

31. Takita, "Tule Lake Pilgrimage," 271, 342. Those I approached for oral history interviews either said no or consented on the grounds that they would remain anonymous.

32. "Status Report of the Oahu Filipino Community Council Task Force on Education, February 9, 1975," n.p., in *Petitions and Recommendations for Equal Educational Opportunity for Filipinos in Hawai'i Public Schools, 1975–1987* (Honolulu: Center for Research on Ethnic Relations, SSRI, UHM, 1996).

33. "In Isle Schools Filipinos Hope to Stem Violence," *Honolulu Star-Bulletin*, 12 June 1975; "To Work on School Tensions, Task Force Organized," ibid., 7 October 1974; Eleanor Nordyke, *The Peopling of Hawai'i*, rev. ed. (Honolulu: University of Hawaii Press, 1989), 80; Jeff Chang, "Lessons of Tolerance: Americanism and the Filipino Affirmative Action Movement in Hawai'i," *Social Process in Hawaii* 37 (1996): 116; Michael Haas, ed., *Multicultural Hawai'i: The Fabric of a Multiethnic Society* (New York: Garland, 1998), 216.

34. Leslie Wilcox, "Filipino Groups Air Concern, Meeting on School Death," *Honolulu Star-Bulletin*, 23 September 1974; Terry McMurray, "2 Injured in Waipahu School Brawl," *Honolulu Advertiser*, 20 September 1975; J. Chang, "Lessons of Tolerance," 117.

35. Tom Kaser, "School Bias Probe Asked by Filipinos," *Honolulu Advertiser*, 19 November 1975; Ken Kobayashi, "New Discrimination Probe," ibid., 3 December 1975.

36. Hildegaard Verploegen, "DOE Affirmative Action Progress is Hit," *Honolulu Star-Bulletin*, 27 May 1983; "Board of Education Citizen Task Force on Affirmative Action for Filipinos," 3 March 1988, in *Petitions and Recommendations for Equal Educational Opportunity for Filipinos.*

37. J. Chang, "Lessons of Tolerance," 118.

38. Tom Coffman, "State Defends Its Hiring Practices," *Honolulu Star-Bulletin*, 22 December 1971, A2.

39. Pub. Law No. 99-603, Title I, §102, 100 Stat. 3359; "Citizen Discrimination and Workplace Laws," US Equal Opportunity Commission, accessed 16 September 2014, www.eeoc.gov/laws/types/nationalorigin.cfm; J. Chang, "Lessons of Tolerance," 122.

40. Hildegaard Verploegen, "75 Percent of Filipino Candidates Fail Teachers Exam," *Honolulu Star-Bulletin*, 10 September 1986, A1; Kay Lynch, "Principals: Trouble with English is Key," *Honolulu Advertiser*, 18 November 1986, A1.

41. Kent Sakoda and Jeff Siegel, *Pidgin Grammar: An Introduction to the Creole Language of Hawai'i* (Honolulu: Bess Press, 2003).

42. The phrase "failed promise of . . . solidarity" is from the title of Neil Foley's book *Quest for Equality: The Failed Promise of Black-Brown Solidarity* (Cambridge, MA: Harvard University Press, 2010).

8

Religion

ADAM LAATS

Academic historians winced when they heard her say it. In her 2011 bid for the presidential nomination, Michele Bachmann insisted that the Founding Fathers had "worked tirelessly" to end slavery. It was simply not true. But Representative Bachmann's statement reflected something other than simple ignorance, simple unawareness of historical fact. It does not fit neatly into what Robert Proctor has defined as the main categories of systematic non-knowledge—"native state," "lost realm," and "strategic ploy" ignorance. Rather, Bachmann's ideas about the ethos of America's past reflected a vibrant and politically powerful counterknowledge bolstered by an array of influential alternative historians and pundits that included David Barton, John Eidsmoe, and Glenn Beck. This counterknowledge has blocked mainstream historical knowledge in a widespread phenomenon we might call *cultural occlusion*. It has encouraged widespread, systematic non-knowledge about US history. As this chapter explores, this contending tradition of counterknowledge has also been developed and disseminated by an influential group of Christian textbook publishers.[1]

Bachmann's statement represented one significant way that Americans do not know US history. But it is not the only common sort of non-knowledge. There is of course simple ignorance, naïve non-knowledge. Many people may simply be unaware of the Founding Fathers' attitudes toward slavery. But historical knowledge is also occluded, blocked by competing knowledge systems. These competing systems do not always represent Bachmann's sort of politically and culturally informed non-knowledge. In many cases, as history-education scholar Sam Wineburg has argued, this occluding is done by popular culture, especially movies that present a vivid but incorrect vision of the past. Other competing historical knowledge systems work in the realm of what

David Lowenthal has called "heritage." School textbooks, museums, public-history sites, and genealogical societies promote a celebratory heritage that crowds out the historical knowledge favored by academic historians.[2]

This chapter explores a different sort of constructed non-knowledge in the field of US history. In the last decades of the twentieth century, conservative evangelical Protestants in the United States argued for a vision of US history strikingly at odds with mainstream academic notions. This evangelical vision found a captive market in the vibrant and growing network of schools poised to teach this version of US history. To serve this subcultural educational niche, a network of Christian-school publishers in the 1970s began to offer textbooks reflecting the knowledge favored by conservative evangelical Protestants. These publishers did not represent a simple monolith. Despite significant differences in content and style, the textbooks produced by these publishers share a common history. Since the 1980s the vision of US history incorporated in these books has become more decisively sectarian, more prominently distinct from mainstream academic understandings. Like Representative Bachmann, readers of these histories are not simply unaware of mainstream academic history; they are reading a history that explicitly disputes the facts and methods of mainstream academic history. Publishers hope that their distinctive Christian history will occlude competing knowledge systems of US history.[3]

New Schools, New Knowledge

Among conservative evangelical Protestants, concern over the content of US history textbooks has been part of a broader concern over the changing nature of education. Beginning in the 1970s, a trickle of new evangelical "Christian day schools" became a flood. During the late 1970s, according to some estimates, roughly two to three of these schools opened per day in the United States. In 1977 the three largest national associations of Christian day schools—the more separatist American Association of Christian Schools (AACS), the more ecumenical Association of Christian Schools International (ACSI), and the Christian Reformed–affiliated Christian Schools International (CSI)—claimed a total of 349,679 students. By 1992 they claimed approximately 4,337 member schools with a total enrollment of 776,649 students, and a 2002 survey identified 4,812 schools belonging to those associations with a total enrollment of 883,066. In 2012 the three leading organizations of Christian schools continued to represent a significant number of schools and students: AACS claimed to enroll just over 100,000 students. the larger ACSI claimed the affiliation of

3,000 schools in the United States and more than 24,000 schools worldwide; and CSI, which had shrunk, claimed to include 475 schools reaching approximately 100,000 students. Of course, these numbers do not represent the exact number of evangelical schools in the United States. Education scholars James C. Carper and Thomas C. Hunt have noted the difficulty of tracking the numbers of dissenting religious schools. Many schools may refrain from joining national organizations for theological, sociological, or financial reasons. Other schools may close after only a few years.[4]

In addition to students in Christian schools, the number of conservative evangelical Protestant homeschoolers has increased significantly in the past three decades. Just as with the numbers of private religious schools, our knowledge about the numbers of such students is inexact. The National Center for Education Statistics notes that the number of homeschooled students reached 1.77 million in 2011–12, with the most common reason given by parents for homeschooling being "a concern about the environment of other schools." Not all of these students, of course, come from conservative religious families, but some evangelical scholars suggest that evangelical Protestants make up 70 percent of homeschooling families.[5]

Given our incomplete knowledge about the numbers of conservative evangelical schools or homeschools, it is impossible to say with certainty how many students may read the textbooks of Christian-school publishers. Moreover, given the broad coalition of denominations and theological attitudes included in the "evangelical" identity, many students in evangelical schools and evangelical families may use textbooks from mainstream academic publishers instead of the textbooks examined in this chapter. Nevertheless, this population, even according to the most conservative estimate, represents a significant segment of the school-age population in the United States. More important than the numbers themselves is the sort of knowledge offered by the publishers who hope to reach these students. Students who read textbooks from self-identified Christian-school publishers read a historical narrative that has grown more distinctly sectarian over time. Though the numbers must remain inexact, the evidence suggests that significant numbers of evangelical parents desire a markedly different type of education for their children.

During the 1970s and 1980s, the boom years of Christian schools, parents chose Christian schools largely to instill in their children a competing way of knowing and a competing body of knowledge. Of course, that was not the only reason they chose Christian schools. One study of school-switching parents in

Philadelphia suggested that parents sent children to new conservative evangelical Protestant schools for a range of reasons, including lack of discipline and weak academics at public schools. But the leading reason given by parents was "Moral values learned in the public schools." Specifically, they worried that their children were being exposed to the ideology of "secular humanism." A similar study from California found that many parents considered public schools the "vehicle of a godless humanistic philosophy." The results of these studies suggest that parents chose Christian schools mainly in order to expose their children to an explicitly subcultural knowledge system. In other words, parents wanted their children to know things that weren't taught in public schools, and they wanted their kids not to know some things that were taught there.[6]

Parents' generic worries about public schooling also motivated evangelical intellectuals in the field of US history. Since the 1970s, evangelical writers have made strident cases for distinctly evangelical interpretations of history. The historical methods of these writers, as well as their conclusions, represent a form of knowing and a body of knowledge distinctly at odds with mainstream academic understandings.

Peter Marshall and David Manuel, for instance, proudly let their research for *The Light and the Glory*, published in 1977, be guided by the Holy Spirit. They interpreted America's early history from a religious viewpoint. For instance, in their telling, Metacom's War in the seventeenth century was "God's blast of judgment," and the Salem witch trials came about as a result of an upsurge in "demonic activity." Marshall and Manuel also asked different questions than did mainstream historians. In their consideration of the causes of the American Revolution, they were bothered by the fact that Romans 13—"Let every soul be subject unto the higher powers. For there is no power but of God: the powers that be are ordained of God"—seemed to condemn any rebellious activity. Marshall and Manuel found their answer in Galatians 5:1: "Stand fast therefore in the liberty wherewith Christ hath made us free, and be not entangled again with the yoke of bondage." Since God gave the colonists freedom, if Americans submitted to British tyranny they would be defying God. By infusing American history with a dose of faith, Marshall and Manuel promised to provide a fuller, truer history than anything mainstream historians could offer.[7]

Ten years later, Tim LaHaye, the fundamentalist coauthor of the blockbuster Left Behind series of apocalyptic novels, argued that the nation had endured a "Deliberate Rape of History." Between 1954 and 1976, LaHaye insisted, a generation of "left-wing scholars for hire" had worked for secular organizations such as the Carnegie Foundation. Such authors had systematically distorted the truth of America's Christian heritage. In order to find the true history of America's founding, readers needed to look to older books, written by those "closest to the events they describe." In *Faith of Our Founding Fathers* (1987) LaHaye worked through the Christian beliefs of the Founding Fathers, demonstrating that "most were deeply religious, all had a great respect for the Christian traditions of the colonies, and all were significantly influenced in their thinking by the Bible, moral values, and their church."[8]

More recently, prominent conservative Protestant voices have complained that, in activist Gary DeMar's words, "we are a nation that has lost its historical memory." DeMar decried the tendency of current American society to promote "modern-day secular assumptions and the normless ideals of multiculturalism, political correctness, and moral relativism." Young people often assumed that the current vision of a wall of separation between church and state had always been America's norm. It had not, DeMar insisted. Instead, the United States had always been a nation with a "rich Christian heritage." DeMar hoped his recitation of professed public Christianity from John Winthrop to Jimmy Carter could help open Americans' eyes to the properly and profoundly religious nature of America's public square.[9]

Religious conservatives in search of this kind of polemic have a shelf of similar histories to choose from. Books such as John Eidsmoe's *Christianity and the Constitution: The Faith of Our Founding Fathers* (1987) and Cynthia Dunbar's *One Nation Under God* (2008) have argued for the essentially religious nature of America's heritage. Public figures such as Glenn Beck, author of *Being George Washington* (2011), and Newt Gingrich, author of *Rediscovering God in America* (2009), have also published well-received histories of early America that argue for a Christian understanding of America's public sphere.[10]

Before we can understand the textbook offerings of leading Christian-school publishers, we need to explore the distinctive understanding of history shared among these conservative evangelical Protestant historians. Fundamentalists have been intensely interested in the intersection of biblical revelation and history. Indeed, as historian Grant Wacker has argued, the world of fundamentalism in the United States can largely be defined as "a world without a modern sense of history." Wacker does not suggest that fundamentalists lack a historical sense. Rather, he argues that among some conservative evangelical

Protestants, "insistence that the method and content of revelation were not a function merely of historical processes . . . came to be known, especially in the United States, as fundamentalism." For American fundamentalists, history does not explain scripture; scripture explains history.[11]

Although many self-identified fundamentalists would be loath to associate themselves with Rousas Rushdoony's Reconstructionist theology, Rushdoony articulated one vision of the meaning of history that resonated among many conservative evangelical Protestants. The problem, as Rushdoony defined it in 1969, was that Christian historians and secular historians alike had succumbed to the error of describing history as "totally the product of impersonal natural forces." Without careful attention to the intellectual requirements of what Rushdoony called "Biblical history," even Christian historians could fall prey to the "cult of 'objective, impartial' scholarship." The true meaning of history, Rushdoony insisted, was an unfolding narrative of "the battle of Christ versus antichrist." In order to teach children in an authentically Christian way, historians must consistently revise history "in terms of a consistent and systematic Biblical faith." The central lesson of history for young people, according to Rushdoony, must be "the Messianic Purpose of Jesus Christ." Rushdoony emphasized the notion that truly Christian education must teach a distinctly Christian vision of history. Instead of merely papering over a naturalistic narrative with scriptural quotations, the flesh and bone of history curricula must be reworked to teach history as an unfolding of divine will across time and space.[12]

The intellectual presuppositions of Christian histories such as those Rushdoony advocated stand in stark contrast to those of mainstream academic historians. Some conservative Christian writers acknowledged this distinction. Ronald Horton, speaking for the faculty of the fundamentalist Bob Jones University (BJU), insisted in 1992 on a "basic difference between Christian and secularist thinking about the human past." Instead of starting with man, the BJU faculty agreed, the study of history must begin with a "Biblical view." In the end, to the BJU faculty this meant that "the facts of history are actions either directed or permitted by God." The purpose of studying, teaching, and learning history was to help students conform to the image of God. For this reason, history should emphasize the pathways to God and the price of disobedience. By presenting stories of good Christians in the past, historians could help readers and students understand the meaning of true faith. By studying the "privation and ultimate destruction" that had resulted from disobedience in the past, Christian students could learn to guide their own actions. In the opinion of the BJU faculty, the highest expression of history was the Bible itself. Indeed, they concluded that "God Himself is a historian." The study of history, in sum, was one of the best ways to help a student "shore up his doctrinal beliefs and reinforce his Christian view of the world." Of course, mainstream academic historians also disagree among themselves about emphasis, goals, and interpretation, but they broadly agree on a vision of history distinct from that of biblical historians such as Horton and Rushdoony.[13]

The intellectual presuppositions of writers of mainstream textbooks have long been a concern of conservative evangelical Protestant activists. In 1923, to cite just one example from the 1920s controversies over the relationship between traditional Protestant religiosity and public schools, Minneapolis fundamentalist leader William Bell Riley blustered that "this war will not abate until teachers and text books that scorn the Christian faith are gone!" In the 1940s, the National Association of Evangelicals began a new campaign to root out "objectionable" textbooks from all levels of education.[14]

Beginning in the 1980s, this perennial concern reappeared as an issue of central interest to many conservative Christians. Evangelical leaders such as Gary Bauer and Mel and Norma Gabler and the conservative Catholic activist Phyllis Schlafly insisted that textbooks gave students a distorted vision of America's past. Conservative evangelical psychologist Paul C. Vitz conducted a study of sixty commonly used history textbooks and concluded that such mainstream textbooks intentionally left out the leading role faith had played in US history. This was more than simple oversight, Vitz insisted. The omission of religion was part of "a general rejection of the entire conservative spectrum of American life. These books omit marriage and the traditional family, along with traditional sex roles, patriotism and free enterprise."[15]

Writers from a variety of conservative Protestant backgrounds agreed that mainstream textbooks ignored the key role of faith in US history. Worse, in the opinion of *Sword of the Lord* editor Curtis Hutson, many books "neglect, deny, and distort" the meanings and influence of religion. Paul Bartz, of the *Bible-Science Newsletter*, agreed in 1988 that "virtually all" mainstream textbooks showed "a strong bias against traditional religion and morals." In mainstream history textbooks, Bartz insisted, the religions of the Hopi and Pueblo Indians were described in deep, spiritual detail, while the religious nature of Protestant America was ignored. Bartz complained that mainstream texts robbed events such as the first Thanksgiving, the Reformation, and the journeys of the Pilgrims of their religious meaning.[16]

In 1989, in the pages of the *New American*, William F. Jasper blasted the "blatant prejudice" of mainstream textbook publishers and authors against conservative evangelical Christianity. He claimed that their books promoted communism, portraying "Saintly Sandinistas," heroic Bolshevists, and kindly, popular Maoists. American history texts, Jasper argued, taught that the United States was a brutal imperialist power, that collectivism and the New Deal had saved the country, and that gun-ownership rights were passé. The problems did not stop with their distorted depiction of American history. Mainstream textbooks taught children to rebel against their parents, to have promiscuous sex, and to denigrate religion. In Jasper's opinion, these texts were the result, not of simple prejudice, but of a "long-term, conscious, concerted effort to subvert, through public school curricula, the hearts and minds of American children."[17]

Worst of all, in the opinion of Baptist writer Dale Fischer in 1987, conservative Christians who complained were ridiculed for their efforts to demand accurate history textbooks. Secularist foes, Fischer warned, called "courageous" conservative parents "uneducated, uncultured . . . advocates of book banning and burning; and other choice characterizations." Such parents had to endure accusations that they planned to impose a "narrow-minded and harebrained brand of fundamentalist fanaticism" on the public school curriculum. But such parents, Fischer insisted, only wanted to give their children an authentic history, one in which the significant role of Christianity in America's history received its due.[18]

Different Knowledge, Different Textbooks

In every case, these evangelical writers have argued for a distinctly evangelical history. They have contested both the conclusions and the methods of mainstream historians. And just as the changes in mainstream academic thinking have been reflected in changing mainstream textbooks, the evangelical desire for more conspicuously "Christian" histories has been reflected in the textbooks produced by evangelical publishers.

The alternate histories offered by conservative evangelical Protestant publishers have not been simply static or monolithic, as some journalists have implied. Just as we could not presume to understand the ideology of mainstream public school history education by only reading the works of leading mainstream scholars such as the philosopher John Dewey, history-education scholar Sam Wineburg, or historian Alan Brinkley, so we must look more closely at differences among Christian-school publishers. Each publisher has promoted an idiosyncratic vision of American history, and these approaches have changed over time. As the wider community of conservative evangelical Protestants demanded textbooks that represented what they saw as an authentically Christian alternative to mainstream textbooks, Christian-school publishers intensified their unique theological interpretations of American history. Such textbooks became more intensely fundamentalist over time. At the same time, however, Christian-school publishers adopted some of the language and ideas of mainstream histories. To begin to pick apart these distinctions, we need to examine more than a single publisher's wares, and more than a single edition of these textbooks.[19]

Two of the leading publishers of conservative evangelical textbooks have been Bob Jones University Press and A Beka Book.[20] Evidence of this domination comes from sources such as anthropologist Alan Peshkin's study of a fundamentalist Protestant school. As Peshkin concluded, the impact of Bob Jones University Press is difficult to overstate among conservative evangelical educators. Beginning in 1974, two BJU faculty members composed their first textbook, *Physical Science for Christian Schools.* Faculty members regularly added to the publishing list until it became a comprehensive curriculum. In 1982 Timothy Keesee and Mark Sidwell published the first edition of BJU Press's *United States History for Christian Schools.*[21]

In the 1970s, A Beka Book also established itself as a leading publisher of Christian-school textbooks. Arlin and Beka Horton had founded the press from their headquarters at Pensacola Christian Academy. At first, in the early 1970s, the Hortons purchased some copyrights from out-of-print textbooks. They added scriptural passages and traditional Protestant religious imagery to the older books and republished them. With a burst of Christian schooling in the 1970s, sales of these books quickly grew. In 1982 A Beka Book published the first edition of *United States History in Christian Perspective: Heritage of Freedom.*[22]

A Beka and BJU certainly did not control the entire market. Smaller publishers such as Alpha Omega Publications, Christian Light Education, Mott Media, Rod and Staff, and Konos offered traditional or explicitly Christian textbooks and curricular materials. Other publishers offered materials tailored to specific denominations. But the combined output of these publishers did not rival that of BJU and A Beka.[23]

After BJU Press and A Beka Book, the third largest conservative evangelical school publisher was Accelerated Christian Education (ACE). Unlike other publishers, ACE did not use a traditional textbook format. Donald Howard founded ACE at his small Christian school in 1970. By its peak in the early 1980s, ACE claimed to serve more than eight thousand Christian schools. Howard bemoaned the state of mainstream school publishing. Most US history books, he warned in 1979, "are socialistic and contain the philosophy of the humanistic left." ACE material, in contrast, "was written by separated Christians with a Biblical philosophy of life with Scriptural content where appropriate."[24]

The content of ACE's curricular materials certainly differs markedly from that of mainstream US history textbooks. To take just one example, the third edition of an eighth-grade US history workbook published by ACE told students the following about the founding of Jamestown: "Captain Smith wisely dropped the communistic system and established Scriptural principles. He said, 'He that will not work, shall not eat, except by sickness he be disabled.' Americans who think we should replace our free-enterprise system with communism should learn a lesson from America's first experiment in communism. The early colonists tried communism and found it didn't work." One of the few academic studies of ACE materials found the content "so skeletal that real understanding of the cause and effect of events seems impossible in most cases." When ACE writers did delve more deeply, the authors of this study found, the information was "distorted or inaccurate."[25]

Leaders of both Bob Jones University Press and A Beka Book agreed with other conservative evangelical Protestant intellectuals that history textbooks for Christian schools must differ markedly from those used in secular or pluralistic institutions.[26] Furthermore, leaders at BJU Press and A Beka Book insisted that their textbooks should reflect both their vision of proper historical thinking and a profoundly biblical approach to the teaching of history. As a result, revised editions of textbooks from both publishers combined changes that brought them closer to mainstream academic history with changes that separated their historical narrative more profoundly from that of mainstream textbooks.

To this end, A Beka's "Buzz" Baker suggested that education in Christian schools must differ radically from that in public schools. "We are not in business to train students for the world," Baker wrote in 1979. "We are in business to train them for the Lord." In history as in all subjects, he argued, any knowledge must be based on biblical precepts. In a true Christian school, every teacher and every textbook ought to begin with the Bible.[27] More recent leaders of A Beka have agreed. In 2011 Troy Shoemaker, Administrator of Pensacola

Christian College, insisted that the Bible must be the starting point of all knowledge, including that in history textbooks. To begin from a purely worldly standpoint, Shoemaker argued, from a standpoint that disregarded God's influence in history, was to misrepresent history.[28]

A Beka leaders have long voiced concern that many purportedly Christian textbooks did not adequately differ from those used in secular or pluralistic schools. Baker warned in 1979 that perhaps the "most dangerous pitfall of all" for Christian schools was to use textbooks that did not radically differ in content from those used in public schools. He warned that some nominally Christian schools actually "harm . . . the cause of Christian education" by clinging to mainstream academic thinking.[29]

Leaders at BJU shared Baker's concern. Though their vision of proper Christian education varied significantly from that of A Beka, BJU leaders also worried that many Christian schools allowed themselves to use secular or pluralistic textbooks used in public schools. In 1979 Elmer Rumminger warned in the pages of the BJU School of Education's magazine that many of the textbooks used in Christian schools were "laced through with this Satanic philosophy" of secular humanism. Similarly, the long-serving dean of the School of Education, Walter Fremont, insisted that truly Christian textbooks must do more than "sugarcoat . . . a humanistic philosophy." In order to present US history in a truly Christian way, the BJU faculty wrote in the early 1990s, textbooks must "emphasize the providential circumstance of [America's] founding and associate its prosperity with its obedience to God."[30]

In light of these pronouncements by leaders at A Beka and BJU, it is not surprising that we find in the content of textbooks from both publishers a steady ratcheting upward of a distinctly and self-consciously fundamentalist vision of American history, even as both publishers revised earlier editions to bring them closer to mainstream histories in other ways. These changes can be seen in different editions of textbooks from both publishers: the 1991 and 2012 editions of BJU Press's *United States History for Christian Schools* and the 1982 and 1996 editions of A Beka Book's *United States History in Christian Perspective: Heritage of Freedom.* I do not claim that my method is exhaustive, but it suggests both the differences between these publishers and the differences between earlier and later editions. In both cases, the textbooks not only satisfy conservative evangelical Protestants' demands to tell a very different story about the first British colony in North America but also show a marked increase in these distinctive historical elements over time. That is, in later editions, both publishers strike a decidedly more conservative tone—more distinct from mainstream history textbooks—in terms of both theology and economics.

In some ways, these books differ dramatically from those on offer from mainstream publishers. Both editions of A Beka's US history textbook deride the religions of American Indians. The early edition notes that Indians "worshiped anything which they could not explain." In the 1996 edition, the author argued that "superstition kept the Indians from working together to develop the land in which they lived." Similarly, both editions suggest, in contrast to the findings of mainstream historians, that pre-Columbian Americans lived in a "sparsely populated wilderness."[31]

The editors' changes brought the newer edition closer in some ways to the understandings of mainstream historians. The newer edition emphasizes that Indians in the land that became the United States came from many different cultures, and it differentiated among Indian groups in ways the earlier edition had not. The first edition explains, "The Indians lived very close to the land; they depended on much of the North American vegetation for their livelihood." The revised edition, in contrast, notes more specifically that "tribes living in the Eastern woodlands depended largely on vegetation for their livelihood." The new edition also includes a full-page map showing pre-Columbian language groups.[32]

Most important, these revisions in the treatment of pre-Columbian American civilizations did not represent a détente with mainstream academic history. To the contrary, each revision marked a notably more biblical, more sectarian vision of US history. The expanded treatment of American Indian cultures includes a profoundly sectarian explanation of the roots of pre-Columbian populations. A large map demonstrates the way pre-Columbian Americans likely migrated from the biblical Tower of Babel. As the authors note in the revised edition, the varieties of American Indian culture probably resulted from the worldwide flood described in the book of Genesis. In the language of the revised edition, "Over 6000 years ago, after the great Flood, God caused the people of the world to speak many different languages and scattered them from the tower of Babel 'upon the face of all the earth' (Gen. 11:9)."[33]

The newer edition often adds language suggesting differences of opinion among historians, a change mainstream academic history educators might applaud. But the discussions among historians always support a sectarian interpretation of the Bible. These biblical allusions are not used merely to add moral weight to the content of the passages but to explain the past in strictly biblical terms. Consider the following passage about the migrations of pre-Columbian populations: "Some historians believe that [Indians] came to North America by way of a land bridge across the Bering Strait, which now separates North America and Asia. These two continents could have been divided in the days of Noah's great-great-great grandson Peleg, whose name means 'divided.' The Bible says that the earth was divided in the days of Peleg (Gen. 10:25)."[34]

In a similar fashion, though both A Beka editions emphasize the providence of God in the timing of European exploration of the Americas, the later edition is more explicit about God's role and more insistent that God's will determined the history of American exploration. The first section of the early edition is titled "The Dawn of the Modern Age: A Nation Born in Due Time." In the newer edition the divine connection is clearer: "Dawn of the Modern Age: The Providence of God." Both editions separate history into three ages: the ancient, the middle, and the modern. The earlier edition notes, "Had God allowed America to be born sooner . . . our history would have been very different. . . . As God allowed it, the New World remained unknown to most Europeans until the Modern Age was dawning in Europe." The newer edition again puts more emphasis on God's direct role in the founding of European America: "In God's perfect timing, America was born with the Modern Age. . . . God, in His wisdom, allowed America to remain hidden until the Modern Age had dawned in Europe."[35]

As the narrative of the A Beka textbooks progresses into the first British colonial experiments, both editions make the biblical nature of the founding of Jamestown clear and explicit. Both editions also point out the lessons to be learned about the nature of capitalism and communism. Just as with the timing and meaning of European exploration and the nature of pre-Columbian civilizations, the later edition makes each of its points about Jamestown clearer and more explicit. Both editions note the biblical nature of John Smith's new rules for the Jamestown colonists. Both emphasize the idea that Smith "established a policy based on Biblical principle that any who would not work should not eat (2 Thess. 3:10)." As with the migration of pre-Columbian Americans, the newer edition suggests that history represents a continuing conversation among historians. In this case, speaking of the struggles of Jamestown, the newer edition reads, "As one historian put it, 'Those who tried to live without working soon found that they must also try that harder thing—to live without eating.'"[36]

The newer edition also emphasizes the cultural and economic lessons to be learned from the Jamestown saga. In the first edition, students read the following passage:

> In essence, the old common-store system was a *communistic economic system*, in which the fruits of one's labors were to be shared with all. The new system was a *private enterprise (capitalistic) system*, in which each provides his own needs. Americans had a lesson very early in their history in the benefits of free enterprise, in which each man is allowed to reap the benefits of his own labor.

In the 1996 edition the passage reads as follows:

> In essence, Jamestown adopted a system of *private enterprise* (capitalism). Under a capitalistic system, individuals are free to make a living and prosper on their own enterprise (initiative). This system provided the incentive necessary for the production of sufficient crops for the colony. Private enterprise would become a cornerstone of America's greatness.

In each edition, authors and editors at A Beka clarified and emphasized the ideas that differed most dramatically from those offered by mainstream educational publishers. At the same time, the new edition moved in the direction of mainstream history education in other areas, by emphasizing the notion of history as an ongoing discussion among historians and by noting the vast diversity among pre-Columbian Americans.[37]

The textbooks developed by BJU Press differed dramatically in tone and content from those of A Beka Book. BJU's textbooks recognized more consistently the diversity of cultures among pre-Columbian Americans, for example. But BJU's textbooks insisted just as ardently as those from A Beka on the biblical meanings of American history. BJU texts also underwent similar double revisions. The later edition moved in some ways closer to mainstream history. In other ways, BJU editors and authors made the sectarian religious meanings of American history more profoundly distinct in later editions.

Both editions of BJU's US history textbook emphasize the diversity and cultural accomplishments of pre-Columbian American civilizations. In the fourth edition, the section on these civilizations is greatly expanded and has a revised structure. The fourth edition both compares and contrasts various European and Indian culture groups. Each "Indian civilization" in the fourth edition is described as "diverse" even within itself. The second edition notes various "tribes" and "culture areas" and states that "American Indians were not backward savages; they built societies of surprising complexity." However, the treatment in the second edition is markedly briefer and less nuanced than that in the fourth edition. Beyond expanding the treatment of pre-Columbian Americans, the fourth BJU edition also includes a section on technological innovations in China in its discussion of "old world" civilization.[38] Both editions stress the biblical nature of John Smith's rules in Jamestown. Both note that "he enforced the Biblical principle of 2 Thessalonians 3:10—that everyone should earn his livelihood through diligent work." Both also make the point that the communal nature of Jamestown's early economy doomed it to failure.[39]

The most striking revision in the BJU textbook comes at the end of the chapter. In the second edition, students are asked to answer review questions about the content of the chapter. The questions are separated into categories: "Terms," "Content," and "Application questions." None of the questions has anything explicitly to do with the Bible or life as a sectarian Christian. In the newer edition the end-of-chapter questions are dramatically changed , asking students to consider the lessons of early American history in explicitly cultural and religious ways. For instance, one "section quiz" asks students, "How did the trials and successes of Jamestown manifest the core values of freedom, individualism, equality, and growth? What contradictions to these values began to develop?" Another question reads: "Referring to Deuteronomy 6:10–15, describe how the Puritans failed to heed the warning that God gave the Israelites. How does this passage apply to your generation?" Similarly, the fourth edition asks students to consider the following: "In light of what you have learned about religion and liberty, use the Bible to respond to the argument that denying homosexuals the right to marry denies them fundamental liberties that heterosexuals possess."[40]

Tellingly, the editors framed these new questions in new ways. Instead of simply prompting students to review the content of a chapter, the new questions were meant to help students in "Making Connections," "Developing History Skills," "Thinking Critically," and "Living as a Christian Citizen." With the exception of this last category, the new language mirrors the language of mainstream history textbooks. Learning history, in the view of mainstream history educators, was not only a matter of reciting facts but also a challenge to engage in what Sam Wineburg has called the "unnatural act" of thinking historically.[41]

Like the A Beka textbooks, the editions from BJU Press changed in two related ways. In some senses, they became more similar to the offerings from mainstream publishers. They expanded their treatment of non-European, non-white agents as essential to understanding US history. They also urged students and teachers to engage in historical thinking beyond rote memorization. But the BJU texts, like those from A Beka, also changed to emphasize the editors' distinctly sectarian vision of American history. The lessons of the past, according to BJU editors, ought to push students to live more completely as Christian citizens.

Cultural Occlusion: What Not to Know and How Not to Know It

How are we to understand the knowledge system advanced by these textbook publishers? Their efforts represent something with which historians have not adequately grappled. In these textbooks we see the deliberate occlusion of mainstream historical knowledge for cultural purposes. The authors and editors knowingly moved their textbook content further away from mainstream historical methods and knowledge. They did so in order to advance a culturally distinct body of knowledge and way of knowing about the past. They wanted students to know history through the lens of biblical hermeneutics. They wanted students to know a history that insisted on the primary role of God's will in determining historical events. We can't know much about the readers of these textbooks by looking at the textbooks alone, but it is clear that the publishers hoped to instill a vision of America's past that would block mainstream historical knowledge. The textbooks examined in this chapter represent an effort to create ignorance about mainstream history by instilling knowledge about a contending approach to historical understanding.

Christian-school publishers are not alone in occluding knowledge about the past. As historians and history-education scholars David Lowenthal, Sam Wineburg, and Robert Proctor have argued, our knowledge of the past is often blocked. Ignorance is often constructed for a variety of reasons. For David Lowenthal, textbooks of all sorts represent "heritage" rather than "history." To his mind, some critics of "heritage" contend that "heritage effaces history's intricate coherence with piecemeal and mendacious celebration, tendering comatose tourists a castrated past." Such critics of "heritage," he argues, focus on the difference between academic visions of historical knowledge and other sorts. Those "heritage" versions that differ from academic histories, these critics insist, are simply "bad history." But Lowenthal makes a more subtle argument. He agreed that history and heritage have different goals. The essence of history is to understand the past on its own terms. In contrast, the essence of heritage, Lowenthal suggests, is a "declaration of faith."[42]

The Christian-school textbooks examined in this chapter fit into this definition of *heritage*. Authors and publishers sought to use history to convince students of theological truths. However, Lowenthal argues that all school histories tend to devolve into heritage rather than history. Even mainstream school histories represent "a saga of civic pieties to be swallowed on faith." As a result, "school history is more heritage than history." In the cases analyzed by Lowenthal, however, the mainstream history textbooks' "heritage" approach to history did not consist in a deliberate replacement of mainstream historical knowledge with a declaration of faith. It is unlikely that mainstream textbook authors and publishers would agree that their work did not represent mainstream historical knowledge. In the case of these Christian-school textbooks, on the other hand, we see a deliberate and explicit replacement of mainstream historical knowledge with knowledge that supports a sectarian interpretation of America's early history. Unlike publishers of mainstream history texts, the makers of these Christian textbooks would not dispute the suggestion that their books were a "declaration of faith." Indeed, over the years, Christian-school textbook publishers and authors worked deliberately and publicly to increase the distinction between the knowledge claims in their books and those of mainstream history textbooks.[43]

History-education scholar Sam Wineburg has also looked at the complex meanings of knowledge and non-knowledge in history education. As Wineburg points out, students do not come to history classes as blank intellectual slates. Most high school students in his study had rich and false knowledge bases about history. Most commonly, students "know" false facts about the past based on their viewing of movies such as *Forrest Gump* and *Schindler's List*. As they confront historical knowledge in school, these students are already full of knowledge, but knowledge derived from what Wineburg calls a "fictionalized past, not the historical event."[44]

In addition, Wineburg argues that students inherit a "collective occlusion." What they know about the past has often been blocked by mainstream cultural tendencies to oversimplify past events. Even when students confront evidence that contradicts their inherited categories, students tend to mold the evidence to fit their occluded understandings of the past. For example, Wineburg showed high school students pictures of demonstrations in support of US policy in Vietnam. Most of the students incorrectly identified the protesters as antiwar. Students had inherited an oversimplified, occluded vision of protests during the Vietnam War as necessarily antiwar. In spite of clearly legible pro-war protest signs, students tended to fit this evidence into their occluded understanding of the past. Students only rarely changed their understanding of the past in light of contradictory evidence.[45]

The Christian-school textbooks examined in this chapter, on the other hand, do not merely accidentally block mainstream knowledge about the past. Rather, they make a deliberate effort to instill a system of knowledge that would serve to occlude mainstream understandings. These textbooks represent a kind of constructed ignorance that is distinct from the occluded knowledge displayed by Wineburg's subjects.

Similarly, these books represent a different sort of constructed ignorance from that studied by Robert Proctor. Proctor studied tobacco companies that sought to buy plausible ignorance. The goal of tobacco executives in the 1960s, Proctor argued, was explicitly to create ignorance of knowledge agreed upon by tobacco executives. That is, tobacco executives knowingly hoped to obscure the public's acceptance of a certain sort of knowledge, to replace correct knowledge with plausible ignorance. If scientists could be persuaded to publish studies that seemed to contradict claims about the health dangers of smoking, tobacco executives could plausibly claim not to know that smoking was dangerous. This "strategic ploy" ignorance represented one sort of constructed non-knowledge.[46]

These textbook publishers engaged in no such strategic ploy. In these cases, the competing knowledge system was seen by publishers, teachers, and parents as superior to those of mainstream academic historians or mainstream school history textbooks. There was no strategic ploy to spread ignorance, but rather a concerted effort to replace mainstream knowledge with knowledge in accord with the beliefs of a cultural minority. In the language of evangelical intellectuals and publishers, the goal was to incorporate a more profoundly biblical understanding of history, an understanding of history distinct in both knowledge and ways of knowing from mainstream history. Each revised edition of these textbooks moved further in this direction from mainstream knowledge. But evangelical intellectuals and pundits never claimed that the ignorance produced by these textbooks was something only for others. Rather, evangelical intellectuals insisted that these sorts of Christian textbooks would bring readers closer to the truth. Textbook authors, editors, and publishers attempted to realize those desires. At least as seen in the small sample of textbooks examined here, Christian-school publishers crafted histories that made knowledge claims more profoundly distinct over time.[47]

To damn these textbooks as incorrect by pointing out the many differences between them and mainstream textbooks is not the point. It is true that these books grew more markedly distinct from mainstream textbooks over time. The writers of these books did not grow more unaware of mainstream textbook content; they did not simply become more ignorant of "real" history. Rather, the publishers worked to improve the ways their textbooks incorporated a distinct knowledge system. In this case, we see a systematic attempt to occlude mainstream knowledge about the past by substituting a culturally distinct body of knowledge and way of knowing. We might call this *cultural occlusion*.

As a result of these efforts, we see a phenomenon in which educated adults confidently evince an understanding of US history starkly at odds with that of mainstream academic historians. When Representative Bachmann proclaimed her faith in the antislavery efforts of America's early leaders, she did not do so out of naïve ignorance. Bachmann likely had learned a story of America's founding distinctly different from the one mainstream historians would tell. Similarly, young people reading history textbooks from Christian-school publishers will not simply miss out on historical knowledge. Rather, the studious among them will learn a history in which God plays a decisive role, a history in which the truths of capitalism trump early collectivism. Their knowledge of US history will not simply be absent. Rather, insofar as they learn anything from their textbooks, their knowledge will be blocked and supplanted by an alternative version of both the facts of the past and their meanings.

NOTES

1. Z. Bryon Wolf, "Michele Bachmann in Overdrive: Her 10 Greatest Hits," *The Note* (blog), ABC News, accessed 27 January 2014, http://abcnews.go.com/blogs/politics/2013/05/michele-bachmann-in-overdrive-her-10-greatest-hits/; Robert N. Proctor, "Agnotology: A Missing Term to Describe the Cultural Production of Ignorance (and Its Study)," in *Agnotology: The Making and Unmaking of Ignorance*, ed. Robert N. Proctor and Londa Schiebinger (Stanford, CA: Stanford University Press, 2008), 3. For examples of works of conservative counterhistory, see, e.g., David Barton, *The Jefferson Lies: Exposing the Myths You've Always Believed about Thomas Jefferson* (Nashville: Thomas Nelson, 2012); John Eidsmoe, *Christianity and the Constitution: The Faith of Our Founding Fathers* (Ada, MI: Baker Academic, 1987); and Glenn Beck, *Being George Washington: The Indispensable Man, As You've Never Seen Him* (New York: Threshold, 2011).

2. Sam Wineburg, "Making (Historical) Sense in the New Millennium," in *Historical Thinking and Other Unnatural Acts: Charting the Future of Teaching the Past* (Philadelphia: Temple University Press, 2001), 232–55; David Lowenthal, *Possessed by the Past: The Heritage Crusade and the Spoils of History* (New York: Free Press, 1996).

3. This effort to produce a compelling new historical understanding is certainly "strategic," but it does not fit neatly into the notion of a "ploy." After all, evangelical intellectuals and publishers herald their new histories as an improvement over mainstream academic history.

4. In the world of conservative evangelical Protestantism, labels can be contentious and difficult to agree upon. In general, conservative evangelical Protestantism in the United States includes several subgroups, including *fundamentalists*, *evangelicals*, *Pentecostals*, and others. As a rough guide, these groups are all united by a theology that insists on an inerrant view of scripture and adherence to traditional theological tenets such as the virgin birth of Jesus and the real, historical Resurrection. The schools called "Christian" schools have sometimes catered to a single denomination, such as Baptist, Christian Reformed, or Presbyterian. More often, though, these schools have opened their doors to Protestants from a variety of conservative backgrounds. In this chapter, I will use the term *evangelical* or *conservative evangelical* to refer to the broadest coalition of conservative evangelical Protestants. Only when necessary or helpful will I make distinctions between *fundamentalist* or *evangelical* identities. I will also follow common usage and call these schools *Christian* or *Christian day* schools instead of insisting on more theologically precise labels. As a general rule, the textbook publishers examined in this chapter came from the "fundamentalist" wing of evangelicalism. As a result, their textbooks are likely most popular among stricter, more separatist schools and churches. For a more detailed look at the theological wranglings among these schools, see Adam Laats, "Forging a Fundamentalist 'One Best System': Struggles over Curriculum and Educational Philosophy for Christian Day Schools, 1970–1989," *History of Education Quarterly* 50 (February 2010): 55–83; S. D. Rose, *Keeping Them Out of the Hands of Satan: Evangelical Schooling in America* (New York: Routledge, 1988), 34; Paul F. Parsons, *Inside America's Christian Schools* (Macon, GA: Mercer University Press, 1987), x; *Characteristics of Private Schools in the United States: Results from the 2001–2002 Private School Universe Survey*, NCES 2005-305) (Washington, DC: US Department of Education, National Center for Education Statistics, 2002), 27; American Association of Christian Schools website, accessed 24 October 2012, http://www.aacs.org/about-us/; Association of Christian Schools International website, accessed 24 October 2012, http://www.acsiglobal.org/; Christian Schools International website, accessed 7 November 2012), http://www.csionline.org/; and James C. Carper and Thomas C. Hunt, *The Dissenting Tradition in American Education* (New York: Peter Lang, 2007), 203–5.

5. US Department of Education, National Center for Education Statistics, *Parent and Family Involvement in Education, from the National Household Education Surveys Program of 2012*, NCES 2013-028, August 2013, http://nces.ed.gov/pubs2013/2013028/findings.asp; Dan Gilgoff, "As Home Schooling Surges, the Evangelical Share Drops," *U.S. News & World Report*, 9 January 2009, http://www.usnews.com/news/religion/articles/2009/01/09/as-home-schooling-surges-the-evangelical-share-drops.

6. Martha E. MacCullough, "Factors Which Led Christian School Parents to Leave Public School" (EdD diss., Temple University, 1984), 114; Peter Stephen Lewis, "Private Education and the Subcultures of Dissent: Alternative/Free Schools (1965–1975) and Christian Fundamentalist Schools (1965–1990)" (PhD diss., Stanford University, 1991), 132.

7. Peter J. Marshall and David Manuel Jr., *The Light and the Glory* (Old Tappan, NJ: Fleming H. Revell, 1977), 146, 254, 223, 236. For their wording of Romans 13 and Galatians 5:1,see 257 and 15, respectively.

8. Tim LaHaye, *Faith of Our Founding Fathers* (Brentwood, TN: Wolgemuth & Hyatt, 1987), 5, 6, 30.

9. Gary DeMar, *America's Christian Heritage* (Nashville: Broadman & Holman, 2003), 83, 7, 81.

10. Conservative pundits have offered glowing reviews of each book. See, e.g., John Mark Reynolds, "*Being George Washington*: A Live Blog Review," *First Things* (blog), 26 November 2011, http://www.firstthings.com/blogs/firstthoughts/2011/11/being-george-washington-a-live-blog-review; Jesse Gordon, "Book Review: Rediscovering God in America," On the Issues, May 2007, http://www.ontheissues.org/Rediscovering_God.htm.

11. Grant Wacker, "The Demise of Biblical Civilization," in *The Bible in America: Essays in Cultural History*, ed. Nathan O. Hatch and Mark A. Noll (New York: Oxford University Press, 1982), 127.

12. Rousas J. Rushdoony, *The Biblical Philosophy of History* (Nutley, NJ: Presbyterian and Reformed Publishing, 1969), 1, 25, 111, 5, 16, 20.

13. Ronald A. Horton, ed., *Christian Education: Its Mandate and Mission* (Greenville, SC: Bob Jones University Press, 1992), 113–15.

14. William Bell Riley, "Reply to University Regents on the Evolutionary Controversy," *Christian Fundamentals in School and Church*, July–September 1923, 51; William M. McCreery, National Association of Evangelicals, form letter to be sent to college presidents, 26 January 1949, Bob Jones University Fundamentalism File, Bob Jones University Library, Greenville, SC.

15. Ann Wharton, "History Texts and Patriotism," *Fundamentalist Journal*, May 1985, 68–69; Paul C. Vitz, "What Can You Do about Textbook Censorship?," *Moody Magazine*, September 1987, 76.

16. Curtis Hutson, "Study of the Public-School History Texts Shows Neglect of Religion," *Sword of the Lord*, 1 June 1984, 2; Paul Bartz, "Eliminating Christianity from School Textbooks," *Bible-Science Newsletter*, June 1988, 3.

17. William F. Jasper, "Public School Students Learn Lies," *End-Times News Digest*, October 1989, reprinted from the *New American*, 28 August 1989, 2–6.

18. Dale Fisher, "Who Are the Bigots & Censors?" *Baptist Bulletin*, March 1987, 11–12.

19. For examples of journalistic assumptions, see, e.g., Thomas Vinciguerra, "Here's the Problem with Emily Dickinson," *New York Times*, 27 November 2005,http://www.nytimes.com/2005/11/27/weekinreview/27vinciguerra.ART.html?pagewanted=all; and "Selections from Fundamentalist School Textbooks," *Free Inquiry*, Fall 1987, 15.

20. Laats, "Forging a Fundamentalist 'One Best System.'"

21. Alan Peshkin, *God's Choice: The Total World of a Fundamentalist Christian School* (Chicago: University of Chicago Press, 1986), 135, 63, 115; Daniel L. Turner, *Standing without Apology: The History of Bob Jones University* (Greenville, SC: Bob Jones University Press, 1997), 265; Timothy Keesee and Mark Sidwell, *United States History for Christian Schools* (Greenville, SC: Bob Jones University Press, 1982).

22. Phillip Smith, provost emeritus, Bob Jones University School of Education, personal communication to author, 5 June 2008; Walter Fremont, "The Christian School Movement Today" (speech, 31 July 1989), audiotape in Bob Jones University archives; A. A. Baker, *The Successful Christian School: Foundational Principles for Starting and Operating a Successful Christian School* (Pensacola, FL: A Beka Book, 1979);

Michael R. Lowman, *United States History in Christian Perspective: Heritage of Freedom*, with Laurel Hicks, George T. Thompson, and the editorial department of A Beka Book Publications (Pensacola, FL: A Beka Book, 1982).

23. "Alpha Omega Publications," *Family Resources* 1 (Spring 1988): 1; "Home School Basic Resource Guide," *Teaching Home*, 1987–88; *Family Resources* 1 (Spring 1988).

24. Gary Coombs, "ACE, An Individualized Approach to Christian Education," *Interest*, September 1978, 9–10; "History and Development of ACE," *CLA Defender* 1, no. 5 (1978): 6, 25–26; *Facts about Accelerated Christian Education* (Lewisville, TX: Accelerated Christian Education, [1982]); Fremont, "Christian School Movement Today"; Donald R. Howard, *Rebirth of Our Nation: The Decline of the West, 1970s; the Christian Educational Reform, 1980s* (Lewisville, TX: Accelerated Christian Education, 1979), 123, 296.

25. *PACE Social Studies 1086* (student workbook) (Lewisville, TX: Accelerated Christian Education, 1998), 15; Dan B. Fleming and Thomas C. Hunt, "The World as Seen by Students in ACE Schools," *Phi Delta Kappan*, March 1987, 523.

26. Because ACE publications come in the form of disposable workbooks rather than traditional textbooks, I have not been able to locate older editions of ACE materials for the purposes of this analysis. Nor has the staff of ACE been willing to make older editions available for study. Walter Schumacher, Accelerated Christian Education Product Advancement Team, personal communication with author, August 28, 2012. As a result, this chapter compares older and newer editions of US history textbooks from BJU Press and A Beka Book.

27. A. A. Baker, *The Successful Christian School* (Pensacola, FL: A Beka Book, 1979), 32, 37.

28. Troy Shoemaker, "Philosophy of Education," A Beka Resources, accessed 25 October 2012, http://www.abeka.com/Resources/Articles/ArticlePDFs/ABBPhilosophyOfChristianEd.pdf.

29. Baker, *Successful Christian School*, 20, 32.

30. Elmer Rumminger, "Secular Humanism in Christian Schools?" *Balance* 1:3 (1979): 1; Walter G. Fremont, "Straight Talk on Traditional Education," *Balance* 2 (May 1982): 1; Horton, *Christian Education*, 121.

31. Lowman, *United States History in Christian Perspective*,1982 ed., 10; Michael R. Lowman, *United States History in Christian Perspective: Heritage of Freedom*, with Laurel Hicks, George T. Thompson, and the editorial department of A Beka Book Publications (Pensacola, FL: A Beka Book, 1996), 7, 5; 1982 ed., 10, 12. For an example of the views of mainstream historians, see William Cronon, *Changes in the Land: Indians, Colonists, and the Ecology of New England*, rev. ed. (New York: Hill & Wang, 2003), 87.

32. Lowman, *United States History in Christian Perspective*, 1982 ed., 12; 1996 ed., 5, 6.

33. Ibid., 1996 ed., 5.

34. Ibid.

35. Ibid., 1982 ed., 2; 1996 ed., 5; 1982 ed., 2; 1996 ed., 5.

36. Ibid., 1996 ed., 27. The 1982 edition reads, "[Smith] established and enforced a policy, based on Biblical principle (II Thess. 3:10), that those who were not willing to work would not eat" (30). These connections between conservative Protestant biblical hermeneutics and procapitalist ideology have a long tradition in American educational activism. See, e.g., Adam Laats, *The Other School Reformers: Conservative Activism in American Education* (Cambridge, MA: Harvard University Press, 2015).

37. Lowman, *United States History in Christian Perspective*, 1982 ed., 31, emphasis in the original; 1996 ed., 28, emphasis in the original.

38. Timothy Keesee and Mark Sidwell, *United States History for Christian Schools*, 4th ed. (Greenville, SC: Bob Jones University Press, 2012), 9–14; Keesee and Sidwell, *United States History for Christian Schools*, 2nd ed. (Greenville, SC: Bob Jones University Press, 1991), 38–39; 4th ed., 3.

39. Ibid., 2nd ed., 13; 4th ed., 27; 2nd ed., 13–14; 4th ed., 25.

40. Ibid., 2nd ed., 36; 4th ed., 29, 43.

41. Ibid., 4th ed., 43; Wineburg, *Historical Thinking and Other Unnatural Acts*.

42. Lowenthal, *Possessed by the Past*, 102, 105, 121.

43. Ibid., 126.

44. Wineburg, "Making (Historical) Sense in the New Millennium," 238–41.

45. Ibid., 242, 247.

46. Proctor, "Agnotology," 9.

47. To draw a finer distinction, there was certainly a "strategic" effort to block mainstream historical knowledge. But it does not fit the historic record to call this a "ploy." Evangelical authors and pundits hoped to give students better history, not worse. Authors hoped that their distinct knowledge and ways of knowing would allow students to remain ignorant—or perhaps innocent—of mainstream historical knowledge. Unlike Proctor's scheming tobacco executives, these textbook publishers did not hope to muddy knowledge only for a duped public. Rather, they hoped to impart an improved knowledge that would simultaneously occlude flawed mainstream historical knowledge. Calling this a "ploy" does not do justice to this benevolent construction of ignorance. Critics might object that the end goal of these publishers remains the intentional creation of ignorance, which does indeed fit in the category of "strategic ploy." One could say that the goal of the textbook writers was to confuse knowledge of history by knowingly introducing alternative sorts of knowledge from the field of theology. However, in my opinion this line of argument underestimates the sincerity of the textbook writers. They were not hoping to peddle what they knew was a worse history in order to shield students from knowledge that might diminish their faith. Rather, they hoped to introduce students to a better history that would bolster their faith. In its broadest terms, this is similar to the sorts of in-group education that every minority group engages in. It only becomes remarkable when, as in this case, the group is a numerically large one that promotes a unique body of knowledge. We need a new term to describe this sort of cultural in-group occlusion of mainstream knowledge with dissenting knowledge.

9

History

DONALD WARREN

Sometime around 1710 four Mohawk chiefs, possibly five, arrived in London under the aegis of the Society for the Propagation of the Gospel in Foreign Parts. They weren't the first Indians to cross the Atlantic for English destinations, and they wouldn't be the last. Still, such voyages were unusual, and a delegation of Native dignitaries to foreign shores even rarer. One wonders what the chiefs thought of their experience and how the English responded to them. In retrospect the moment appears inconsequential, but at the time it teemed with educative potential begging for conjecture and analysis. For the Mohawks, the journey alone, back and forth across an unknown ocean, must have amazed. The English too entered unfamiliar territory when confronted by their guests' physical bearing, regalia, language, mores, and spirituality. Either party may have seen savages, royalty, or confusing blends of habits. Granted, the Mohawks had more experience with such interactions than most of their foreign hosts.

Historian Lawrence Cremin mentioned this astonishing cultural encounter as a brief aside in his three-volume study *American Education*. He chose not to pursue the matter probably because surviving written descriptions lacked context, consistency, and reliable origins. On other topics, however, he willingly and insightfully guessed. In the end, the Mohawk meeting slipped from Euro-American recorded memory. Segments of it may yet be recoverable within Mohawk and Iroquois oral traditions, the kinds of sources Cremin cited in general references to American Indian pasts.[1]

What piqued his curiosity instead was why colonial sectarian evangelism among Indians faltered. With regard to the Iroquois, he relied on accounts produced by a missionary at the time who offered explanations: mistreatment of Indians by land speculators, disreputable behavior by white military personnel, and misrepresentation of English motives by Dutch merchants seeking advantages in their trade with the widely dominant Iroquois. Cremin added an implication. The missionary "could scarcely carry on his work in the face of this larger and more powerful education." Discordant education and discordant pedagogies surfaced as interpretative themes in his ambitious survey. Summarizing colonial experience, Cremin posted a binary framework for analyzing Indian and European cultural conflicts, with "the more powerful education of tribal folkways on the one hand and white exploitation on the other." A fair fight didn't ensue. Rather, "the vast majority of the Indians were formed by tribal values for a way of life that was at best marginal to the social mainstream, and at worst crumbling." Cremin found education within the dynamics of Native cultures and judged it deficient. By the American Revolution, Indian "ways" were long, insurgent struggles against European institutions and strategies of teaching and learning—churches, families locked within social-class hierarchies, property ownership, monotheistic moral imperatives, schools and colleges, modes of written communication and recordkeeping, nascent hegemonic momentums of the invading economy, and related technologies.

This chapter explores methodological and narrative effects of what can be classified in Cremin's universe as a war on education: serial confrontations, often bloody, among quite different and contested practices of cultural continuity and change. Looking back from what would become the United States, he didn't ask whose social mainstream qualified as normative during the colonial period, apparently assuming the answer was known. Not pressing this question has deleted topics from the menu of research interests enticing education historians. Most significantly, it deflected curiosity from the educative processes and institutions active among Indigenous Americans before and after contact with white intruders. The chapter develops a preliminary examination of the roles played by historians of education in such historiographical omissions and circumnavigations. These and related missteps, committed wittingly or unwittingly, have left us intellectually poorer, or, as some have begun to describe it, with a socially constructed ignorance about our collective past. With regard to North American Native peoples, the errors have coalesced to form an arc of knowledge lost, forgotten, or devalued. It threatens to confine us within a habit-forming ignorance in the same way a trusted lens colors historical conscience and consciousness. To break free, historians might ask who is "we" in historical narratives. One can imagine a Native historian responding to Cremin's orientation as inventive folly that locates contact, and Euro-American schooling along with it, as the colonial-era intersection where history ends. An Indian might propose instead that since then education historians have inhabited a posthistorical mind-set wherein repetitions remain abstract and fail to delineate events fully, much less edify. The following pages cite Indian scholars and those who rely on them to assess the damage of the distortions within mainstream literature.[2]

This chapter aspires to depict education history within what might be indelicately termed a broader realm of inquiry. It focuses on certain education history practitioners with the aim of relocating their subfield to history's center stage. In recent years insecurities over the specialization's relevance and narrow audience have prompted attention to policy issues. Needed, some insist, are public, op-ed contributions to discussions on race, gender, testing, teacher competence, and other matters affecting what students know and can do. Historians, we hear, are equipped to advance our understanding of these issues by explaining what has worked well or not so well in the past. At minimum, the tactic garners attention if not respect. This chapter suggests an additional, not an alternative, orientation, namely, that education history's benefits originate from and depend upon diverse research designs. Its essential tasks become historical diagnoses of cultures forming or in distress, probes with the potential to generate richer, sharper questions about what schools and teachers do and why they matter. History begins with historians but doesn't have to end with them, and shouldn't.

Other Conceptual Foundations

As if speaking to the Lawrence Cremins of the field, American Indian scholars and specialists have suggested that education historians need sharp eyes and ears ready to hear. The suggestion may seem odd. It's meant collegially to aid detection of often muted and submerged noises of cultures in motion. The advice has far-reaching applications. Ill-prepared education historians can miss David Van Reybrouck's *Congo*, Amanda Vaill's *Hotel Florida*, and Theodore Rosengarten's *All God's Dangers* as deeply nuanced reconstructions of learning processes. Van Reybrouck configures the long, turbulent histories of the Congolese from the pre- to the postcolonial era, determined to find Indigenous perspectives on lessons wrought from European and US imperialism. Vaill uses a much frequented hotel in Madrid to filter experiences of the Spanish

Civil War, asking variations on a single question: What did participants learn from the conflict? By contrast, Rosengarten focuses on one man, a black Alabama sharecropper who labored across the twentieth century, and produces a study of cultures fixed, fluid, and ironically integrated. In one critic's view, *All God's Dangers* is the book that "best explains the [US] South." It won a National Book Award in 1974 and then vanished from scholarly and popular discussion, including the education canon, where it never found a home in the first place.[3]

Education historians might also overlook Thomas Piketty's magisterial *Capital in the Twenty-First Century*, which, like Daniel T. Rodgers's *Age of Fracture*, scans the history of economics for clues to the ways and means of the education of the public. Similarly, Werner Sollors's *Temptation of Despair* looks for the public emerging in Germany after World War II. Despair becomes an educational bulwark in the same way that inequality functions as an educating institution in Piketty's analysis, and social fragmentation surfaces as a vehicle of widespread teaching and learning in Rodgers's work. Save for Rosengarten and Rodgers, none of these authors pay much attention to schools or other Euro-American pedagogical institutions; all treat education as the fraught process of cultural formation.[4]

They have methodological similarities as well. The authors merge research designs of various disciplines and rely on extensive resources ranging across archaeological digs, commercial motion pictures, correspondence, documentary films, fiction, government records, news reports, oral history, photojournalism, and statistics. They neither disparage precolonial times as prehistory nor otherwise elevate western European culture as the preeminent arbiter of ideas, values, explanations, and meanings. They consider oral traditions and storytelling historically useful and in some cases necessary. There's no presumed progressive arc driving the inquiries and explicitly denying the fantasized ameliorative powers of time. Things don't necessarily get better. Not all wounds heal, and those that do may leave barely visible scars. Preferences, like other actions, are choices that have consequences, whose roots the authors intend to find. They select words carefully, demonstrating the importance of history writing. Prose, energized by verbs, becomes the third front of their history projects, the others being method and topic. Although intimately joined, as in a string ensemble, each plays a unique role. By these means, but not expressly, the authors push education history onto less trafficked landscapes. Their work implies the field is again due for revision, the research agenda needing new formats and sensibilities, as Native scholars have advised.

What's required in this revision is an expedition into sources, published ones primarily, synthesizing works not usually linked to education history. The changes move from a definition of historiography as not only literature review but also, more basically, critique of research methods. The task is to join historians on their hunt for sources and examine how they reach their findings. Doing so raises a debatable proposition: Education history's intrinsic and most inviting contributions begin as research methods. Hypothetically, any topic can be entertained, and reach important destinations, if methods are driven by accumulating evidence of when, where, and by what means individuals and groups have learned, for good or ill. The approach allows historians a functioning definition of education from which to begin and proceed.

American Indian histories offer particularly valuable analytical levers for getting at questions of substance, method, and narrative. Collectively, they provide prismatic case studies of what happens methodologically when education historians attempt to cleanse their work of ethnocentrism and similar predispositions. Such efforts inevitably fall short, given durable habits of mind, but the research process itself benefits. It can bloom profusely. Education history can satisfy a basic standard. It must be educative along the path toward discovery of a particular relevant topic. In the education history literature, conceptual and evidentiary problems stem from historians' settling on topics in Native education at the outset. A revealing premature selection elevates white-imposed boarding and reservation schools at the expense of older Indigenous traditions of teaching and learning. This chapter argues that the choice has been unnecessary and misguided. Its conclusions apply to indigenous histories of other nations, but the probability is merely hinted.[5]

From several angles, Bernard Bailyn must bear some of the blame. His problematic definition of education has cast a long shadow over Native Americans' perspectives and experiences and over others' ability to observe both their own and different cultural patterns and values. He presented his now famous paper on the history of US education in October 1959 as part of a conference series hosted by the Institute of Early American History and Culture at Williamsburg, Virginia. Bailyn defined education "as the entire process by which a culture transmits itself across the generations." One problem stems from the presence of multiple cultures in colonial America, while another may be seen in the potential for Bailyn's definition to simplify the dynamics of confrontations among different cultures. If the cultures are prioritized, even tacitly, their changes likewise can be demoted without attribution. Burden falls on when, how, and by whom a culture, as well as the notion of culture itself, is defined. Bailyn brought to the conference a more restricted agenda to draw battle lines among those who study education history. The broad construction, he argued, enabled historians to see "great variations" as "schools and universities fade in significance next to other social agencies," to see "education in its elaborate, intricate involvements with the rest of society" and "its shifting functions, meanings, and purposes."[6]

At the time, Bailyn's definition received both warm and cool receptions for moving beyond the confines of formal pedagogy. Many celebrated an open, more empirically grounded version of education history. It liberated those in the field from unquestioning advocacy of schools, colleges, and similar institutions. Critics, on the other hand, feared the new definition signaled a retreat from battles over publicly financed and controlled common schooling. They also warned that Bailyn's hypothesis couldn't produce written histories distinguishable from general history. The history of education would disappear as a specialization. Almost at once, Lawrence Cremin set out to prove the naysayers wrong, subsequently devising more detailed, if malleable, definitions of *education* to inform his comprehensive survey. Other revisionist trends ensued, but in the twenty-first century self-identified historians of education routinely work from Bailyn's assumptions and definition and Cremin's examples. Nevertheless, education historians seem to have returned to a schooling focus, as some have found in the pages of the *History of Education Quarterly*, the *American Educational History Journal*, and conference programs of the History of Education Society, the Organization of Educational Historians, and the Southern History of Education Society—all US-based associations. Work emanating from other countries may limn a more intricate contemporary portrait of the field.[7]

Cremin refined and augmented Bailyn's critique of education history but left it basically unchallenged. Furthermore, as Sol Cohen observed in detail, Cremin's *American Education* trilogy proved Bailyn's critics right. A coherent, fulsome survey of US education broadly defined couldn't be written with all the inclusiveness they had envisioned. To obtain a manageable working sample of teaching and learning, they defaulted to preemptive Eurocentric benchmarks. Both Cremin and Bailyn aimed for comprehensive hypotheses and explanations yet glossed disputes revealing irreconcilable cultural differences. They tended not to examine taxation and funding controversies as indicators of education's priority status. Both tethered their histories to the expansion of European styles of schooling and their financing before the Revolutionary War.[8]

From Cremin we learn of Americans' arrays of educational practices but little of their intense, unresolved competitions. According to Bailyn, the most notable departure from European influences during colonial times was the enlisting of education for homegrown social purposes. From their wilderness experience as settlers on the Atlantic coast, Americans concluded that deliberately fashioned group teaching and learning, compared, say, with apprenticeships, could encourage much-needed social cohesion. They survived and eventually flourished not as individuals and families alone but as a union. Organized schooling brought efficient fuel and cultural glue to the enterprise, and the march toward a new nation commenced. Granted, the trek had ups and downs, although they were correctable within a generally sure-footed directive. War on divergent education was fought to achieve unity, and as predictably as in a fight to determine who will appear in the mirror when you look into it, the newcomers won.

Early on, Richard Storr warned of the chilling effects of "teleology in reverse," reconstructing education's past through the rose-tinted glasses of reigning presentism. He worried about educative phenomena that would be missed, while documenting preestablished definitions and remained deeply skeptical of methods that omitted serendipitous, accidental, and experiential learning. With tacit reference to Bailyn and Cremin's definitional approach, Storr saw an inadequate, even crippling research design. Knowing education's appearance in advance, they could confidently locate it in trolling the past. The conceptual collapse of education as a single-purposed, legislated, and easily detected endeavor entered at the onset of research. Only tainted findings could follow. Education historians have tended to ignore Storr's advice, as several critics have observed.[9]

Historians who underestimate Bailyn and Cremin are bound to find themselves embarrassed later. They are celebrated exemplars of a cultural perspective that lifts teaching and learning, variously discerned, onto history's center stage. Yet, if education is understood as the process by which people struggle to survive, grow, and form society, their critique of education history fails at other critical junctures. As the story goes, part of the attraction to Bailyn and Cremin stems from the contrast between their work and that of Ellwood P. Cubberley. Understandably, Cubberley became a personalized target of their analyses and an easy mark for later criticism because of his overt biases against women, selected immigrant populations, people of color, and farmers. No competent historian, they complained, treats the past as "the present writ small." The problem, however, was that Bailyn and Cremin enacted their own methodological predilections. Reclaiming the past as distinctly different from the present, a foreign territory, merely restates a western European abstraction of time as linear and progressive.

British novelist Julian Barnes helps readers perceive the more twisted chronologies charting people's lives, where past and present mingle to project a future and memory functions as more of a solvent than a fixative. In lived chronologies people forget and alter past events, as philosophers and sociologists of knowledge have observed. Time and memory neither recur purely nor cohere upwardly. Insistently situational, they emerge as tools for organizing personal and communal experiences, constructed realities that come and go in relation to equally mobile signposts and other triggers. Euro-American historians have battled over these complexities for centuries. If only intuitively, we've grasped revision as the lifeblood of our discipline and sometimes welcomed it. We write and rewrite and perform our detective work tentatively, acknowledging, if grudgingly, that sooner or later others may offer different interpretations. Barnes evades the ambiguities and dysfunctions of memory by confessing them through a fictional hero revisiting his personal past.[10]

In Eurocentric fashion Barnes omits an alternative perspective on memory as a regularly exercised ceremonial muscle. That's how American Indians, and arguably other societies, form oral traditions, the amended and amendable versions of their pasts. These aren't stolid orthodoxies but processes affirming and updating history's contours, details, and meanings. Oral traditions too, it turns out, accommodate revisionist necessities. Finally, although the Bailyn-Cremin criticism of education history as celebratory and evangelical sounds wise and empirically attractive, it crumbles upon inspection. If education is a cultural asset, arguably the necessary one as they propose, then finding evidence of its traces occasions historians' warm greetings and their wonder, or it should, if they intend to understand how societies form and fade. Education history portends the depth and breadth of experiments in cultural change and continuity, including those telling failed ones, inevitably happened upon after the fact. Indigenous Americans often enjoy using a closely woven but organic filter in retelling their stories.

To capture the diverse, typically subtle phenomena of cultural formation, education historians need a different conceptual net than the one Bailyn and Cremin posit. It may be broader, although it doesn't have to be, as Rosengarten shows, but it must be finely meshed to catch encoded nuances and unanticipated details. Equally necessary, it must be open-ended, admittedly unfinished, and permitted to grow. In *The Barbarous Years*, his latest addition to the multivolume *The Peopling of British North America*, Bailyn reveals his famous aphorism as finite and skewed. Education remains the process of cultural development across generations, but his model is western European, biased by colonial and mercantilist values against other, apparently lower-order, cultures. Does education perform its magic in all cultures? Are some more educative than others? Instead of providing a reliably useful analytical framework, Bailyn leaves us in a conceptual cul-de-sac.[11]

To back out of it, consider a different working hypothesis. The most pervasive, embedded, corrosive, and effective agent of education in the United States during the eighteenth and nineteenth centuries was slavery. As an institution, it taught Americans how to see and measure their world and their own worth. It functioned as the lodestar of their moral compass, infused public and scholarly rhetoric, and molded unremarkably, but overtly, thinking about literacy, religion and theology, commerce, science, technological advancement, and art. It was literally, although not ideally, the American common school. It received direct and indirect public subsidies and political ratification. Slavery's educative force encompassed the nation, far beyond southern regions, and embraced blacks and whites, not merely the owners and the enslaved. It infected policies, territorial, state, and federal constitutions and laws, public entertainments, and accepted common sense. Popular views on women, American Indians, and immigrants, particularly Irish Catholics, eastern and southern Europeans, Asians of all nationalities, and Spanish and Portuguese speakers, hardened along the way. Some of these consigned strangers predated the arrival of other western Europeans and are incorrectly classified as immigrants, except from the perspectives of Native Americans.[12]

Slavery's hegemony didn't end with the Emancipation Proclamation or two years later at Appomattox Court House but persisted legally, by common practice and consent, and insidiously, as Douglas A. Blackmon documents in *Slavery by Another Name*. Recent research on the history of slavery, the Civil War, and Reconstruction anticipates and reinforces Blackmon's penetrating study. The contributions frame afresh questions of where and how Americans learned, proposing slavery as their principal educative institution perhaps beyond the nineteenth century, a conclusion W. E. B. Du Bois reached long ago and Blackmon confirms. Recent events—in Ferguson, Missouri; Cleveland, Ohio; Staten Island, New York; Baltimore, Maryland; and Charleston, South Carolina—underscore the ongoing and as yet still raw psyche of slavery for whites and blacks alike, as well as other people of color. There, slavery remains visible in the durable teachings of Jim Crow, poverty, bigotry, and political power maldistributions and vacuums. Drawing from methodological and conceptual innovations, queries about the cultural reach of slavery in the United States propose a truism. Historiographical scopes in education depend on scholars' abilities to reconceive the search for relevant phenomena. None of this fertile work has been produced by professed education historians, although Cremin anticipated the necessary reorientation: "Slavery itself became a vast system of education—or one might better term it miseducation."[13]

If, as in the example of slavery, research methods are fundamental in the history of education, a question rises in importance and urgency: How do education historians begin and proceed? Advice can be found in recent histories of American Indian tribes and nations and reconsiderations of older—often much older—contributions. Like all intellectual domains, stretching from pure science to popular culture, education history benefits from new and revised findings, but now it's especially clear that the investigations' necessary precondition and consort is another way of thinking. The times seem ripe for a thorough rebuke of the field via methodological initiatives. Piketty's critique of economics offers a model that's at once rigorously and systematically historical, empirical, literary, multidisciplinary, and educative. Although not discussed here, his approach could prove useful as education historians ponder escapes from delimiting assumptions and intellectual habits.

At War with Education

As much if not more than the study of slavery, American Indian histories hold great promise for giving education historians direction, recasting our understanding of substantive and methodological concerns, and examining the social construction of ignorance. These are significant claims, and to make good on them requires foundation.

Bailyn defined education as the processes involved in the transmission of culture from one generation to the next. His *Education in the Forming of American Society* was part of a series of conferences and subsequent publications intended to identify worthwhile topics emerging historians should pay attention to as they looked ahead. Note the practical goals of the initiative: "The primary aim of the series is to serve the needs of graduate students and those directing their studies, and thereby to foster better research." Unrecorded was a tacit purpose to send historical research in certain directions and not in others. The Bailyn volume stated the aim explicitly, to "provide a new and challenging perspective for the study of early American education, indeed for a reassessment of the history of education in the United States down to the present." Was the aim, then, to cement history's substantive relevance to contemporary education policy or, separating methods from findings and committing simultaneously the sins of anachronism and contextual isolation, to enforce a particular way of doing history?[14]

Taking cues from an ongoing program in the 1950s funded by the Ford Foundation, Bailyn intended to decouple "the present interpretation of education in American history" from "its leading characteristic . . . its separateness as a branch of history, its detachment from the mainstream of historical research, writing, and teaching." Education history was an embarrassment, "a distinct tributary," its origins traceable to a marriage of convenience with professional teacher preparation at the turn of the twentieth century. (Similar complaints were issued from psychology and sociology, both maturing disciplines during the same period.) Bailyn envisioned a change of venue to one where education historians would ply their trade within the canons, if not under the authority, of history departments. He reckoned undeterred by higher education's protocols, values, and policies. The proposed marriage failed to be consummated. If a Bailyn revolution was in the making, its trajectory fell short. An alignment of education history with university history departments assumes the latter know how to frame and pursue educational queries and are inclined to do so. Until recent decades that assumption has proved to be optimistic, and contrary to typical history department reward systems. Within a few decades, education history usually still could be found in professional academic units and retained its selective institutional focus. Looking back, we can detect a more basic problem with the key concept of culture. It forged the foundation of Bailyn's thinking about change, continuity, and societal development, and that foundation was essentially Eurocentric, ill prepared to admit or sustain studies of different cultures.

The problem becomes acute in investigations of American Indians before or independent of their Euro-American contacts. If education is the means by which a culture transmits itself across time, we have to ask: taking into account tribal and national distinctions, where and how do we find specific manifestations of American Indian education? If we label the instances of traditional Native learning "informal," only to differentiate bureaucratized Euro-American schooling, education historians risk devaluing or subverting educative purposes and effects. Prospects for more culturally inclusive approaches in teacher preparation and education research generally might go unnoticed or remain invisible. In the same vein, slave education, owner learning within the culture of bondage, multilateral exertions of agency development, and education as lived experience forging group identity and differentiated resilience all intersect here as well. In the 1950s and moving toward century's end, such calibrated phenomena were the foster children of US historical and education research, typically unseen by many leading practitioners. Yet even these marginalized resources bear insights on interacting cultural strangers and the intersections' national implications.[15]

Before Bailyn gave his paper, William N. Fenton, an anthropologist committed to the emerging field known then and still today as ethnohistory, joined the Needs and Opportunities project at a conference in February 1953. Fenton's paper was meant "to encourage a broadening of historical studies into fields where relatively little original research has been carried on or where new approaches to old problems challenge investigators." Four years passed before the book appeared, even though Fenton's paper was second in the sequence of conferences. Other commitments consumed his time; nevertheless, his contribution seems to have been troubled. Fenton declined to document his paper in formats familiar to historians, and his list of published and primary sources struck conference organizers as abbreviated. They invited other scholars to expand it. The difficulties apparently surfaced from the start with Fenton's conference presentation itself. Here's how he began: "It is in the spirit of Dekanisora, speaker of the Five [later Six] Nations, who called on Governor Spotswood here at Williamsburg [VA] supposedly before 1720, that I have come to your fire to polish the 'Chain of Friendship' between the ethnologists and the historians."[16]

Fenton offered wampum (in the form of his paper) as the bond of friendship attesting words spoken and urged the audience of mostly professional historians "to wipe away tears for those who have gone the long trail since our last meeting," to open their ears for better hearing, and "to clear your throats of any bitterness . . . between us that you may reply later with a clear mind."

Trained "conventionally" as an ethnologist, Fenton also recalled he had been initiated into adulthood by the "old men of the Six Nations." He was no stranger to distrustful collaborations between anthropologists and historians, having been engaged for years with the preparation of historical ethnologists. He then admitted his opening was ritualistic "in the manner of the old Iroquois orators." Its format could be found in the records of numerous Indian negotiations with British and French representatives, who had routinely reported feeling stymied by the delay in turning to the business at hand. The reactions mystify. Euro-Americans of the day (and subsequently) typically began conferences with prayers and other invocations. Fearing that his 1953 audience would be similarly restive and ethnocentric, Fenton underscored an elementary point. Given the topic—the history of Indian and white relations in what became the United States—both groups needed to be understood in terms of their own cultural markers, ideally with respect for differences. (Presciently, the hardcover imprint of his book carried an image of a peace pipe.)[17]

The Iroquois routinely attended meetings that promised significant outcomes bearing wampum, a specifically crafted or selected gift representing the event. They began with speeches honoring the dead and the unended past, admonishing attendees, themselves included, to listen carefully to one another, and urging all to lay aside acrimony and bitterness. Not a tribe but rather the confederation of the Cayuga, Mohawk, Oneida, Onondaga, Seneca, and Tuscarora Nations, the Iroquois governed themselves by consensus, their path to transcending their own diversity. Oratory was a necessary skill, and they were famous for it among colonists and Indians alike. It was the means by which they persuaded others and themselves of productive actions and resolutions of internal and "foreign" conflict. Oratory advanced cross-generational teaching and learning. Its centrality to Iroquois education suggests a component of what Euro-Americans label the essential, or core, curriculum.[18]

Following the ritual opening and related history lessons, Fenton turned to the conference theme in an effort to complicate what he feared could become bland and boring issues. His presentation immediately posed historiographical problems with regard to Indian life before European contact. To reconstruct this past, he cautioned, historians required access to oral traditions, with more than mere nods to archaeology's and anthropology's research tools and findings. Lacking them, uninitiated historians could not grasp generally or in detail the trajectories of cultural change and continuity Indians had instigated across millennia. Not all their societies had developed in similar ways and at similar paces; some had devolved or disappeared completely. Tribal differences had remained, and all had been in motion. Some had become farmers; others, abetted by horses and guns, had ranged farther afield as hunters and gatherers. Their spirit lives—their theologies, rituals, and cosmologies—had modified. Unless historians paid attention to the changes, they risked committing their guild's fundamental error, writing Indian histories without Indian sources. In assessing relations between whites and Native Americans before 1830, historians needed to know Indians from Indian perspectives.[19]

Fortunately, Fenton promised, historians could find dizzying arrays of relevant materials. Written in English, French, Spanish, or other European languages, they lay scattered across the country in massive and local collections and in European libraries, museums, and archives. Great numbers were warehoused as US government documents, and others were held by state historical societies. Although recorded in a linear medium Indians generally rejected as inaccurate, the materials nonetheless opened doors onto pasts not otherwise accessible through prized historical methods. Surviving captivity memoirs by whites, a popular genre of the colonial and early national periods, and even fiction proved informative once contextually triangulated to help frame research questions. With difficult labor, oral traditions could be accessed and situated. They constituted, after all, organic, growing bodies of evidence, not anything resembling museum relics. But be forewarned: they typically held closely guarded social and ceremonial secrets not meant to be shared with outsiders. Depending on their lineage, historians and anthropologists could anticipate silence or misdirection in response to some of their queries.

Fenton concluded with a research agenda: (1) the supplanting of Indians in New England; (2) the role of the Six Nations in founding the republic; (3) biographies of Natives, whites, and their societies; and (4) as a beginning, the connections among leading actors in emergent narratives, each rooted within Indian and white frameworks: missionary, ethnologist, educator, and patriot. It was a short list that also outlined a tangled web of complex, transdisciplinary, and in some instances impossible work to be done. The supporting bibliography, not written entirely or principally by Fenton, provided citation details. The section on "Missions and Education" exposed an originating problem. It posited a false assumption that should have troubled many historians, namely, that Euro-Americans through evangelical initiatives had brought education to Indians. Here a discordant note sounded in sharp contrast to Fenton's emphasis on Indians' long (and diverse) traditions of teaching and learning to guide cultural change and form their societies. The list and categorization of sources revealed, for all who cared to see, that the Euro-American concept of education was terrifyingly, maybe incurably, embedded and ignorant, that its true purpose was imperialistic, or both.[20]

Bailyn gave his paper on early American education six years after Fenton's presentation. It was published the following year with the intent of disturbing often trod areas of study overgrown with "an excess of writing along certain lines and almost undue clarity of direction." It identified Ellwood P. Cubberley as a major culprit in the commission of historiographical errors. At stake was serious business, a reconceptualization of early American life as one dictated by an unimaginable, unforgiving, and impersonal (one might say anonymous) future, and not by a familiar, well-rehearsed past. In part because of inadequate attention to the colonial period, education historians had missed this context, in which "learning—the purposeful acquisition not merely of technical skills but of new ways of thinking and behaving—was essential."[21]

On this point, Bailyn channeled Benjamin Franklin, pioneering exemplar of transformations in the inherited pattern of education among his Euro-American contemporaries. Replacing stable cross-generational family status, the nature and persistence of servitude, and restricted opportunities for careers, "new devices for self-improvement and education" had appeared. Colonists had relaxed and amended expected family traditions and social structures as they invented wilderness communities and struggled to survive in them. Sectarian religion had provided mooring of sorts, a trusted anchor. For Bailyn, that explained why they had launched missions to convert, school, and otherwise civilize the "savages" who surrounded and outnumbered them. Understandably, the newcomers might have wanted to temper, if not reverse, their own lack of control over their future, but no, Native schooling had been more clearly a hostile and hegemonic act, a usurpation of sovereignty. Social life in the colonies had required unity, and the Indians had definitely been different and eventually hostile. The colonists' outreach had exuded "bland piety" and "hypocrisy" but also "sincerity." The effort had failed. "The English settlers, insensitive, inflexible, and righteous," despite their need to adapt to strange and unwelcoming lands, had proved to be "atrocious anthropologists," habitual practitioners of a feral ethnocentrism.[22]

In *Education in the Forming of American Society* Bailyn credited white newcomers with introducing "the problem of group relations in a society of divergent cultures," adding "a new dimension to the social role of education." The "self-conscious, deliberate, aggressive use of education . . . spread throughout an increasingly heterogeneous society and came to be accepted as a normal form of educational effort." It was to be enforced as ointment for the "unstable, highly mobile, and heterogeneous society" of colonial America. Schooling had become "an instrument of deliberate group action." The "transformation of education, turning on the great axles of society—family, church, community, and the economy—had become clear before the end of the colonial period." Note that the transformation grew from improvisation and generational breakdown. Experiments were under way. Unfortunately for Indians, the English missionary campaigns were both evangelical and more essentially propagandistic, experiments of a different order. Consider the contrasting anthropological and archaeological evidence of an inconvenient truth. The Powhatans, other Algonquians, and the Iroquois had long understood and practiced education as social intention, and now also facing grave uncertainty, they too tested new norms of educational effort. Bailyn does not recount their side of the story.[23]

As promised, his essay delivered a hypothetical history, identifying fundamental needs and opportunities for study, but it slanted the perspective. Closely reasoned and documented, it nonetheless privileged a narrative that tilted the playing field on which envisioned investigations were to be conducted, admitting "it may well prove to be wrong or misleading." In the essay on sources, Bailyn acknowledged the publication by "Fenton and others" but dwelt on the mission and education section of its bibliography, the details of which, he thought, "need not be repeated here." Arguably, he should have dug deeper, especially into Fenton's layered conceptions of culture drawn from outside the history discipline at the time. In doing so, he would have found nonimperialistic examples of education as the process of cultural change and continuity or at minimum framed rudimentary questions opening them to inquiry.[24]

Several decades later Bailyn reconfirms the bias in *The Barbarous Years*, subtitled *The Conflict of Civilizations, 1600–1675*. The implications become explicit when he reconstructs seventeenth-century British encounters with Powhatans and other Algonquins without recourse to sources providing Indian perspectives on the invasions of their lands. The historiographical error could have been avoided, given the accumulating new research on North American Indians, much of it coming from Native scholars working in anthropology, archaeology, demography, ethnohistory, plant biology, and US history. Merle Curti's social and intellectual history of human nature in Euro-American thought also would have helped fashion more complex and dynamic depictions of Indigenous and white confrontations.[25]

Unlike in Bailyn's earlier work, emphasis in *The Barbarous Years* falls on events through which diplomatic and common courtesy extended by the English and Algonquian tribes degenerated quickly into barbarity within both societies. Pointedly, he asks why the English became savages. One reviewer, a former Bailyn doctoral student at Harvard, judged the question faulty. First, it was prematurely formed, surfacing in Bailyn's original plan for the multivolume series published before the study was far advanced. Second, partly for this reason, the sources cited almost three decades later tended not to take advantage of recent Native studies. The comment warrants an underline. American Indian historical literature has grown exponentially over the past thirty years, with Native scholars contributing significantly to the enterprise. Critical acclaim, sales, and prestigious awards have accompanied the new productivity. Charles Mann complains, somewhat unfairly, that Bailyn's premise commits the sins of anachronism and historicism, given the undeniable presence of Native inhabitants who met European invaders. His general title notwithstanding, Bailyn knows his study concerns "the repeopling" of North America's Eastern Seaboard.[26]

Bailyn intends to construct narratives of the colonists, a heterogeneous mix of European ethnicities, not English alone, and of Indians ("The Americans"), but his interpretation and the sources supporting it suggest grounds for skepticism. His secondary materials in history and anthropology tend to be old and narrowly selective. He dispenses rather abruptly with contested findings, for example, estimates of precontact Indigenous populations. This particular dispute feeds other numbers, notably the mortality rates traceable to imported European diseases, matters of distinct consternation and fury among Native peoples. Worth emphasis, he also avoids searching critiques of European missionary impulses. He displays careful attention to the lifestyles and values of specific tribes, relying heavily on testimony from outsiders, but does not acknowledge limitations of this historiographical approach. His attraction to the British story, despite recognition that the Spanish had come earlier, hints at racial, ethnic, and linguistic predispositions. With lyrical sophistication, he portrays Indians as exotic and doomed, a commonplace but backward-reading depiction of Native peoples as perceived by white outsiders. It tacitly assumes the clearest understanding of Indians is as conquered peoples.

Bailyn misses opportunities in his subsequent investigations to amend, update, and otherwise extend his hypothesized linkage of education and culture in *Education in the Forming of American Society.* The omission casts doubt on the earlier work as merely a learned restatement of the Cubberley thesis: schooling in the United States developed amid serial battles and opponents as deepening socialization. Does Lawrence Cremin's take on historiographical sins of education history also warrant reassessments? An irony looms. Whatever failures of application occurred, Bailyn and Cremin bequeathed a lens that can be refined for detecting education in the past. Look undeterred for tracks left by diverse cultural formations. Combatants in the war over education commenced by colonists four centuries ago now include historians.

Breaking Ignorance

American Indian histories portray intricate understandings of ignorance made and undone. They rely on unusual, often literal, digging. In the process nuances and narrative prototypes surface. Tragedy, heroism, comedy, and irony intersect richly and can't be untangled. Two hypotheses, now securely established, offer starting points. First, American Indian histories, whatever the particular topic and whether case studies or surveys, qualify as educational narratives. They reconstruct cultures on the move and the risks such experimentation runs. When we begin by positing Native peoples as educated, their unfolding histories acquire regenerating explanatory power, upending inclinations to structure them as merely accounts of victims. Second, filling blanks in US history with Native content, if only to respond to agnotological insights, has far-reaching effects, reordering the basic narrative, its beginnings, twists, and pivotal turns.

Jack Brink illustrates these uncommon and fundamental contributions. Drawing principally from archaeology and historical anthropology, he focuses on "one of the truly remarkable accomplishments in human history. It is the story of an unheralded, unassuming, almost anonymous group of people who hunted for a living." There is no Great Wall of China at the center of his study. No Stonehenge, Egyptian pyramids, or even Greek and Roman civilizations with their monumental ruins and tattered documents vie for his spotlight. Rather, he wants to understand how Indigenous people of what became North America managed to survive by hunting buffalo, the largest mammal on the continent and a fearsome beast. He stumbles upon an anomaly. Single kills occurred, but far more impressively, so too did massive slaughters of herds, remnants of which archaeologists have unearthed. How did these ancient peoples, hunting in groups on foot, lacking horses, gunpowder, wheels, or metal tools and weapons, accomplish such feats? Brink's answer is education, cross-generational teaching and learning driven by needs for food, spiritual roots, and communication systems linking diverse tribes. There was nothing informal about the process. Recovered archaeologically, this was intentional education propelled toward social formation.[27]

Recent investigations on the Comanches, Sioux, and other dominant Plains nations in the eighteenth and nineteenth centuries detail the dangers of bison hunting and the educational processes Indians institutionalized to minimize the risks. Brink moves the calendar back by several thousands of years, finding evidence that these ancients studied buffalo, their reflexes, migration patterns, the effects of climates and seasons on their weight and fur, and the different textures of their flesh and fat content (males were not preferred), and plotted their forays accordingly. Early on, they learned that buffalo could be stampeded. Instigating a run near a cliff or outcropping brought untold bounty of meat, hides for clothing, bones for tools, and inedible body parts for twines and containers. A buffalo jump resulted in food and supplies lasting for months and occasioned communal celebrations. The benefits accrued from an organic bison-hunting curriculum. It grew as learning accumulated, assigned distinct roles to women, men, and children, and bestowed sacred meanings on some cliffs over others. The shared information brought different tribes of hunters to the same site repeatedly over the years. Not just any cliff would do. Some had divine blessings. For the most part, Brink's aged monument to human ingenuity lies buried in stories, ceremonies, dances, and songs, appearing piecemeal as artifacts found at archaeological digs. Brink connects them imaginatively.

Benjamin Franklin confronted a similar curriculum problem. He wanted to advance the argument for practical over classical learning. Instead of Latin, Greek, the finite spectrum of metaphors implanted by recitations on dead civilizations, and preparations for the ministry, he favored science, mathematics, and other studies intended to prepare young men for useful professions like medicine, law, business, and engineering. As a favorite rhetorical weapon, he wielded a quotation from the Iroquois. Pouring salt onto wounded white pride, he fondly reported on a 1744 meeting between the government of Virginia and the Six Nations. The Indians were told that if they would send some of their boys to Virginia, the English would give them an education at the William and Mary College. An Iroquois orator replied:

> We know that you highly esteem the kind of learning taught in those colleges, and that the maintenance of our young men, while with you, would be very expensive to you. We are convinced therefore, that you mean to do us good by your proposal, and we thank you heartily. But you who are wise must know that different nations have different conceptions of things; and you will therefore not take it amiss, if our ideas of this kind of education happen not to be the same with yours. We have had some experience of it; several of our young people were formerly brought up at the colleges of the northern provinces; they were instructed in all of your sciences; but when they came back to us they were bad runners; ignorant of every means of living in the woods; unable to bear either cold or hunger; knew neither how to build a cabin, take a deer, or kill an enemy; spoke our language imperfectly; were therefore neither fit for hunters, warriors, or counsellors; they were totally good for nothing. We are, however, not the less obliged by your kind offer, though we decline accepting it; and to show our grateful sense of it, if the gentlemen of Virginia will send us a dozen of their sons, we will take great care of their education, instruct them in all we know, and make men of them.[28]

A notorious wit, Franklin may have missed the unnamed orator's dry irony, a trope famously employed in Iroquois speeches, but he captured the essence of the cross-cultural encounter. Contrasting conceptions of education emerge and harden among and within societies. To survive, or better yet to flourish, they experiment and risk, casting curriculum into a future beyond discordant pedagogies and currently imagined binary options. Franklin grasped the high stakes being wagered and remained pessimistic about how many contemporary colonists shared his sense of urgency. He thought goading them might help.

Similarly, American Indian pasts draw education historians onto forbidding land where boundaries separating past, present, and future remain stubbornly murky. We can test how they fare by reviewing survey history textbooks routinely marketed for students in teacher preparation programs. Those published before 2000 tend to omit Indians entirely or offer superficial general accounts. Exceptions to this pattern and a few surprises can be found, typically in textbooks issued in multiple editions across the years, some well into the 1990s. The widely adopted survey by R. Freeman Butts and Lawrence A. Cremin grouped references to American Indians with those citing generalized experiences of racial and ethnic minorities and women. Writing in the early 1950s, the authors traced the fits and starts of efforts to achieve schooling equality in the United States, surely an anticipatory theme at the time. Fueled by his commitment to public education as a major beneficial force in US history and reservations about the accuracy and utility of subsequent revisionist trends, Butts went farther in his single-authored textbook of 1978. He retained the tragic mixture of ignorance and dashed hope that accompanied attempts to impose ethnocentric schooling on Indians, inserted details on selected institutions and their justifications, and worried about the future:

> [Mounting] attendance figures, of course, did not solve the problems of education for Indians. If anything, they simply brought them into the mainstream of educational problems: rural oriented youth transferred to an urban environment that was more alien and hostile than the reservation and more contemptuous of poor vocational skills; a curriculum that paid little attention to Indian heritage; teachers who had little training, understanding, or appreciation of the special need for respect and dignity of a people who were being asked to move skillfully between the traditional culture of a tribal folk society and a modern urban civilization, and be able to find values in both.[29]

Other authors also wrote of Native Americans as unique mirrors reflecting systemic failures in teaching and learning and the necessity of weighing cultural variables in education history and policy. S. Alexander Rippa endorsed one Native leader's challenge: "The question is not how you can Americanize us but how we can Americanize you. The first thing we want to teach you is that, in the American way of life, each man has respect for his brother's vision." Rippa underlined the point: "Still proud and defiant, Native American leaders today are struggling to conserve an ancient and priceless heritage." Editions of his textbook appeared from 1967 through the 1990s. The interpretation and tone remained consistent as he emended content with fresh details.[30]

These traits became more prominent and extensive in textbooks revised or published after 2000. A consensus took shape. White-enforced schooling of Natives delivered miseducation to both Indian and white societies. It disparaged vibrant cultures, promoted a false homogeneity while reinforcing racial divisions, and slammed doors to opportunity and equality. Respectful with regard to enlightened education policy and American Indian histories, recent textbooks also perpetuate analytical quarantines. The authors may acknowledge treasuries of cultural assets but tend to leave them untapped beyond broad references. Details on named tribes and nations tend to be missing, tacit disparagement of the constructive influences of cultural diversity. Intentions aside, these surveys emphasize the heartbreaking story of victimization at the expense of agency. Indians suffered Euro-American schooling, and by implication education, suppressing instrumental roles in their own cultural formation. Benefiting or irreparably damaged, in the long run they acquiesced to assimilation. Pride, defiance, Indigenous insights and values, and ancient, priceless heritage sank to subterranean levels.[31]

Textbooks by John Rury, Wayne Urban and Jennings Wagoner, and Joseph Watras offer exceptions that can guide narrative critiques and revisions. Watras begins with globalization, setting the world stage as the appropriate context for understanding American education history. He then proceeds chronologically, which means he takes the uncommon step in such surveys of first attending to Spanish colonization. He emphasizes Florida, the Southwest, and the West before turning to New England and the Puritans. Watras sketches the distinguishing characteristics of Spanish and British rule on the continent, integrating the different but nonetheless disastrous effects of each on original inhabitants. In later sections, he acknowledges persistent shortcomings of nineteenth-century training institutions for Natives and briefly notes legislation intended to reorient programs of the US Bureau of Indian Affairs (BIA) toward Indigenous interests. Hints of reversals in a long string of failures began appearing in the twentieth century. Watras's approach corrects the counterfactual claim that the United States developed from eastern bases and British hegemony. No such unilateral growth pattern occurred, as his inventive search for sources demonstrates. Notably, specific Indian societies rarely enter the analysis, omissions that weaken general explanations of Indigenous resistance to Euro-American invasions.[32]

Urban and Wagoner also deliver a chronological account but move from a different foundation. Their survey begins with "pre-colonial America" and the Native peoples who lived there. Multiple disciplinary sources support their reconstructions of Indigenous education prior to the arrival of Europeans. Central to Indian survival, processes of teaching and learning spanned the generations, compelling the acquisition of typically gender-specific skills in hunting, fishing, tool and weapon production, military preparedness, farming, hide preservation, and, depending on climates, ecological manipulations. With distinct cultural identities, languages, and spirit lives, various tribes coexisted and interacted. The evidence tells us they learned from one another, often peacefully.

Tribes' cultures, social organizations, and religious ceremonies changed over the millennia, although evolution describes the process imprecisely. Choices were made, with risks in tow. Urban and Wagoner emphasize a survivalist curriculum, but some of their sources additionally depict expansive versions of the practical curriculum. The first Americans also included artists, craft makers, singers, dancers, healers, and, yes, historians, the tellers of tales. Numerous societies left behind monuments to themselves, mounds, some quite tall and ornately sculpted, constructed mysteriously without the aid of wheels, draft animals, or metal tools. In the context of Indigenous histories, Urban and Wagoner introduce Spanish, French, Italian, and Dutch colonialism. This is one way they challenge the canard about British and New England dominance of US development. Unfortunately, particularly with regard to New Spain, the detours weaken their hypothesized Indigenous foundation of American education, move it aside, and leave Spanish influence underreported. The authors acknowledge the persistence of "education offered within the tribe" and its competitions with white-imposed schooling but leave its energies and effects unassessed. A tacit image forms of postcontact Native cultures as stoic and static. Urban and Wagoner give us a provocative beginning toward a needed fresh conception of American education history. Left for others is the weaving of these thematic strands into a more intricate theoretical framework.[33]

That's the difficult assignment John Rury undertakes. He looks for episodes of social change in American history and the ways education (sometimes labeled "schooling") contributed to the dynamics, blending evidentiary practices of two often separated genres—social and intellectual history. Rury's text marshals connections among ideas, economic forces, demographic trends, population migrations, racial and gender predilections, and institutional formation, among other variables, giving rise to an "emerging national culture, embracing the values of individualism and self-determination." In this context American Indians, who long ago melded tribal and individual orientations, presented a thorny problem of assimilation.[34]

Historically, Native Americans did not clamor for access to Euro-American schooling, resisting the importunities of Richard Henry Pratt, other white reformers of Native societies, and the BIA generally. They wanted sovereignty, not equality, which meant at minimum respect and unwavering enforcement of negotiated treaties. They "had their own traditions of socialization, and doubtless preferred to teach the children their own values, customs, and beliefs." Amounting to "one of the largest experiments in deliberate cultural revision in the nation's history," the reforms failed. "The cultural identities of indigenous children, it turned out, were considerably stronger than Pratt and other educators had imagined." Rury sees an object lesson in the outcome. Reformers had stretched schooling beyond its capacities to achieve social change. Other variables, "cultural chauvinism, abetted by racism and nationalism, blinded them to the enormity of the task they had undertaken and the larger problems associated with assimilation in American life." Paternalism and intolerance worked their will. Rury knows such educative forces were in play from the outset of the colonial era, when "contact between two vastly different social and cultural traditions marked what was perhaps the first great process of social change in what is today described as American history." The problem of how to tell this story accurately, without ignorance, persists.[35]

On lands incorporated as the contiguous United States, contacts among Indian peoples, white invaders, and migrants from the south and the north occurred in tidal waves of settlers, missionaries, itinerant hunters, and military personnel across more than four centuries, some predating Columbus's voyages. Like that with Mexico, the long Canadian border remained relatively porous throughout, an openness compromised periodically by the Russian and US acquisitions of Alaska and fluctuating policies of governments up and down the continent. Neither demarcation constrained travel by Indigenous Americans and never had. After the 1880s, US expansion into Pacific and Caribbean islands, importantly Hawaii, changed the targets of empire building but not its rationale. What had happened previously proved to be merely a dress rehearsal of a Euro-American script many years in the writing. Advantaged by new research, we can see now the US newcomers and their governments were neither prepared for nor inclined to accept the cultural legitimacy of the people occupying lands they wanted. As a way to safeguard the forced acquisitions, the declared war on Native education made sense. Other interpretations of longer duration are available for consideration—conquest, genocide, misguided assimilation—but they leave much unexplained. Why did American Indians typically resist white strategies and tactics, favoring instead calculated preferences for their own cultural formation? How have so many tribes and nations managed to survive against overwhelming military, economic, and political forces and pandemic diseases? The second question implies they are winning the war. If so, how?[36]

Answers to these three leading questions can be found in resilient Native cultures long practiced in change and continuity. Outsiders know more about them than previously, and evidence continues to accumulate. Newcomers to the United States during the post-Columbian era encountered culturally heterogeneous societies, some blended and formally allied, others distinctive, even warlike. All were in process. Their oral traditions and sources reconstructed by archaeologists and historical anthropologists tell us they adapted cross-generational education to survive threats and seize opportunities. Adrea Lawrence perceptively names it "epic learning." They molded their cultures as disruptions occurred, reactivating habits of struggle. Education enabled accumulations of cultural assets, what Jonathan Lear, in his study of the Crow Nation, labels "radical hope," pervasive devastation notwithstanding.[37]

Other examples abound. The Arikara Indians, long established agricultural pioneers, adapted corn horticulture and related religious ceremonies on a pre-Columbian, millennium-long migration from south Texas to North Dakota. Along the way, they engineered corn for diverse soils and climates, introduced the crop to Plains tribes (before Euro-American contact), and acquired a reputation as entrepreneurs by organizing gatherings of the nations to exchange hides, tools, crafts, dances, songs, and stories. William Fenton's high regard for the Iroquois has been widely shared. Charles Mann judges the confederacy to have been North America's most influential polity north of the Rio Grande River at the beginning of the eighteenth century. The Sioux attained similar status in the upper Plains a century later. The Comanches operated primarily in the lower Plains, the Southwest, and northern Mexico during this period. Recent studies conclude that they had become the mightiest empire on North American soil by the mid-eighteenth century, eclipsing the North American designs of the world's superpowers in cultural sway and military capacity. The list is only illustrative of the many ways American Indians displayed distinct cultural strengths as educational attainments.[38]

These tribes' histories can liberate education historians to think again about their field and to weigh expansions. For one, the studies multiply the available instances of cultural continuity and adaptation, and of education, that occurred throughout the American continents outside those acknowledged within Euro-American frameworks. They serve to deprovincialize the detection of teaching and learning, releasing curiosity from cultural blinders. For another, they draw upon sources and methods that can inform research on other topics in education history and education research. Multidisciplinary and comparative approaches seem essential, but Indian histories warn us they can't be taken at face value. Anthropology, for example, has its own history of bias and curtailed interests to contend with. Buyers need to beware. Scholars in the social sciences generally and in the humanities have proved noticeably inept in posing and pursuing questions about teaching and learning phenomena. Their primary interests lie elsewhere, a preoccupation helping to explain why so many ignore or devalue Native scientific achievements and other indicators of advanced culture. In conducting contrarian investigations, education historians contribute not only to their own specialization but also to the literatures of other disciplines. Language can help with the liberation. As repositories of unique metaphors, symbols, and modes of analysis, American Indian histories offer diverse ways of thinking, knowing, and writing about education's pasts, including that long period of declared war when ignorance was made and broken.[39]

NOTES

I thank Ben Warren for entering and revising multiple versions of the manuscript, constructing and correcting citations, reading it all with fine critical sensibilities, and telling me exactly what he thought. Others helped significantly in similar ways, in writing and face to face: A. J. Angulo, Sara Clark, Abigail Gundlach-Graham, Jeremy Jernigan, Adrea Lawrence, Daniel Perlstein, Charles A. Tesconi Jr., Wayne Urban, and Christian Ydesen.

1. Lawrence A. Cremin, *American Education: The Colonial Experience, 1607–1783* (New York: Harper & Row, 1970), 136, 349–50; Samuel G. Drake, *Biography and History of the Indians of North America, from Its First Discovery* (Boston: Benjamin B. Mussey, 1851), 509–12.

2. Cremin, *American Education: The Colonial Experience*, 349–50, 354.

3. Vine Deloria Jr., *We Talk, You Listen: New Tribes, New Turf* (New York: Macmillan, 1970); Vine Deloria Jr. and Daniel R. Wildcat, *Power and Place: Indian Education in America* (Golden, CO: Fulcrum, 2001); Raymond J. DeMallie, "'These Have No Ears': Narrative and the Ethnohistorical Method," *Ethnohistory* 40 (Autumn 1993): 515–38; Donald L. Fixico, *The American Indian Mind in a Linear World: American Indian Studies and Traditional Knowledge* (New York: Routledge, 2003); K. Tsianina Lomawaima, "Tribal Sovereigns: Reframing Research in American Indian Education," *Harvard Educational Review* 70 (Spring 2000): 1–23; Devon A. Mihesuah, ed., *Natives and Academics: Researching and Writing about American Indians* (Lincoln: University of Nebraska Press, 1998); David Van Reybrouck, *Congo: The Epic History of a People*, trans. Sam Garrett (New York: HarperCollins, 2014); Amanda Vaill, *Hotel Florida: Truth, Love, and Death in the Spanish Civil War* (New York: Farrar, Straus & Giroux, 2014); Theodore Rosengarten, *All God's Dangers: The Life of Nate Shaw* (Chicago: University of Chicago Press, 1974); Dwight Garner, "Lost in Literary History: A Tale of Courage in the South," *New York Times*, 19 April 2014, sec. C.

4. Thomas Piketty, *Capital in the Twenty-First Century*, trans. Arthur Goldhammer (Cambridge, MA: Belknap Press of Harvard University Press, 2014); Daniel T. Rodgers, *Age of Fracture* (Cambridge, MA: Belknap Press of Harvard University Press, 2011); Werner Sollors, *The Temptation of Despair: Tales of the 1940s* (Cambridge, MA: Belknap Press of Harvard University Press, 2014).

5. William N. Fenton, *American Indian and White Relations to 1830: Needs and Opportunities for Study* (Chapel Hill: University of North Carolina Press, 1957); Takashi

Fujitani, *Splendid Monarchy: Power and Pageantry in Modern Japan* (Berkeley: University of California Press, 1998); Sven Beckert, *Empire of Cotton: A Global History* (New York: Knopf, 2014).

6. Bernard Bailyn, *Education in the Forming of American Society* (New York: Vintage Books, 1960), 14; Jack Brink, *Imagining Head-Smashed-In: Aboriginal Buffalo Hunting on the Northern Plains* (Edmonton, AB: AU Press, Athabasca University, 2008), xii; Lawrence A. Cremin, *The Wonderful World of Ellwood Patterson Cubberley* (New York: Teachers College Press, 1965); Cremin, *American Education: The Colonial Experience*; Cremin, *American Education: The National Experience, 1783–1876* (New York: Harper & Row, 1982); Cremin, *American Education: The Metropolitan Experience, 1876–1980* (New York: Harper & Row, 1988); Milton Gaither, *American Educational History Revisited: A Critique of Progress* (New York: Teachers College Press, 2003); Gaither, "History of North American Education, 15,000 BCE to 1491," *History of Education Quarterly* 54 (August 2014): 323–48. For evolving details of Cremin's definition of *education*, see the preface and introduction to each volume of his trilogy in the order of publication.

7. Gaither, *American Educational History Revisited*; Gaither, "History of North American Education, 15,000 BCE to 1491."

8. Sol Cohen, *Challenging Orthodoxies: Toward a New Cultural History of Education* (New York: Peter Lang, 1999); Joseph Watras, *A History of American Education* (Boston: Pearson/Allyn & Bacon, 2008).

9. Richard J. Storr, "The Role of Education in American History: A Memorandum for the Committee Advising the Fund for the Advancement of Education in Regard to the Subject," *Harvard Educational Review* 46 (August 1976): 334; Storr, "The Education of History: Some Impressions," ibid. 31 (Spring 1961): 124–35; Yesenia Lucia Cervera, "Negotiating the History of Education: How the Histories of Indigenous Education Expand the Field," *History of Education Quarterly* 54 (August 2014): 362–83; Joshua B. Garrison, "Ontogeny Recapitulates Savagery: The Evolution of G. Stanley Hall's Adolescent" (PhD diss., Indiana University, 2006); Glenn P. Lauzon, *Civic Learning through Agricultural Improvement: Bringing "The Loom and the Anvil into Proximity with the Plow"* (Charlotte, NC: Information Age, 2011); Adrea Lawrence, *Lessons from an Indian Day School: Negotiating Colonization in Northern New Mexico, 1902–1907* (Lawrence: University Press of Kansas, 2011).

10. Ellwood P. Cubberley, *Public Education in the United States: A Study and Interpretation of American Educational History* (Boston: Houghton Mifflin, 1919); Cubberley, *Public Education in the United States: A Study and Interpretation of American Educational History*, rev. ed. (Boston: Houghton Mifflin, 1934); Bailyn, *Education in the Forming of American Society*, 9; Julian Barnes, *The Sense of an Ending* (New York: Vintage Books, 2011).

11. Bernard Bailyn, *The Peopling of British North America: An Introduction* (New York: Knopf, 1986); Bernard Bailyn, *The Barbarous Years: The Peopling of British North America; The Conflict of Civilizations, 1600–1675* (New York: Knopf, 2012).

12. William Cronon, *Changes in the Land: Indians, Colonists, and the Ecology of New England* (New York: Hill & Wang, 2003); Andrés Reséndez, *A Land So Strange: The Epic Journey of Cabeza De Vaca; The Extraordinary Tale of a Shipwrecked Spaniard Who Walked across America in the Sixteenth Century* (New York: Basic Books, 2007); Jonathan D. Steigman, *"La Florida Del Inca" and the Struggle for Social Equality in Colonial Spanish America* (Tuscaloosa: University of Alabama Press, 2005); John Willinsky, *Learning to Divide the World: Education at Empire's End* (Minneapolis: University of Minnesota

Press, 1998); Edward E. Baptist, *The Half Has Never Been Told: Slavery and the Making of American Capitalism* (New York: Basic Books, 2014); Beckert, *Empire of Cotton*; Ira Berlin, *Generations of Captivity: A History of African-American Slaves* (Cambridge, MA: Belknap Press of Harvard University Press, 2003).

13. Douglas A. Blackmon, *Slavery by Another Name: The Re-Enslavement of Black People in America from the Civil War to World War II* (New York: Doubleday, 2008); Daniel Carpenter and Colin D. Moore, "When Canvassers Became Activists: Antislavery Petitioning and the Political Mobilization of American Women," *American Political Science Review* 108 (August 2014): 479–98; Drew Gilpin Faust, *This Republic of Suffering: Death and the American Civil War* (New York: Knopf, 2008); Annette Gordon-Reed, *The Hemingses of Monticello: An American Family* (New York: Norton, 2008); Steven Hahn, *A Nation under Our Feet: Black Political Struggles in the Rural South, from Slavery to the Great Migration* (Cambridge, MA: Belknap Press of Harvard University Press, 2003); Hahn, *The Political Worlds of Slavery and Freedom* (Cambridge, MA: Harvard University Press, 2009); Edward P. Jones, *The Known World* (New York: Amistad, 2003); Louis Menand, *The Metaphysical Club: A History of Ideas in America* (New York: Farrar, Straus & Giroux, 2001); Bertram Wyatt-Brown, *Southern Honor: Ethics and Behavior in the Old South* (New York: Oxford University Press, 1982); W. E. B. Du Bois, *Black Reconstruction in America: An Essay toward a History of the Part Which Black Folk Played in the Attempt to Reconstruct Democracy in America, 1860–1880* (Cleveland, OH: World, 1962); Du Bois, *The Souls of Black Folk* (New York: Modern Library, 2003); Cremin, *American Education: The Colonial Experience*, 411; Ta-Nehisi Coates, *Between the World and Me* (New York: Spiegel & Grau, 2015).

14. Fenton, *American Indian and White Relations to 1830*, ix; Bailyn, *Education in the Forming of American Society*; Fenton, *American Indian and White Relations to 1830*, vi; Gaither, *American Educational History Revisited.*

15. Paul Buck et al., *The Role of Education in American History* (New York: Fund for the Advancement of Education, 1957); Bailyn, *Education in the Forming of American Society.*

16. Fenton, *American Indian and White Relations to 1830*, 5.

17. Ibid., v, 3.

18. DeMallie, "These Have No Ears"; Barbara Alice Mann, *Iroquoian Women: The Gantowisas* (New York: Peter Lang, 2004); Charles C. Mann, *1491: New Revelations of the Americas before Columbus* (New York: Knopf, 2005).

19. Colin G. Calloway, *First Peoples: A Documentary Survey of American Indian History* (Boston: Bedford / St. Martin's, 2008); Preston Holder, *The Hoe and the Horse on the Plains: A Study of Cultural Development among North American Indians* (Lincoln: University of Nebraska Press, 1970); Roy Willard Meyer, *The Village Indians of the Upper Missouri: The Mandans, Hidatsas, and Arikaras* (Lincoln: University of Nebraska Press, 1977); Jeffrey Ostler, *The Plains Sioux and U.S. Colonialism from Lewis and Clark to Wounded Knee* (New York: Cambridge University Press, 2004).

20. Fenton, *American Indian and White Relations to 1830*, 109–13.

21. Bailyn, *Education in the Forming of American Society*, 36, 37; Cubberley, *Public Education in the United States* (1919); Cubberley, *Public Education in the United States* (1934). See also Cohen, *Challenging Orthodoxies.*

22. Bailyn, *Education in the Forming of American Society*, 37.

23. Ibid.

24. Ibid., 5, 66.

25. James Axtell, *The Rise and Fall of the Powhatan Empire: Indians in Seventeenth-Century Virginia* (Williamsburg, VA: Colonial Williamsburg Foundation, 1995); Fenton,

American Indian and White Relations to 1830; Paula Gunn Allen, *Pocahontas: Medicine Woman, Spy, Entrepreneur, Diplomat* (San Francisco: HarperSanFrancisco, 2003); Merle Curti, *Human Nature in American Thought: A History* (Madison: University of Wisconsin Press, 1980); Bailyn, *Education in the Forming of American Society*, 38–39, 41, 45.

26. Fred Anderson, "America: 'Into the Heart of Darkness,'" *New York Review of Books* 60 (4 April 2013): 55–57; Bailyn, *Peopling of British North America*; David Wallace Adams, *Education for Extinction: American Indians and the Boarding School Experience, 1875–1928* (Lawrence: University Press of Kansas, 1995), xiii; Allen, *Pocahontas*; Axtell, *Rise and Fall of the Powhatan Empire*; James Axtell, *Natives and Newcomers: The Cultural Origins of North America* (New York: Oxford University Press, 2001); Calloway, *First Peoples*; Cervera, "Negotiating the History of Education"; Brenda J. Child, *Holding Our World Together: Ojibwe Women and the Survival of Community* (New York: Viking, 2012); Deloria and Wildcat, *Power and Place*; Vine Deloria Jr., *God Is Red: A Native View of Religion* (Golden, CO: Fulcrum, 2003); Fixico, *American Indian Mind in a Linear World*; Donald A. Grinde and Bruce E. Johansen, *Exemplar of Liberty: Native America and the Evolution of Democracy* (Los Angeles: American Indian Studies Center, University of California, Los Angeles, 1991); Pekka Hämäläinen, *The Comanche Empire* (New Haven, CT: Yale University Press, 2008); Kuunux Teerit Kroupa, "Through Arikara Eyes: History of Education as Spiritual Renewal and Cultural Evolution" (paper, annual meeting of the American Educational Research Association, Denver, 2010); Kroupa, "Education as Arikara Spiritual Renewal and Cultural Evolution," *History of Education Quarterly* 54 (August 2014): 303–22; Lawrence, *Lessons from an Indian Day School*; Adrea Lawrence, "Epic Learning in an Indian Pueblo: A Framework of Studying Multigenerational Learning in the History of Education," *History of Education Quarterly* 54 (August 2014): 286–302; Lomawaima, "Tribal Sovereigns"; C. Mann, *1491: New Revelations*, 375, n95; Jerald T. Milanich and Charles M. Hudson, *Hernando De Soto and the Indians of Florida* (Gainesville: University Press of Florida, 1993); Ostler, *Plains Sioux and U.S. Colonialism*; Anthony F. C. Wallace, *Jefferson and the Indians: The Tragic Fate of the First Americans* (Cambridge, MA: Belknap Press of Harvard University Press, 1999). For his reflections on historiography and culture, see Bernard Bailyn, *Sometimes an Art: Nine Essays on History* (New York: Knopf, 2015).

27. Brink, *Imagining Head-Smashed-In*, xii.

28. Reprinted Iroquois orator, quoted in Drake, *Biography and History of the Indians of North America*, 41; Quincy A. Kuehner, *A Philosophy of Education, Based on Sources* (New York, Prentice-Hall, 1935), 276–77.

29. R. Freeman Butts and Lawrence A. Cremin, *A History of Education in American Culture* (New York: Henry Holt, 1953); R. Freeman Butts, *Public Education in the United States: From Revolution to Reform* (New York: Holt, Rinehart & Winston, 1978), 340.

30. S. Alexander Rippa, *Education in a Free Society: An American History*, 8th ed. (White Plains, NY: Longman, 1997), 322; H. Warren Button and Eugene F. Provenzo, *History of Education and Culture in America* (Englewood Cliffs, NJ: Prentice-Hall, 1983); Robert L. Church and Michael W. Sedlak, *Education in the United States: An Interpretive History* (New York: Free Press, 1976); Sheldon S. Cohen, *A History of Colonial Education, 1607–1776* (New York: John Wiley & Sons, 1974); Newton Edwards and Herman G. Richey, *The School in the American Social Order* (Boston: Houghton Mifflin, 1963); Michael B. Katz, *Reconstructing American Education* (Cambridge, MA: Harvard University Press, 1987); Adolphe E. Meyer, *An Educational History of the American People* (New York: McGraw-Hill, 1967); David B. Tyack, *The One Best System: A History of American Urban Education* (Cambridge, MA: Harvard University Press, 1974).

31. See John D. Pulliam and James J. Van Patten, *History of Education in America*, 9th ed. (Upper Saddle River, NJ: Pearson Education, 2007); Joel H. Spring, *American Education*, 15th ed. (New York: McGraw-Hill, 2011); L. Dean Webb, Arlene Metha, and K. Forbis Jordan, *Foundations of American Education*, 5th ed. (Upper Saddle River, NJ: Pearson Education, 2007).

32. Watras, *History of American Education*.

33. Wayne J. Urban and Jennings L. Wagoner Jr., *American Education: A History*, 4th ed. (New York: Routledge, 2009), 192.

34. John L. Rury, *Education and Social Change: Themes in the History of American Schooling*, 2nd ed. (Mahwah, NJ: L. Erlbaum Associates, 2005), 56.

35. Ibid., 49–50, 128, 129, 28.

36. Axtell, *Natives and Newcomers*; David Ewing Duncan, *Hernando De Soto: A Savage Quest in the Americas* (New York: Crown, 1995); Charles M. Hudson, *Knights of Spain, Warriors of the Sun: Hernando De Soto and the South's Ancient Chiefdoms* (Athens: University of Georgia Press, 1997); C. Mann, *1491: New Revelations*; Reséndez, *A Land So Strange*; A. J. Angulo, *Empire and Education: A History of Greed and Goodwill from the War of 1898 to the War on Terror* (New York: Palgrave Macmillan, 2012); Julian Go, *American Empire and the Politics of Meaning: Elite Political Cultures in the Philippines and Puerto Rico during U.S. Colonialism* (Durham, NC: Duke University Press, 2008); Go, *Patterns of Empire: The British and American Empires, 1688 to the Present* (New York: Cambridge University Press, 2011); Walter Nugent, *Habits of Empire: A History of American Expansion* (New York: Vintage Books, 2009); K. Tsianina Lomawaima and Teresa L. McCarty, *"To Remain an Indian": Lessons in Democracy from a Century of Native American Education* (New York: Teachers College Press, 2006).

37. Calloway, *First Peoples*; Hämäläinen, *Comanche Empire*; Ostler, *Plains Sioux and U.S. Colonialism*; Bernard W. Sheehan, *Savagism and Civility: Indians and Englishmen in Colonial Virginia* (New York: Cambridge University Press, 1980); David E. Whisnant, *All That Is Native and Fine: The Politics of Culture in an American Region* (Chapel Hill: University of North Carolina Press, 1983); Lawrence, "Epic Learning in an Indian Pueblo"; Jonathan Lear, *Radical Hope: Ethics in the Face of Cultural Devastation* (Cambridge, MA: Harvard University Press, 2006).

38. Holder, *Hoe and the Horse on the Plains*; Kroupa, "Through Arikara Eyes"; Kroupa, "Education as Arikara Spiritual Renewal and Cultural Evolution"; R. Meyer, *Village Indians of the Upper Missouri*; Ostler, *Plains Sioux and U.S. Colonialism*; Hämäläinen, *Comanche Empire*; Thomas W. Kavanagh, *Comanche Ethnography: Field Notes of E. Adamson Hoebel, Waldo R. Wedel, Gustav G. Carlson, and Robert H. Lowie* (Lincoln: University of Nebraska Press in cooperation with the American Indian Studies Research Institute, Indiana University Bloomington, 2008); Kavanagh, *The Comanches: A History, 1706–1875* (Lincoln: University of Nebraska Press in cooperation with the American Indian Studies Research Institute, Indiana University Bloomington, 1996); Philipp Meyer, *The Son* (New York: HarperCollins, 2013).

39. See, e.g., W. W. Newcomb Jr., *The Indians of Texas: From Prehistoric to Modern Times* (Austin: University of Texas Press, 1961); Deloria, *We Talk, You Listen*; Kroupa, "Through Arikara Eyes"; Kroupa, "Education as Arikara Spiritual Renewal and Cultural Evolution"; C. Mann, *1491: New Revelations*; David Delgado Shorter, *We Will Dance Our Truth: Yaqui History in Yoeme Performances* (Lincoln: University of Nebraska Press, 2009).

PART III / Nationalizing and Globalizing Ignorance

10

US

LISA JARVINEN

In February 1901 Mark Twain's famous essay titled "To the Person Sitting in Darkness" appeared in the *North American Review* and was widely distributed by Anti-Imperialist Leagues, which had emerged in response to US territorial expansion following the War of 1898. Twain satirized the claim that overseas missionaries and European imperialists were delivering those cursed to sit in darkness to the light of civilization. The narrator imagines how such a person might respond to news of a rapacious American minister in China squeezing local communities for indemnities for every Christian convert killed during the Boxer Rebellion. He asks whether the customer for the "Blessings-of-Civilization Trust" might not become a little suspicious of the bill of goods, given the news coming out of China and South Africa at the height of European imperialism. He wonders whether the person sitting in darkness might still swallow the story in view of the ongoing Philippine-American War being fought by Philippine revolutionaries against the United States, which had acquired the islands by treaty from Spain at the end of the War of 1898. The narrator worries that "the Person Sitting in Darkness is almost sure to say: 'There is something curious about this—curious and unaccountable.' "[1]

Twain's denunciation of the imperial endeavor takes feigned ignorance as its central trope. His biting sarcasm suggests little doubt that the person who allegedly sits in darkness—the colonial subject—is really much deceived by his supposed deliverers. If there is a veil of ignorance to be stripped away, Twain argues, it is that which covers the eyes of the American public, a public he implicates in the essay with an inclusive "we." This "we" implies a willful blindness vis-à-vis the entry of the United States into the company of the imperialist powers. The narrator concludes that "for the sake of the Business we must persuade [the person sitting in darkness] to look at the Philippine matter in another and healthier way. We must arrange his opinions for him." Twain's essay speaks, in short, to the strategic production of ignorance with the intent to deceive.[2]

Would the United States recognize itself as imperialist? In a nation that had constituted itself as a republic through a revolutionary war against an imperial power, the concept seemed anathema. To Americans such as Twain who looked askance at the US intervention in the Philippines and elsewhere, the question of the legitimacy of these actions could be posed simply: republic or empire? To those who supported the acquisition of dependent territories, the question and contrast of terms proved troublesome. Implied in the belief in republican values was the simultaneous fear of their corruption. Rome taught that republics were fragile and that imperial ambition sowed the seeds of destruction. At the moment when the United States most clearly was engaging in imperialism abroad, those who advocated the US acquisition of the Philippines as a permanent colony shied away from using the word *imperialism* unless they articulated a positive definition for it. But the fierce opposition of those who styled themselves anti-imperialists ultimately made even this impossible. In response, annexationists denied the word *imperialism* by ridiculing the charge and implying that its use by their political opponents was dangerously disloyal.[3]

This chapter explores the origins and function of empire denial in the period immediately following the War of 1898 and argues that the deliberate rejection or redefinition of the words *empire*, *imperialism*, and *colony* by annexationists constituted a form of public education. Indeed, the debate over whether US actions in the Philippines were imperialist, and if so, what that might mean for the United States, frequently invoked education and learning as key tropes. Participants repeatedly called on the "lessons of history" to support their ideas about the nature of republics and empires. Annexationists extolled the virtues of a tutelary relationship for Filipinos, who were characterized as childlike dependents, while those opposed argued that education and freedom were inextricable. Mark Twain included "education" as an important item on the satirical list of goods being sold to the person sitting in darkness. As with Twain's hypothetical person, ignorance too was a frequent theme. While much was made of the supposed ignorance of Filipinos, it was commonly accepted that Americans knew little to nothing about the Philippines. In political speeches, editorial cartoons, and newspaper opinion pieces, authors explicitly sought to teach lessons grounded in American history and culture about the meaning of imperialism.[4]

Political speech that aimed to educate, or miseducate, Americans about the meaning of imperialism as practiced by the United States suggests the country's domestic culture underwent a transformation in the wake of 1898. Here I refer specifically to what Robert Beisner once dubbed "the 'semantic' issue" surrounding the use of the words *empire* and *imperialism*. This case of empire denial at the semantic level offers a slightly different kind of analog to recent research from historians of science about how powerful interests have deliberately sowed doubt and confusion over questions that could be resolved by scientific inquiry. (Smoking tobacco either does or does not cause cancer and other physical ills; climate change either is or is not occurring.) The question whether the United States is or ever has been an imperial power certainly intersects with the social construction of ignorance, but its resolution ultimately belongs to the realm of rhetoric and language. To answer the question, one first needs to define *empire* and *imperialism*. The conditions or practices that fit given definitions of those words are largely a matter of nomenclature. By this I do not mean to imply that it does not matter whether these terms are used—far from it. As Ian Tyrell has argued in regard to American historians' perennial avoidance of the term, "How empire is remembered and forgotten needs to be a subject of analysis in its own right." This chapter responds to that call for research by examining how these words originally acquired a toxicity for those in the United States that complicated their adoption as analytic terms useful for understanding the country's history. It identifies the ways in which annexationists and anti-imperialists explicitly employed the rhetoric of education, teaching, and learning to bolster their attempts to define the terms *empire* and *imperialism*. It also shows how each side accused the other of miseducating the public.[5]

Within the United States, the question whether the country is or ever has been an empire has proven difficult to answer for more than a century. Many scholars have argued for understanding the United States as an imperial power, although their arguments repeatedly have been countered by others suggesting that the term *imperial* was misapplied and that its use in describing the United States distorted our ability to understand the unique nature of American power and influence. This debate tends to resurface most intensely when the United States engages in unilateral military actions abroad. In such periods, scholarly arguments about the historical nature of US expansion and hegemony quickly spill over into broader public discussions in which employing, rejecting, or repurposing the terms *empire* and *imperialism* become ways to criticize or endorse the role of the United States in the world. In the past decade, since the interventions in Iraq and Afghanistan, the notion that the United States is now and has been an imperialist power has become more broadly accepted in both scholarly and public discourse, but this has by no means eliminated the perennial question.[6]

Before turning to a study of the uses of the words *imperialism* and *empire* during the time period, it is important to note that the debate over imperialism that burned most intensely from late 1898 through 1902 cannot stand in for much larger discussions of whether the United States has ever been an imperial power. Historians, political scientists, and cultural theorists employ a broad range of definitions of *empire* and *imperialism* that distinguish among formal, informal, cultural, commercial, trade, and liberal empires, among others. While a commonly accepted definition of *empire* is the exercise of power by one country over other countries or territories, the nature of the relationships between those entities and the ways in which power is exercised are sources of considerable disagreement when it comes to determining what qualifies as empire. This has been a particular difficulty in relation to the United States because of the varied ways the nation has established itself as the world's leading power.

While many events in the aftermath of the War of 1898 have influenced how *empire* and *imperialism* have resounded in the United States, one can nevertheless hear echoes of the original debates about the aptness of those terms. The debate over the acquisition of the Philippines—under US control from 1898 through 1946—is particularly interesting in this respect; it is the one episode in American historiography on which there is the least disagreement that it constituted imperialism. Even in the case of the Philippines, many scholars argue that the United States fulfilled its stated intention of preparing the country for self-government and independence by gradually increasing autonomy, and that this undermines any claims of exploitative colonialism. In the period between 1898 and 1902, however, this course of action had not yet been decided, and the possibility that the Philippines and other territories would remain colonies was an open question. For this reason, and in view of the longer-term polemics over the question of American imperialism, it is well worth examining closely how those who argued most forcefully for subduing the Philippine insurrection and retaining these and other territories as permanent possessions denied that these actions constituted empire-building and imperialism.[7]

These debates echoed the political jousting in Britain over the term *imperialism*, which had coalesced around the liberal and illiberal aspects of colonialism and economic and military hegemony exercised by the great powers of the time period. By the late nineteenth century, *empire* and *imperialism* had taken on positive qualities in Britain in relation to its own practices, although earlier negative connotations would reemerge with the start of the Boer War in 1899. The identification of the terms with partisan politics transcended discussions internal to the United States. This would continue to be the case throughout the twentieth century, especially as US imperialism became a core tenet of anti-American thought in the Soviet Union and among the Latin American left. Nevertheless, at the turn of the twentieth century the concept of empire held the possibility of either positive or negative definition.[8]

The heated discussions over whether the United States should retain the colonies that Spain had ceded through the Treaty of Paris—Cuba, Puerto Rico, the Philippines, and Guam—marked the first time that Americans used the word *imperialism* to describe the United States in a sustained way. In the political debate that ensued, politicians and commentators attempted to define *imperialism*. Those who opposed the occupation of the islands and the denial of citizenship rights or coequal sovereignty to their peoples called themselves *anti-imperialists*. Those who supported annexation—even at the cost of a brutal war of counterinsurgency in the Philippines—frequently refused the label *imperialist*. The strategic ploys annexationists used to create ignorance about *imperialism* as an appropriate descriptor for US action in the Philippines are worthy of note because of their lasting resonance. In his historiographical review essay of American historians' use of empire as a category of analysis, Paul Kramer points out the strangeness of a situation that has made "imperial self-consciousness the litmus test of empire" and ceded "conceptual control to the historical actors themselves." In an ironic twist, although the anti-imperialists failed to halt the annexation of the Philippines as a colony and other expansionist policies, they were largely successful in exploiting and expanding the negative connotations of the words *imperialism* and *empire*. Those who advocated for expansion and colonialism had to take great care to either avoid or redefine what those words meant and thereby create doubt about whether the United States was becoming an imperial power even as the country was in the midst of acquiring colonies.[9]

For their part, the anti-imperialists of the late 1890s and early twentieth century worked to persuade and teach others that there was no difference between imperialism and colonialism. At a minimum they opposed the annexation and colonial rule of the Philippines, although many similarly opposed American policies in Hawaii, Puerto Rico, and Cuba. They objected principally to what they saw as violations of the American ideals, expressed in the Declaration of Independence and the Constitution, of popular sovereignty and the right of self-government. They tended to oppose republicanism with imperialism, which, in the words of Erving Winslow, the secretary of the New England Anti-Imperialist League, "implies the exercise of autocratic power, secrecy, and brutality." Along with their moral and legal objections to annexation, they frequently decried militarism and occasionally influence by financial interests, given that those who favored expansion often cited the need for increased markets abroad for American goods. For the anti-imperialists, "the undeviatingly true lesson of all history" was that aggression and militarism led directly to imperialism, and imperialism led to the conversion of free people into serfs.[10]

The anti-imperialist movement first came to national attention in June 1898, when a small group of Bostonians established the public information American Anti-Imperialist League. They founded this organization in order to inform the public about and to create opposition to US annexation of the Philippines. It also had a broader goal of preventing what they saw as a general tendency toward colonialism. The movement gained strength over the next two years as numerous local chapters of the national league were formed and other, similar organizations adopted anti-imperialist positions. Anti-imperialists came from across the political spectrum, although the most well known and most influential were those known as Mugwumps (former Republicans who had left their party in the 1880s to take a stand against corruption), who tended to be drawn from the ranks of traditional northeastern elites and included men such as Charles Francis Adams and Carl Schurz. The Mugwumps were joined by some dissident Republicans, such as George Frisby Hoar, and many Democrats, Populists, and others, including writer Mark Twain, industrialist Andrew Carnegie, and labor leader Samuel Gompers. Annexationists, on the other hand, were likely to be Republicans who supported the policies of the William McKinley administration and that of his successor, Theodore Roosevelt.[11]

In 1900 William Jennings Bryan, the Democratic candidate for president, ran as an anti-imperialist, although the anti-imperialists were divided over his candidacy because of other policy views he held. His loss to McKinley represented a blow to the movement. The ongoing war in the Philippines soon supplied a new issue to the anti-imperialists when reports of American atrocities began to appear in the press, but the commission convened to investigate the situation was dominated by pro-annexationist Republicans. By 1902, when the war came to an end and Congress passed the Philippine Organic Act, which established a civil government on the island (although it remained subject to the American government), anti-imperialism began to lose steam as a national cause. Although the Anti-Imperialist League continued in existence through 1920, the anti-imperialist movement of 1898–1902 is often considered a failure because its adherents did not achieve their goals of halting the legal annexation and subjugation of territories outside the United States. This view has recently been challenged by historian Michael Patrick Cullinane, who argues that anti-imperialist dissent had the effect of limiting what was politically possible, especially as the movement turned toward denunciations of human rights violations not only by American troops in the Philippines but also in other colonial settings. Cullinane emphasizes the impact of this dissent on legal and political speech and connects this to the successes of the movement through his discussion of how anti-imperialists articulated the central concept of liberty by defining the concept and word empire in opposition to it. Here, I point to the ways in which those who supported annexation and colonization ultimately attempted to avoid the words *imperialism*, *empire*, and *colonies* so that they could pursue their policies.[12]

When faced with accusations of imperialism, annexationists often defined their goals in terms of wanting to create a benevolent and fundamentally liberal regime of short-term tutelage. They also drew upon earlier uses of the term that were understood to be largely figurative, as in the phrase *commercial empire* or Jefferson's *empire of liberty*. Frequently, if somewhat ambivalently, the annexationists defended colonialism in the island territories by arguing that it was merely a continuation of continental expansion. For them, the lesson of history taught by the continental growth of the United States was that if acquisition of the islands constituted illegitimate expansion, then so, too, did the Louisiana Purchase and the Mexican cessions. In 1898 one speaker at a local event noted, "If it be imperialism to favor territorial expansion, then imperialism began with Jefferson. . . . [It] continued with Monroe. . . . It includes among its votaries Andrew Jackson . . . and makes an imperialist of Ulysses S. Grant." Albert Beveridge, the political figure of the era most frequently associated with the advocacy of imperialism, framed the issue similarly. A stalwart Republican, Beveridge had presidential ambitions. Even prior to winning his seat in the Senate, where he would represent Indiana from 1899 through 1910, he worked to make his mark on American foreign policy. During his 1898 electoral campaign, Beveridge first attracted widespread attention with his phenomenally successful "March of the Flag" speech, delivered at the Indiana Republican convention on September 16, 1898, and subsequently widely reprinted by the party. In it, Beveridge made the case for retaining and governing the island possessions without the consent of their peoples on the basis that they were not capable of self-government. He argued that it was the God-given destiny of the American people as Anglo-Saxons to expand in the world, and he directly compared this expansion to that of contemporary European powers. In his repeated praise of "the march of the flag," Beveridge linked continental expansion to overseas expansion and pointed out that ruling over the Indians of the Plains or the Mexicans of the New Mexico Territory without their consent was no different than what was presently proposed for the islands.[13]

This earlier expansion had required civilizing Native Americans and people of Mexican descent. Rule in the islands would likewise entail an educational effort before independence could be bestowed. Indeed, teaching became an important metaphor for reaffirming the American value of self-government by an enlightened citizenry and was used both by annexationists as justifications for a tutelary (i.e., colonial) relationship and by the anti-imperialists to denounce the hypocrisy of using coercion while preaching consent. President McKinley told a Boston audience in 1899 that in the Philippines, "God gave us the light to see our duty," which meant submitting the Filipinos to "the guiding hand . . . the uplifting education, not of their American masters, but of their American emancipators." McKinley put education in the context of Manifest Destiny and the missionary work typical of the nineteenth-century. Beveridge urged Congress to retain the Philippines, arguing that self-government was "a method of liberty . . . acquired only after centuries of study and struggle and experiment and instruction. . . . It is the degree which crowns the graduate of liberty, not the name of liberty's infant class, who have not yet mastered the alphabet of freedom." How long might this take? Theodore Roosevelt judged that it would take at least two generations, given the "density of ignorance" from which he claimed 90 percent of Filipinos suffered.[14]

Education was not, of course, merely a metaphor for the colonial relationship. In Cuba, Puerto Rico, and the Philippines the occupation governments introduced sweeping educational reforms and imposed new school systems superficially modeled on those in the United States. Frequently, the American administrators chose curricula drawn from those used by the Bureau of Indian Affairs, the Hampton and Tuskegee Institutes, and other such industrial- or agricultural-skill-based programs, which were usually intended for groups within American society who had a subordinate status and were believed to be mostly incapable of higher education or self-government. (Twain pointed to such gaps between stated intention and actual practice when he speculated that the person sitting in darkness might notice that there was "a difference in the two brands" of civilizing education being sold in exchange for "his blood and tears and land and liberty.") What education supervised by the United States would mean for the peoples of the new island possessions has been well chronicled elsewhere. In the debate over imperialism, the claim that offering education justified the dependent status of the islands provoked strong responses from the anti-imperialists, who saw them as contradictory. That republican practice required an educated citizenry was a fundamental American belief, and any claim to educate therefore implied a commitment to liberty.[15]

Charles Francis Adams doubted that the American occupiers truly meant to provide education to the peoples of the islands. He cited British precedent when he stated in a book on imperialism and the American tradition that "Hindostan would furnish another highly suggestive example of the educational effects of 'tutelage' on a race. After a century and a half of that British 'tutelage,' what progress has India made towards fitness for self-government? Is the end in sight?" Adams repeatedly put quotation marks around *tutelage* to imply that what the word really meant was imperial control disguised as liberal benevolence. Andrew Carnegie believed that a free education created the desire for self-government. Also arguing from the British experience in India, he warned that "education makes rebels," and later he asked, "With what face shall we hang in the school-houses of the Philippines the Declaration of our own Independence, and yet deny independence to them? . . . Are we . . . to teach rebellion in our books, yet to stamp it out with our swords?" American policymakers would have a difficult time resolving such conundrums. Those who argued in favor of tutelage for people considered unprepared for self-government understood the process of "teaching" this skill in two senses: implementing educational programs designed to produce Americanization and

Figure 10.1. School Begins. Uncle Sam (to his new class in civilization): "Now, children, you've got to learn these lessons whether you want to or not! But just take a look at the class ahead of you, and remember that, in a little while, you will feel as glad to be here as they are!" Louis Dalrymple, "School Begins" (illustration), *Puck* 44, no. 1142 (25 January 1899), http://www.loc.gov/pictures/item/2012647459/.

teaching by example; in other words, people in the territories would learn about American good government by being subjected to it.[16]

These themes come through clearly in an editorial cartoon of the time supporting annexationism (figure 10.1). The artist depicts US relations with insular territories as a literal classroom in which Uncle Sam is teaching a class of children who represent the different territories and peoples still considered dependent. The school building is a classically American one-room schoolhouse of the type that had served rural communities and was by the 1890s already a nostalgic icon associated with American liberty and democracy. In this metaphorical rendering of the relationship between the United States and its outlying lands, the advanced class of pupils, who are neatly dressed and obediently studying their textbooks, are the territories that had recently become states or would soon. The newest class of students is made up of four grumpy, restive black children representing the four island acquisitions: the Philippines, Hawai'i, Puerto Rico, and Cuba. An African American boy is washing the classroom windows; a Native American child sits dozing near the door with an ABC primer held upside down; and a Chinese child stands just outside the door, unsure whether to enter. The lesson of civilization that Uncle Sam imparts is spelled out on the blackboard. It states that England governed its colonies without their consent for the good of civilization and that the United States will in turn govern its territories without their consent until such time as they are capable of self-government. The cartoon is typically, and correctly, read as a clear expression of the racial thinking that contributed to American expansion, but one might also see another lesson in this cartoon. While the pupils of the class are ostensibly the peoples of the territories and racial minorities, the audience for the cartoon was the domestic reading public. The purpose, in other words, was to teach the meaning of American imperialism to *Americans* as a tutelary relationship that might require force—Uncle Sam tells his students in the caption that "you've got to learn these lessons whether you want them or not"—but was ultimately for the good of those so subjugated.[17]

In making such arguments highlighting the benevolence of American rule over the islands acquired from Spain, annexationists at first were not unwilling to characterize the relationship as explicitly "imperial." During the summer and fall of 1898, in the period between when the United States signed an armistice with Spain in July and the adoption of the Treaty of Paris, which ceded the islands to US possession in December, annexationists could still use the words *imperialism* and *empire* in a positive sense, although their opponents had already begun to use them to denounce the prospect of the United States exercising sovereignty over the island territories (including the recently annexed Hawaii).[18]

There was a tentativeness as Americans considered the outcomes of the war. One newspaper in June 1898 titled an article on McKinley's deliberations over his party's policies "'Imperialism': the new word used for the territorial expansion people." In July, as the war was winding down, a *New York Times* journalist speculated that some would use the Fourth of July as an occasion to raise their concerns about imperialism, although "just what it means is left to the imagination." A related article the same day on the possible need for a new cabinet position for a "colonial secretary" stated that the conclusion of the war would surely lead to the acquisition of colonies, "or of what would be called colonies if they were held by any other great power." In an issue that announced the Spanish surrender at Santiago in mid-July, a Grand Rapids newspaper included a piece on the financial benefits of colonialism, an article proclaiming Puerto

Rico the "next stop" on the triumphant march, and another that declared "God's Time Has Come: American Imperialism Must Be in Spite of Ourselves." This article summarized a sermon given at a local church in which the pastor defined American imperialism positively as a means of educating the nations of the world in the principles of popular sovereignty. Reflecting a position common to supporters of expansion, the article flatly rejected the notion that American imperialism could or would be militaristic or despotic. Albert Beveridge, in "The March of the Flag," as in other speeches he gave during 1898, employed the words *empire* and *imperialism* positively in the sense of an extension of American ideals and leadership. He labeled Americans "a people imperial by virtue of their power, by right of their institutions, by authority of their Heaven-directed purpose." In a speech shortly after the declaration of war against Spain in April 1898, Beveridge praised Americans' "imperial intellects" and looked forward to the "Republic's imperial career." Later that summer, he appealed to America's "imperial destiny."[19]

Still, the word imperialism raised questions. In the fall of 1898, at the banquet of the California Society of the Sons of the Revolution, the president of the organization noted in his opening remarks that "'American Imperialism' is certainly something new to America. We have been accustomed to associate that word—Imperialism—with thrones and kingdoms. I am sure we are all desirous of being informed how we can assimilate and adjust it to American interests and to Republican institutions." The evening's speakers, some opposed and some in favor of holding the islands as colonies, all sought to situate that experience in the larger sweep of American history. In Boston, an anti-imperialist reverend preached a Thanksgiving Day sermon that warned against those who had tried to give a positive definition to the word *imperialism*. He rejected both the word—"I abominate it"—and the practice of territorial expansion, which led to an emphasis on materialism rather than idealism. In his view, it was a dangerous word that should be rejected, because "correct thinking requires correct terminology."[20]

Unfortunately for the reverend, annexationists quickly learned to modify their terminology in ways that would allow them to express ignorance of any imperial intent, although this had little effect on the policies they proposed and supported. By the fall of 1898 the question of annexation had become a burning political issue, as the United States would soon have to negotiate final treaty terms with Spain. It was also an election year, and anti-imperialists had forcefully been making their case against expansion and colonialism. For these reasons, in spite of attempts by annexationists to positively repurpose it, the word *imperialism* quickly acquired a set of scare quotes in their responses to opposition to territorial annexation. Even though Beveridge had used the term *imperial* in positive constructions in his "March of the Flag" speech, one of his correspondents who wrote to congratulate him on the success of the speech referred to former concerns he had had about "so called 'imperialism.'" Beveridge's speech had helped persuade him that these concerns were unfounded. A Philadelphia newspaper editor exhorted his audience to "call it imperialism if you will" but reminded them that "American imperialism" is "the higher and nobler Imperialism that voices the sovereign power of this nation [and is] interwoven with the destiny of our great free government." Other annexationists attempted to reclaim the use of the word *imperialism* by disputing that it was to be feared. Already in June 1898 one author who contested former president—and member of the Anti-Imperialist League—Grover Cleveland's characterization of expansion as imperialism declared that "in its present guise this word appears as a miasm [*sic*] arising from the horrid field of war, and some of us, not knowing what it means, run from it in terror under the impression that we have seen some hobgoblin bent on our destruction." In November 1898 a bishop of the Protestant Episcopal Church speaking in Albany urged Americans to take on imperialism as a duty of their race and stated that he was "not frightened by this new coined word 'imperialism' which may mean much of good or much of evil."[21]

In this political contest over the meaning of the word, annexationists sometimes explicitly addressed the negative definitions of *imperialism* used by their opponents and sought to replace them with positive characterizations. James C. Fernald begins his 1899 book *The Imperial Republic* by invoking "the ghost of imperialism" as a literally rendered specter, with "antique helmet, spear, and shield," embodying Americans' fears of despotic militarism and oppression in a representation of the Roman republic, which had degenerated into an empire. Fernald ultimately embraces the term *imperialism* once he has redefined it in the American case to mean the expansion of the republic and the extension of human freedom. In a chapter on the need for establishing an office of colonial affairs, he deals forthrightly with the term *colony* as well by noting that Americans resist the use of this word simply because the nation had its origin as British colonies and suggests that using *dependency* or *territory* would overcome this difficulty. Similarly, John L. McAtee, judge of the Supreme Court of the Territory of Oklahoma, delivered a speech that was later reprinted as a book titled *For the Treaty of Paris and American "Imperialism."* In it, McAtee sets the word off in quotation marks for most of the text until he arrives at his conclusion, in which he urges his audience to "admit that this government and people is imperial . . . but imperial, with discriminations. . . . It is imperial in its free institutions . . . and the liberality with which it extends those institutions. . . . This country is imperial in its uplifting of the race by education and enlightenment." Beveridge takes the semantic issue head on in "The Star of Empire," a speech delivered on September 25, 1900, at the start of the Republican campaign. He marks off the word *empire* with quotation marks at several points to imply its use as a word meaning "autocratic oppression" by those he labels "the Opposition" before recasting it positively as the empire of Jefferson and Washington. Under this "westward star of empire," the nation was expanding "liberty and law . . . commerce and communication . . . social order and the Gospel of our Lord." While Beveridge no longer uses the word *imperial* or *imperialism*, he argues by analogy that Americans are as right to hold the Philippines, Puerto Rico, and Cuba (whose independence he declares should never have been promised) as the British are to hold India, and the Boers, the Transvaal.[22]

Particularly important for understanding the rhetorical denial of empire are the ways in which annexationists negatively characterized the charge of imperialism leveled at them by their opponents. They accused the anti-imperialists of political opportunism and rejected the label of imperialism simply on the grounds that it was pejorative and therefore could not possibly apply to the actions of the United States. More seriously, the most fervent annexationists argued that the anti-imperialists were disloyal and, in an appeal to consequences, that their accusations of imperialism were undermining the war effort and costing American lives. Anti-imperialists were depicted as fools, turncoats, weaklings, and nagging old women. The nagging had some effect, however. Annexationists became so averse to the term *imperialism* that they ultimately only used it when it was clearly marked off as ironic. By 1899 they frequently referred to the anti-imperialist faction simply as "the opposition" or as "anti-expansionists" so as not to allow that there was any imperialism to oppose.[23]

As the Philippine-American War heated up over the course of 1899 and the presidential election of 1900 approached, public debates over the use of the term *imperialism* sharpened. Annexationists became increasingly likely to flatly contradict the charge, to belittle it with sarcasm, or to exaggerate it into a caricature. One dubbed it a "queer kind of imperialism" that would give Cuba its independence or refuse to occupy territory in China. A Republican senator reported that while on a trip to Britain, he had been asked what all the noise in the American press about imperialism meant. He had replied, "I can't tell you what an imperialist in our country is. I never saw one . . . I think that with us the species is extinct." Increasingly, as the charge of imperialism began to stick, annexationists came to deeply resent the pejorative connotations. They frequently counterattacked by implying that the anti-imperialists were the ones who were anti-American in their speech and in their actions. In 1900 Secretary of War Elihu Root denounced the charge as "the cheapest and most threadbare of the demagogue's stock." Beveridge articulated this tactic most clearly in his speech titled "Our Philippine Policy," in which he supported the bill authorizing the formal annexation of the Philippines. (In his original draft he had written and emphatically underlined that they should be annexed "permanently.")[24]

Beveridge won election to the United States Senate in 1899 partly on the strength of his proexpansionist sentiments. Prior to taking his seat in the Senate, he took the bold step of traveling to the Philippines to learn about conditions there firsthand. As much as the ignorance of the Filipinos was widely assumed, Americans were quite candid in describing their own ignorance about the island archipelago. Historian Samuel Flagg Bemis probably put it best: "Before the war there had not been the slightest demand for the acquisition of the Philippine Islands. The average American citizen could not have told you whether Filipinos were Far Eastern aborigines or a species of tropical nuts." When Beveridge returned, he avoided speaking to reporters and did not reveal the results of his trip to anyone other than President McKinley until the occasion of Beveridge's speech to the Senate on January 9, 1900. It was a deliberate strategy to drum up interest and to highlight his possession of expert knowledge of the situation, which no other American politician could match. During the speech, Beveridge displayed a small red notebook that he had used in the Philippines to record his observations and discussions with military men, foreign business or plantation owners, and a few local leaders. Beveridge had distinguished himself on his trip by traveling to the front lines of the fighting between American soldiers and Filipino insurgents and developing a friendship with Major General Henry Lawton, who died in battle just a few weeks prior to Beveridge's speech. A highly skilled orator, Beveridge proved able to combine his claim of hard-won, firsthand knowledge with an emotive delivery that derived from his acquaintance with soldiers who had lost their lives in the line of duty. His speech became a lesson in patriotism. In it he argued not only that retaining the Philippines was in the interest of the United States and the Filipinos themselves but also that failing to support this cause was to betray the troops deployed there.[25]

Beveridge began by stating that he was prompted to give the speech by the many requests he had received from members of Congress for information about conditions in Asia and by his deep concern that some of what was being said in the Senate was "costing the lives of American soldiers." He elaborated later in the speech by explaining that news of American opposition to the war in the Philippines had given the Filipino insurgents hope that they might achieve independence, and therefore they continued to fight. In Beveridge's view, this mistake was the same as that made during the Indian Wars, which he claimed might have cost fewer lives of "soldiers and settlers" had they been fought more decisively and continuously. Beveridge explained to his audience that Filipinos did not understand the concept of free speech and therefore could not comprehend a political debate in the United States over Philippines policy that seemed to include "treasonable criticism" by a minority party. After this lesson in comparative civics, Beveridge informed his listeners that these criticisms of the war effort had aided the insurgents "more than climate, arms and battle." He admonished "those whose voices in America have cheered those misguided natives to shoot our soldiers down, that the blood of those dead and wounded boys is on their hands."[26]

If Beveridge had previously used the words *imperialism* and *empire* to characterize what he was in fact proposing in his Philippines speech (that the United States should retain the islands as permanent colonies), he avoided those words entirely here except by implication. Beveridge's Senate colleague John C. Spooner of Wisconsin made the point directly. He linked what he described as the "fictitious" issue of imperialism to continued Filipino resistance and the consequent deaths of American soldiers. He cited in his speech a letter from Emilio Aguinaldo to one of his generals in which Aguinaldo expressed the hope that the "honorable people of North America" would vote out "the imperialistic party" and end the war by granting the Philippines independence. Spooner stated that although he did not believe that those who opposed the war or supported Bryan's candidacy intended to encourage the insurgents in their resistance, he was giving notice that "good citizenship" demanded that they cease encouraging the enemy when the "question of imperialism was not here then, [is] not here now, and [is] not to be here." Elihu Root likewise insinuated the disloyalty of the anti-imperialists in a campaign speech in which he stated that he would not "say that the men who are encouraging the Filipino soldiers here are traitors to their country. They do not know what it is they do." Theodore Roosevelt was less cautious when he referred to the anti-imperialists as "simply unhung traitors" in a letter to Leonard Wood. After Root's speech, the *New York Times* declared that the anti-imperialists had "no competence to instruct the American people, but only to misguide them." The lesson was clear: crying imperialism was foolhardy at best, treasonous at worst, and in any case misleading.[27]

The anti-imperialists were acutely aware of the significance of the semantic issue. Bryan tried to pin responsibility on the annexationists and said that if there was to be a colonial policy, the Republicans should state it clearly:

> The word expansion would describe the acquisition of territory to be populated by homogenous people and to be carved into states like those now in existence. An empire suggests variety in race and diversity in government. The Imperialists do not desire to clothe the Filipino people with all the rights and privileges of American citizenship; they want to exercise sovereignty over an alien race and they expect to rule the new subjects on a theory entirely at variance with constitutional government.[28]

Senator George Turner of Washington spoke directly to the ways in which language expressed key constitutional principles. He denounced the phrase *benevolent assimilation*—used by the McKinley administration to characterize its Philippines policy—as meaning "the exact opposite of what it means" outside "the Presidential lexicography." Turner was particularly concerned about the substitution of *territory* for *colony*. While those who supported "expansion" could blithely suggest that the substitution of terms was simply a matter of local preference given America's own colonial past, Turner pointed out that it had consequences, as the constitutional clauses regulating the acquisition and government of territories were clearly articulated and framed in a way that precluded an expectation that these would be held as colonies. Had the framers of the Constitution meant to provide for colonies, Turner argued, they surely would have done so explicitly by including the word itself. Turner proved prescient, as it would take the US Supreme Court more than a decade to work out the legal implications of what the Constitution permitted as far as territorial administration. The series of decisions that would later be known as the Insular Cases ultimately allowed for the islands to be governed as dependencies but required considerable semantic wrangling over the meaning of such words as *territory* and *citizen*.[29]

The knowledge that they were losing this semantic battle frustrated the anti-imperialists. They had been highly effective, however, in their negative characterizations of the words *imperialism* and *empire*. After 1900 the likelihood of the terms being used as positive descriptors of American policy toward the insular territories was almost nil. The words had become inextricably linked with purely partisan politics. Furthermore, after the capture of Filipino rebel leader Emilio Aguinaldo in early 1901, US forces began to gain the upper hand, and more features of civilian governance were established across the islands. Annexationists caricatured and exaggerated anti-imperialists' fears that endless militarism and oppression would result from the occupation of the islands in ways that allowed these fears to be easily dismissed as the bulk of the fighting in the Philippines wound down in 1902. That same year, an Omaha newspaper sympathetic to anti-imperialism reprinted a reader's letter that sought to draw back the "verbal veil" over the word *imperialism* and reveal the "philological feats of the imperialists." These included redefining the word as expansion, absorption, assimilation, pacification, extension of civilization, duty, providence, fate, supporting the administration, industrial supremacy, and letting "the flag stay put." However perceptive the observation, it merely chronicled the fact that by this time the philologists of expansion had carried the day.[30]

An extraordinary facet of this war of words over the terms *empire* and *imperialism* was the intense emotionalism it generated within the United States then and has continued to generate since. Observers have remarked with bemusement American reactions to the suggestion that theirs is an imperialist nation. In 1965, during another era in which US involvement overseas was creating intense controversy, an article by Henry Fairlie titled "A Cheer for American Imperialism" appeared in the *New York Times* headed by the photograph of a model school built in San Salvador, the capital city of El Salvador, under the auspices of John F. Kennedy's Alliance for Progress program. Fairlie, a well-known journalist in both his native Britain and the United States, which he dearly loved and often wrote of, took up the question whether America was an empire. To him the answer was self-evident, but he recognized that it troubled most Americans:

> Their nation began its life in rebellion against a colonial power, and consequently the mere suggestion that they may themselves be an empire is immedi-

> ately taken as an accusation. "It is that word. It is just that word," says one American with whom I have discussed the question over and over again. Call it, then, by another name—though none is so short and convenient—but the fact will remain.[31]

More recently, British historian Niall Ferguson popularized the phrase *empire denial* to draw attention to the fact that the United States needed to fully assume its responsibilities as a superpower. In the wake of the US wars in Iraq and Afghanistan, Ferguson vigorously advocated for the nation to recognize and embrace its role as an imperial power—in his view, a necessary stabilizing force in an often chaotic and dangerous world. This concept of imperial denial has also been used by other scholars, several pointing out that the peoples of other, previous empires had similarly denied that the label "imperialist" correctly characterized their societies.[32]

The use of *empire* and *imperialism* as useful and appropriate conceptual tools for understanding American history and culture has become far more common and acceptable in scholarship produced in the United States over the last two decades, but this development has by no means eliminated the polemic. The debate has persisted for so long that one now hears a distinct note of fatigue in the scholarly literature that takes up the question. One historian writes, "Is the United States an empire? Only in the United States could such a question even be asked." Another declared that the publication of a recent book on the outcomes of the American experience in the Philippines "should end any discussion as to whether the category 'empire' applies to the United States," although the need to make such a declaration suggests that that outcome seemed unlikely.[33]

There is persistent doubt in the United States that the terms *empire* and *imperialism* can be anything other than polemical given the "unsavoury baggage that 'empire' hauls in its wake." Those who use the terms as analytical concepts frequently preface their use with an emphatic statement that the terms do not imply judgment and should not be considered pejorative. Authors of works on American imperialism have stated the intention to remain "impartial," declared that *empire* should not be considered "an epithet," and affirmed that the term was being used in a "nonpejorative, nonideological, nonjudgmental sense." Some speculate that the debate has become counterproductive. Paul Kramer makes the case for understanding American history through the lens of the "imperial" as a mode of analysis but suggests eschewing the category "empire," as it leads to "mostly unproductive" comparisons with other empires. In his book-length consideration of the traits of empires, which is a meditation on the ascendancy of US power, historian Charles Maier chooses to avoid stating what he concludes about the United States and empire. He makes this choice because he has "found such assertions so polarizing that readers never get past the definition." This choice frustrated reviewers who admired Maier's scholarship but faulted the imprecision of its conclusions, one suggesting that given the book's implied argument, failing to use the "e-word" amounted to waffling.[34]

Some American historians have objected to both *empire* and *imperialism* as lacking sufficient precision to be useful analytic lenses. In response to New Left historians of the 1960s and 1970s who identified persistent imperialist tendencies in American foreign policy stemming from economic interests, historian James A. Field Jr. once described this interpretation as flawed by a "one-size-fits-all interpretation, which suggests if only we repeat, like John Randolph's whippoorwill, 'empire, empire, empire,' everything . . . will be explained." Field recognized that in his own essay on American expansion in the 1890s he had "tried very hard not to use the word [*imperialism*] outside of quotation marks or paraphrase" because he preferred to analyze and describe the events in question rather than to fit them to what he saw as a predetermined label. Field was not alone in this insistence on arguing inductively. Other historians of empire have proposed using a nominal criterion to decide that "empires are states that call themselves empires" and that this self-designation should count for more than a definition applied "arbitrarily." More recently, Jeremy Suri has termed *empire* a "catchall" label that obfuscates more than it reveals about complicated and contradictory tendencies in American history. He finds that relying on this label "distracts from the real motivations and actions that constitute American foreign policy [and] implies the wrong solution for those frustrated with the nation's actions abroad," although he concedes that from certain vantage points—such as that of Central America—it is quite possible that the United States has appeared to be an imperial power. Nevertheless, he argues that a better understanding of America's role in the world is as a nation builder. In a recent monograph titled *American Umpire* that characterizes the United States' role in the world as an arbitrator, Elizabeth Cobbs Hoffman agrees with Suri that to define the United States as an empire is a mischaracterization and one with serious consequences given that imperialism is "a phenomenon more widely reviled today than wife beating." These consequences include damage to American morale at home and to the United States' reputation abroad. Hoffman suggests that academics who have incorrectly applied the term *imperialist* to the United States have contributed to the deaths of Americans at the hands of extremists and terrorists, a claim that she links to the teaching of undergraduates (including international students) at American universities, who hear the United States described as an empire or an imperial power.[35]

These more recent discussions reflect larger historiographical debates about the United States in the world. Nevertheless, their themes of concern over the consequences of the terms *empire* and *imperialism* to the United States and the sense that these terms are important to the education of the broader public are consistent with the ways in which the debate over these words developed. During the first domestic debate over whether American actions abroad constituted imperialism, the semantic issue became tightly linked to particular political positions and thereby became subject to polarizing rhetoric. In this way, the terms acquired an emotional valence when used within the United States that has persisted and has complicated their use as analytical concepts. To deny the relevance of the terms *empire*, *imperialism*, and *colony* in regard to US acquisition, subjugation, and occupation of the island territories in the wake of the War of 1898 was to deny that such expansionism violated core American ideals. At the turn of the century, agnotological efforts at redefining and later denying the validity of the terms became the means to square the fact of American imperialism with the need to assert the rightness of American idealism.[36]

At the close of the Republican campaign leading up to the midterm elections of 1902, in which the Republican party would maintain its control of Congress, Albert Beveridge gave a speech declaring anti-imperialism to have been nothing but fear mongering and slander:

> The people know that where American soldiers went, peace followed; that justice followed peace; that industry followed justice; and that education kept step with both. The people know that the sentinels America has posted all over the island are American teachers; the fortresses America has erected are American school-houses; the military roads America has erected are American highways of commerce . . . [and] so, "Imperialism" is dead at the hand of truth.[37]

As in Twain's famous quip about his own reported demise, Beveridge was premature in declaring the death of the word he sets off in quotation marks, but he was largely correct in suggesting that a lesson had been learned. Dissociating American actions abroad, whether military, economic, or cultural, from the language of imperialism would remain a consistent feature of American political and historiographical discourse.

NOTES

1. Mark Twain, "To the Person Sitting in Darkness," *North American Review* 172 (February 1901): 165, 170.

2. Ibid., 170; Robert N. Proctor, "Agnotology: A Missing Term to Describe the Cultural Production of Ignorance (and Its Study)," in *Agnotology: The Making and Unmaking of Ignorance*, ed. Robert N. Proctor and Londa Schiebinger (Stanford, CA: Stanford University Press, 2008), 8.

3. William Jennings Bryan, *Republic or Empire? The Philippine Question* (Chicago: Independence, 1899). For a thorough treatment of the impact of Roman history and classical learning on the formation of the American political system, see Mortimer N. S. Sellers, *American Republicanism: Roman Ideology in the American Constitution* (New York: New York University Press, 1994).

4. Both annexationists and anti-imperialists showed great concern for what the "lessons of history" offered the United States vis-à-vis its acquisition of the Philippines. Noted Orientalist William Elliott Griffis, for instance, wrote in a book on policy of his concerns about US Americans' lack of experience with colonial administration and advised caution about allowing political infighting to blind them to the lessons of history that could be learned from the British and Dutch colonial endeavors. Historian and president of the University of Wisconsin Charles Kendall Adams echoed this theme in a contemporary article titled "Colonies and Dependencies," which took the lessons of history that could be learned as its main theme. The trope was so common that the editors of one journal wrote in regard to the responses to their position on the Philippines that "our readers have before them two interpretations of the lessons of history, ours and that of our correspondents. We leave them to weigh for themselves these two interpretations and to decide between them." William Elliott Griffis, *America in the East* (New York: A. S. Barnes, 1899), 231–32; Charles Kendall Adams, "Colonies and Other Dependencies," *Forum* 28 (March–August 1899): 33; Correspondence, *Outlook* 66 (September–December 1900): 186; Twain, "To the Person Sitting in Darkness," 165.

5. Both American studies scholars and historians have emphasized the ways in which imperialism transformed American culture at home. Some relevant examples include Amy Kaplan and Donald Pease, *Cultures of United States Imperialism* (Durham, NC: Duke University Press, 1993); Christine G. Appy, *Cold War Constructions: The Political Culture of United States Imperialism* (Amherst: University of Massachusetts Press, 2000); Mary Renda, *Taking Haiti: Military Occupation and the Culture of U.S. Imperialism, 1915–1940* (Chapel Hill: University of North Carolina Press, 2003); Kristin Hoganson, *Consumer's Imperium: The Global Production of American Domesticity* (Chapel Hill: University of North Carolina Press, 2010); Robert Beisner, "Comments" in AHR Forum on James A. Field, "American Imperialism: The Worst Chapter in Any Book," *American Historical Review* 83 (June 1978): 673; Ian Tyrell, "Empire in American History," in *Colonial Crucible: Empire in the Making of the Modern American State*, ed. Alfred McCoy and Francisco Scarano (Madison: University of Wisconsin Press, 2009),

544. As William Appleman Williams famously stated in 1955, "One of the central themes of American historiography is that there is no American Empire." Williams, "The Frontier Thesis in American History," *Pacific Historical Review* 24 (1955): 379.

6. The first such period, discussed below, came after 1898. Since then, discussions of American imperialism were particularly fierce during the Cold War and following the 9/11 attacks that led the United States to intervene in Iraq and Afghanistan. Beginning in the late 1950s, William Appleman Williams was the most influential voice arguing for an interpretation of American history through the lens of imperialism in view of American actions during the Cold War. See esp. Williams's *Tragedy of American Diplomacy* (Cleveland, OH: World, 1959); and Walter LaFeber, *The New Empire: An Interpretation of American Expansion, 1860–1898* (Ithaca, NY: Cornell University Press 1963). For a direct comparison of the politics of the War of 1898 and those of the Vietnam War, see Robert Beisner, "1898 and 1968: The Anti-Imperialists and the Doves," *Political Science Quarterly* 85 (June 1970): 187–216. For a succinct summary of the scholarly back and forth over the significance of "American imperialism" from this period, see Field, "American Imperialism: The Worst Chapter in Almost Any Book," *ibid.*, 644–68; Walter LaFeber and Robert Beisner, Comments, ibid., 669–78; and James A. Field, "Reply," ibid., 679–83. For a particularly relevant example of debates over the term in regards to contemporary policy, see the debate between Robert Kagan and Niall Ferguson titled "Is the United States an Empire?," *AEI Newsletter* 1 September 2003, http://www.aei.org/article/foreign-and-defense-policy/is-the-united-states-an-empire-article-1/. For overviews of the recent historiography of American empire, see Paul K. MacDonald, "Those Who Forget Historiography Are Doomed to Republish It: Empire, Imperialism and Contemporary Debates about American Power," *Review of International Studies* 35 (January 2009): 45–67; Dane Kennedy, "Essay and Reflection: On the American Empire from a British Imperial Perspective," *International History Review* 29 (March 2007): 83–108.

7. Citing this state of affairs, William Appleman Williams remarked, "Most historians will admit, if pressed, that the United States once had an empire. They promptly insist that it was given away." The classic statement of this latter position was made in 1936 by Samuel Flagg Bemis, who described the acquisition of the Philippines and other island possessions as "the great aberration." Bemis, *A Diplomatic History of the United States* (New York: Henry Holt, 1936), 463–78. See also Eric Rauchway's discussion of how this historiographical construction evolved over time in "The Global Emergence of the United States, 1867–1900," in *A Companion to 19th Century America*, ed. William Barney (Malden, MA: Wiley Blackwell, 2008), 104–20.

8. Richard Koebner and Helmut Dan Schmidt, *Imperialism: The Story and Significance of a Political Word* (Cambridge: Cambridge University Press, 1964), 221–49. The association of the use of *imperialism* as an analytical term and Marxist thought became entrenched especially as a result of the thesis advanced by Vladimir Lenin in his influential work *Imperialism, the Highest Stage of Capitalism* (1916).

9. As scholars of the loosely defined anti-imperialist movement of the late nineteenth century make clear, anti-imperialist thought preceded the War of 1898 and had its roots in opposition to much of the US continental expansion. However, the polemical term of discussion was *Manifest Destiny*, not *imperialism*. E. Berkeley Tompkins, *Anti-Imperialism in the United States: The Great Debate, 1880–1920* (Philadelphia: University of Pennsylvania Press, 1970), 7–10. See also Michael Patrick Cullinane, *Liberty and American Anti-Imperialism, 1898–1909* (New York: Palgrave, 2012); and Jim Zwick, *Confronting Imperialism: Essays on Mark Twain and the*

Anti-Imperialist League (West Conshohocken, PA: Infinity, 2007). On connections between continental and overseas expansion and anti-imperialism, see Frank Schumacher, "American Way of Empire: National Tradition and Transatlantic Adaptation in America's Search for Imperial Identity, 1898–1910," *Bulletin of the German Historical Institute* 31 (2002): 35–50; and Paul Kramer, "Power and Connection: Imperial Histories of the United States in the World," *American Historical Review* 116 (December 2011): 1359.

10. Beisner, "1898 and 1968," 188; Fabian Hilfrich, *Debating American Exceptionalism: Empire and Democracy in the Wake of the Spanish-American War* (New York: Palgrave Macmillan, 2012), 7; Erving Winslow, "The Anti-Imperialist Faith," *North American Review* 175 (December 1902): 811–18; G. D. Herron, editorial, *Social Forum* 1 (June 1899): 28.

11. Stuart Creighton Miller, *Benevolent Assimilation: The American Conquest of the Philippines, 1899–1903* (New Haven, CT: Yale University Press, 1982), 104–10.

12. Cullinane, *Liberty and American Anti-Imperialism*, 1–10, 180–81.

13. In September 1898 an article insisted that the supposedly opposed definitions of militaristic and benevolent imperialism actually had a great deal in common and that the only "true American imperialism" could be that of trade and fair dealing. "Two Kinds of Imperialism," reprinted from the *Yale Review* in the *Harrisburg (PA) Patriot*, 18 September 1898, 2. This appraisal harked back to earlier usages, such as in Washington Irving's 1836 history *Astoria*, in which he describes the western parts of the country as a "commercial empire beyond the mountains, 'peopled by free and independent Americans.'" Irving, *Astoria* (Philadelphia: Casey, Lea & Blanchard, 1836), 261. *Banquet of the California Society, Sons of the American Revolution: Anniversary of the surrender of Cornwallis at Yorktown, Va., October 19, 1898, California Hotel, San Francisco* (San Francisco, CA: The Society, 1898), 23.

14. William McKinley, "Speech at Home Market Club Dinner, Boston, February 16, 1899," in *Speeches and Addresses of William McKinley* (New York: Doubleday, 1900), 189–91; Albert J. Beveridge, "Our Philippine Policy" (delivered in the US Senate, 9 January 1900), in *The Meaning of the Times and Other Speeches* (Indianapolis: Bobs-Merrill, 1908), 72; Theodore Roosevelt and William Howard Taft, *The Philippines* (New York: Outlook, 1902), 15.

15. Erwin H. Epstein, "The Peril of Paternalism: The Imposition of Education on Cuba by the United States," *American Journal of Education* 96 (November 1987): 1–23; Pablo Navarro, "Acculturation under Duress: The Puerto Rican Experience at the Carlisle Indian Industrial School, 1898–1918," *Centro: Journal of the Center for Puerto Rican Studies* 16 (2006); Hazel McFerson, *Mixed Blessing: The Impact of the American Colonial Experience on Politics and Society in the Philippines* (Westport, CT: Greenwood, 2002), 92–94; Anne Paulet, "To Change the World: The Use of American Indian Education in the Philippines," *History of Education Quarterly* 47 (May 2007): 173–202; K. Tsianina Lomawaima, "Estelle Reel, Superintendent of Indian Schools, 1898–1910: Politics, Curriculum and Land," *Journal of American Indian Education* 35 (May 1996); Frank Guridy, *Forging Diaspora: Afro-Cubans and African Americans in a World of Empire and Jim Crow* (Chapel Hill: University of North Carolina Press, 2010), 1–17; Twain, "To the Person Sitting in Darkness," 166. Important examples of work on the intersection of education and empire include Noah Sobe, ed., *American Post-Conflict Educational Reform: From the Spanish-American War to Iraq* (New York: Palgrave, 2009); A. J. Angulo, *Empire and Education: A History of Greed and Goodwill from 1898 to the War on Terror* (New York: Palgrave, 2012); "Education," pt. 3 of McCoy and Scarano, *Colonial Crucible*, 131–92.

16. Charles Francis Adams, *Imperialism and the Tracks of Our Forefathers* (Boston: Dana Estes and Company, 1899), 34; Andrew Carnegie, "Distant Possessions: The Parting of the Ways," in *The Gospel of Wealth* (NY: The Century Company, 1901), 159–60.

17. Jonathan Zimmerman, *Small Wonder: The Little Red Schoolhouse in History and Memory* (New Haven, CT: Yale University Press, 2009), 57–96. See also, for instance, Solsiree del Moral, *Negotiating Empire: The Cultural Politics of Schools in Puerto Rico, 1898–1952* (Madison: University of Wisconsin Press, 2013), 37–40.

18. An article from June 1898 stating that "if we are to pursue what may be called imperialism . . . we must clearly understand its import" is a good example of the tentative nature with which the word was being used in regard to the United States at the time. "Imperialism," *Wilkes-Barre (PA) Times*, 9 June 1898, 4. Democrats were already seizing on the notion of imperialism as a means to denounce annexation. See, e.g., "Debate on Annexation: Representative Bland Declares the United States is on the Verge of Imperialism," *Baltimore Sun*, 14 June 1898, 7.

19. *Minneapolis Journal*, 17 June 1898, 8; "The Talk of the Capital: Vague Dread of Imperialism Voiced in Whispers," "Chance for a Colonial Secretary," *New York Times*, 3 July 1898; "God's Time Has Come: American Imperialism Must Be in Spite of Ourselves," "Colonial Possessions," and "Porto Rico the Next Stop," all in *Grand Rapids Herald*, 18 July 1898, 4. For other examples of positive definitions of the word *imperialism*, see "American Imperialism," by Col. A. K. McClure, in *Frank Leslie's Popular Monthly*, reprinted in *Phoenix Weekly Herald*, 3 November 1898), 43; and "American Imperialism," *Santa Fe New Mexican*, 13 December 1898, 2. Beveridge, *Meaning of the Times*, 42, 45, 47, 87.

20. *Sons of the American Revolution*, 18; George C. Lorimer, "Extent of Empire Does Not Make Us a World Power," in Bryan, *Republic or Empire?*, 701.

21. Henry D. Estabrook to Albert Beveridge, 23 September 1898, Correspondence, A–J, Box 120, Albert J. Beveridge Papers, Manuscripts Division, Library of Congress; A. K. McClure, "Our Flag in Foreign Lands," *Frank Leslie's Popular Monthly* 47 (November 1898): 4; "Imperialism," *Sioux City Journal*, 23 June 1898, 10; "Bishop Doane on the War," *New York Times*, 18 November 1898.

22. James C. Fernald, *The Imperial Republic* (New York: Funk & Wagnalls, 1899), 7–20, 164; John McAtee, *For the Treaty of Paris and American "Imperialism"* (Guthrie, OK: State Capital Printing, 1900), 50–52; Albert J. Beveridge, "The Star of Empire," in *Meaning of the Times*, 118–43. In another speech from late 1899, Beveridge states that "imperialism is not the word for our vast work . . . imperialism, as used by the opposers of national progress, is a word to frighten the faint of heart, and so is powerless with the fearless American people." "For the greater republic, not for imperialism: an address delivered by Hon. Albert J. Beveridge, at the Union League of Philadelphia, February 15, 1899," reprinted in *Orations from Homer to McKinley*, ed. Mayo W. Hazeltine, vol. 25 (New York: Collier & Son, 1902), 11012.

23. These characterizations are apparent in numerous editorial cartoons. See, e.g., "Our Busy Old Women," *Puck*, 22 March 1899. http://www.loc.gov/pictures/item/2012647407/; "The Dream of the Anti-Expansionist," ibid., 19 April 1899, http://www.loc.gov/pictures/item/2012647415/; and "A Powerful Argument against Imperialism," *Judge Magazine*, 11 August 1900, http://upload.wikimedia.org/wikipedia/commons/4/49/Judge_August_11_1900_Bryan_Against_American_Imperialism.jpg.

24. "A Queer Kind of Imperialism," *Grand Rapids (MI) Herald*, 17 August 1900, 4; "M'Comas Talks," *Wilkes-Barre (PA) Times*, 5 September 1900, 6; *Speech by the Hon.*

Elihu Root, Secretary of War, Canton, Ohio, October 24, 1900 (Washington, DC, 1900), 8; John Braeman, *Albert Beveridge: American Nationalist* (Chicago: University of Chicago Press, 1971), 44.

25. Bemis, *Diplomatic History*, 469; Claude G. Bowers, *Beveridge and the Progressive Era* (New York: Literary Guild, 1932), 97–122. In 1899 Charles Morris published *Our Island Empire: A Hand-book of Cuba, Porto Rico, Hawaii, and the Philippine Islands*, with the stated intention of fulfilling Americans' "desire to know" about these new possessions (Philadelphia: J. P. Lippincott, 1899), x.

26. Beveridge, "Our Philippines Policy," *Meaning of the Times*, 58, 71.

27. John C. Spooner, *The Philippine tariff: speech . . . in the senate of the United States, Friday, February 21, 1902*. Washington: [s.n.], 1902, 32–33. See also Spooner, "Imperialism: A Forced and Fictitious Issue: Government of the Philippine Islands," United States Senate, 22–24 May 1900; *Speech by the Hon. Elihu Root*, 25; Theodore Roosevelt to Leonard Wood, 17 April 1901, quoted in Beisner, "1898 and 1968," 189; "Mr. Root's Crushing Answer to the Anti-imperialists," *New York Times*, 26 October 1900.

28. William Jennings Bryan, "Jackson Day Speech at Chicago," in *Republic or Empire?*, 23.

29. William Jennings Bryan, "The Paralyzing Influence of Imperialism," in *Official Proceedings of the Democratic National Convention Held in Kansas City, Mo., July 4, 5 and 6, 1900* (Chicago, 1900), 205–27; George Turner, "Our Ship of State," in Bryan, *Republic or Empire?*, 564–65. For a full discussion of this legal history, see Christina Duffy Burnett and Burke Marshall, eds., *Foreign in a Domestic Sense: Puerto Rico, American Expansion, and the Constitution* (Durham, NC: Duke University Press, 2001).

30. "Little Left of Imperialism," *Philadelphia Inquirer*, 30 March 1901, 8; "A Woman's Views of Imperialism," *Omaha World Herald*, 3 August 1902, 4.

31. Henry Fairlie, "A Cheer for American Imperialism," *New York Times*, 11 July 1965.

32. Robin Winks, "The American Struggle with Imperialism: How Words Frighten," in *The American Identity: Fusion and Fragmentation*, ed. Rob Kroes (Amsterdam: Amerika Instituut, 1980), 173; Niall Ferguson, "America: An Empire in Denial," *Chronicle of Higher Education*, 28 March 2003, http://chronicle.com/article/America-an-Empire in Denial/29867. Ian Tyrell has written that "ignorance of the United States' deeply entrenched imperial experiences has been common not only in the work of that well-known historian and philosopher Donald Rumsfeld but across the historical profession and in related disciplines." Tyrell, "Empire in American History," 544–45. John Darwin opines that "it is pointless to debate whether America should be seen as an imperial power: the case has been made." *After Tamerlane: The Rise and Fall of Global Empires, 1400–2000* (New York: Bloomsbury, 2008), 482. Michael Mann argues that empire denial was common to previous empires as well. Mann, "The Recent Intensification of Military and Economic Imperialism: Are They Connected?" in *Sociology and Empire: The Imperial Entanglements of a Discipline*, ed. George Steinmetz (Durham, NC: Duke University Press, 2013). See also Priya Satia, "Inter-war Agnotology: Empire, Democracy and the Production of Ignorance," in *Brave New World: Imperial and Democratic Nation-Building in Britain between the Wars*, ed. Laura Beers and Geraint Thomas (London: Institute of Historical Research, 2011), 209–26; and David Armitage, *The Ideological Origins of the British Empire* (Cambridge: Cambridge University Press, 2000).

33. Eric Foner, review of *The Dominion of War: Empire and Conflict in North America, 1500–2000*, by Fred Anderson and Andrew Clayton, *London Review of Books*, 19 May 2005, http://www.ericfoner.com/reviews/051905lrb.html. See also José Anazagasty,

review of *Colonial Crucible: Empire in the Making of the Modern American State*, ed. Alfred W. McCoy and Francisco A. Scarano, *Centro Journal* 22 (2010): 320; and Marilyn Young, review of *Colonial Crucible: Empire in the Making of the Modern American State*, ed. Alfred W. McCoy and Francisco A. Scarano, *Pacific Historical Review* 79 (August 2010): 491.

34. Frank Ninkovich, *The United States and Imperialism* (Hoboken, NJ: Wiley-Blackwell, 2001), 7; McCoy and Scarano, *Colonial Crucible*, 4; Mary Ann Heiss, "The Evolution of the Imperial Idea and U.S. National Identity," *Diplomatic History* 26 (Fall 2002): 513; Martin Walker, quoted in Macdonald, "Those Who Forget Historiography Are Doomed to Republish It," 46; Kramer, "Power and Connection," 1350; Charles S. Maier, *Among Empires: American Ascendancy and Its Predecessors* (Cambridge, MA: Harvard University Press, 2006), 3; Andrew Bacevich, "Commentary, Charles Maier *Among Empires*," H-Diplo Roundtable, ed. Thomas Maddux, 2006, http://www.h-net.org/~diplo/roundtables/PDF/Bacevich-MaierRoundtable.pdf. See also Michael H. Hunt, "Commentary," http://h-diplo.org/roundtables/PDF/Hunt-MaierRoundtable.pdf and Anna K. Nelson, "Commentary," http://h-diplo.org/roundtables/PDF/Nelson-MaierRoundtable.pdf.

35. Macdonald, "Those Who Forget Historiography Are Doomed to Republish It," 45. Joseph Esherick, Hasan Kaya, and Eric Van Young, introduction to *Empire to Nation: Historical Perspectives on the Making of the Modern World*, ed. Esherick, Kaya, and Van Young (Lanham, MD: Rowman & Littlefield, 2006), 6. Paul Walker, "Legacies of Nation Building in the Soviet Successor States," in ibid., 303. See also Joseph Esherick, "Afterword: The Return of Empire?" in ibid., 373–86, in which he concludes that the term *empire* does not apply to the United States. Jeremy Suri, "The Elusive Character of American Global Power," in McCoy and Scarano, *Colonial Crucible*, 524. Suri, *Liberty's Surest Guardian: Rebuilding Nations after War from the Founders to Obama* (New York: Simon & Schuster, 2012); Suri, "Is America Really an Empire?," *Salon*, 26 October 2011, http://www.salon.com/2011/10/26/is_america_really_an_empire/. Elizabeth Cobbs Hoffman, *American Umpire* (Cambridge, MA: Harvard University Press, 2013), 19, 336. In a related opinion piece, Hoffman argues that the mischaracterization of the United States as an imperial power was a factor in the Boston Marathon bombing. "Terrorism: Is U.S. imperialism inviting it?," *San Jose Mercury News*, http://www.mercurynews.com/ci_23159588/terrorism-is-u-s-imperialism-inviting-it.

36. Winks, "American Struggle with Imperialism," 146–47.

37. Albert J. Beveridge, "Positive and Negative in Politics," in *Meaning of the Times*, 202.

11

Germany

LISA PINE

Having been arrested on November 11, 1923, after the failure of the "beer hall putsch," Adolf Hitler was imprisoned at Landsberg Castle, in a "large, airy, and comfortably furnished" cell, where he dictated an account of his life and worldview (*Weltanschauung*) to Emile Maurice and Rudolf Hess in 1924. The manuscript he produced was published as *Mein Kampf* (*My Struggle*) in 1925. In it, Hitler firmly stated his beliefs on a wide variety of topics, including education. Even at this early date, education formed a key component of Hitler's designs for the future of the German nation. For example, he stated that "from all the innumerable great names in German history, the greatest must be picked out and introduced to the youth so persistently that they become pillars of an unshakeable national sentiment." Hitler's ideas on education came to underpin the Nazis' goals for transforming German society once they came to power. In a speech to representatives of the press on March 15, 1933, Josef Goebbels, minister for propaganda and popular enlightenment, asserted that "it is not enough to reconcile people more or less to our regime, to move them towards a position of neutrality towards us, we would rather work on people until they are addicted to us."[1]

This chapter explores the centrality of education to the Nazi regime, highlighting the social construction of ignorance in Germany during the period from 1933 to 1945. It examines both the perpetuation of ignorance by the Nazi administration through the use of censorship, propaganda, and disinformation and the efforts of the Nazi leadership at various types of miseducation of German children in schools, which were designed to generate unquestioning faith and belief in the Nazi system. As such, it contributes to the study of ignorance and its social construction by examining the policies of the National

Socialist government, which disinformed the German population in a variety of ways. Recent scholarship has begun to explore the active engineering of ignorance, the kind of ignorance that can be "made, maintained and manipulated" as a "strategic ploy" or "active construct." This was undoubtedly the case on the part of the Nazi regime. The creation and preservation of ignorance by the Nazi government covered a wide range of aspects of German life, many of which were interrelated. It was purposely orchestrated to deter any large-scale resistance or even nonconformity to the regime and its ideology.[2]

The focus of this chapter is on three key areas: the promotion of Nazi ideology and the perpetuation of myths; the censorship of artistic and cultural life; and the control of the content of school textbooks. The first section of the chapter illustrates how the Nazi regime utilized miseducation or disinformation in order to promote its ideological imperatives of "racial purity" and anti-Semitism. In addition, it shows how the Nazi administration employed disinformation to legitimate its racial hygiene policies to "improve" the quality and quantity of the German population and how it sought to promote the national community (*Volksgemeinschaft*) and nationalism more generally. Using the myth surrounding the *Führer* (leader), and his infallibility and the perpetuation of the myth of the omniscient and omnipresent secret state police, the Gestapo, as key examples, it illustrates how the Nazi regime perpetuated (through the use of propaganda) all types of myths that were accepted by the German population for much of the Third Reich. The second section of the chapter considers how the Nazi government clamped down upon and censored entire realms of artistic and cultural life that did not conform to its attitudes about German art and German literature. The final section of the chapter examines the Nazi censorship of school textbooks in order to create homogeneity in what was taught. An analysis of these three broad areas illuminates the determined use of disinformation and miseducation by the Nazi government in Germany between 1933 and 1945.

The Promotion of Nazi Ideology and the Perpetuation of Myths

Nazi ideological imperatives lay at the heart of the regime's attempts both to construct its ideal society and to promote its worldview to the German national comrades (*Volksgenossen*). The regime used film and other media to spread its message about anti-Semitism and the eradication of the "unfit." Similarly, it employed such channels of information and knowledge to advocate its nationalist ideology and to perpetuate myths to persuade and instruct its population. Racial purity, the concept of the master race (*Herrenvolk*), and anti-Semitism were core aspects of Nazi ideology. The regime built upon and created popular mistrust toward the Jews by means of propaganda—in particular, through the use of posters, as well as *Der Stürmer* (The stormer), an anti-Semitic journal published under the aegis of Julius Streicher, the regional leader (*Gauleiter*) of Franconia. The front cover of the September 1943 issue, for example, featured a photograph of a Jew with the caption "Satan." In addition, in 1940 a trilogy of films about the Jews was screened—*Die Rothschilds* (The Rothschilds), *Jud Süss* (Jew Süss), and *Der ewige Jude* (The eternal Jew). These films sought to instruct the German population about the "threat" posed by the Jews. The first of these, *Die Rothschilds*, was an attempt to explain the rise to power and wealth of the Rothschild family and the emergence of the "Jewish British plutocracy." It revealed the "historical fact" that Jewish financiers had profited from the death of German soldiers. In doing so, it rationalized the extermination of the Jews as expressed in Hitler's *Mein Kampf* and in his speeches. In *Die Rothschilds*, the Jews were portrayed as a racial and economic threat.[3]

Veit Harlan's *Jud Süss* showed the inherent rootlessness of the Jewish people and their ability to assimilate into any society. Süss, the elegant and fashionable Jewish lawyer, personified "the Jew in disguise." Film historian Eric Rentschler has shown that in "constructing a malevolent other, Nazi propagandists insisted that they were serving the public good by revealing the Jew's true face." The rest of the Jews in the film were portrayed as dirty, hook-nosed, and physically repellent. They represented "authentic Jewry." In contrast, the true "Aryan" prototypes were depicted, such as, for example, Dorothea, the classic German maiden, played by Christina Söderbaum. This film unquestionably contributed to the anti-Semitism already prevalent in Germany, for it brought together archetypes and themes that created the desired antipathy toward the Jews under the guise of entertainment and was a great box office success. Rentschler notes that "it not only confirmed existing prejudices; it agitated, militated, and called for action." This was intentional engineering on the part of the National Socialist government and its propaganda machine.[4]

In its efforts to "educate" the German population, Fritz Hippler's *Der ewige Jude* covered the entire gamut of Nazi allegations against the Jews. It was one of the most virulent and effective propaganda films ever made. Much of its strength lay in its pretense of documentary objectivity. Whereas *Jud Süss* was a period costume feature whose message was nevertheless crystal clear, *Der ewige Jude* claimed to be a documentary film about world Jewry. The substance and message of the film very much reflected the idea that the Jew had cunningly assimilated himself into European society. Furthermore, the association of Jews with rats led the audience to believe that they were disease bearers and subhuman. Statistics were produced to show that Jews figured predominantly in the world of crime, out of all proportion to their number in the population, but no source was given for the figures. The film showed images of Jews with beards, long hair, and skullcaps. Each shot faded into one of the same man "disguised" in European clothing. In this way, the Jews were portrayed as an almost invisible threat to the health of the "Aryan" race. By showing pictures of Jews in prominent positions of world power—for example, the financial houses of Jews such as the Rothschilds, the Warburgs, and the Montefiores—the film played on the myth of the "international Jewish conspiracy." It also exacerbated sentiments of jealousy, indignation, and resentment by showing the disproportionate number of Jews in the upper echelons of society during the Weimar Republic and their corresponding disproportionate absence from the more menial strata. Finally, after this great profusion of images and rhetoric, the viewer was confronted with the slaughterhouse scene—anti-Semitic propaganda at its most extreme. The inflammatory scenes were immediately followed by shots of Hitler's Reichstag speech of January 30, 1939, in which he prophesied that a forthcoming war would bring about "the annihilation of the Jewish race in Europe." A sequence of blonde "Aryan" stereotypes concluded the film. These three films were intended to teach a particular message about the Jews to the German population at a time when the Nazi leadership sought increasingly drastic "solutions" to "the Jewish Question." They formed a central component of the efforts of the Nazi government to demonize the Jews and to assert a whole array of allegations against the Jews that misinformed the German populace.

In addition, the Nazi government employed disinformation to legitimate its racial hygiene policies to "improve" the quality and quantity of the German population. To this end, the regime introduced eugenic measures, including compulsory sterilization. Policies directed against the physically and mentally ill eventually culminated in their mass murder (Operation T4) during the war. Nazi educational and propaganda materials showed the high costs of welfare for people the regime considered to be "inferior" or even "unworthy of life," money that could be saved and spent instead on more "worthy" or "valuable" members of the German national community. In October 1939 Viktor Brack, one of the creators of the T4 scheme, commissioned Hermann Schweninger to make propaganda films on the subject of "euthanasia." These too were educative in nature. The aim was to juxtapose the expense put into maintaining "ballast existences" to the limited therapeutic results achieved. In 1935, 21 million people watched these films stigmatizing the mentally ill, and this number rose to 40 million in 1939. They contrasted the most shocking cases of children and adults with physical disabilities and mental illnesses with the luxuriously appointed, expensive and modern asylums with attractive gardens and grounds in order to appeal to the "healthy" instincts of the German nation and to demonstrate to ordinary Germans that this money could be better spent on more deserving causes. After 1940 a different approach was taken to achieve this end: the production of a feature film instead of documentaries. The feature film *Ich klage an* (I accuse), on the subject of euthanasia, was released in August 1941. Eighteen million people saw the film. Nazi reports compiled by the Security Service of the SS (Schutzstaffeln), the Nazi elite organization led by Heinrich Himmler, in January 1942 suggested that the film had been "favourably received and discussed" and "enthusiastically received." The film showed a medical professor named Heyt, whose young wife had multiple sclerosis. He killed her with a morphine overdose and was subsequently tried for murder. The aim of the film was to show that the law should be changed in order to permit "mercy killings," and clearly audiences were encouraged to take this view. Goebbels was astute in his judgment that messages could be effectively taught through the medium of feature films, not just through overtly propagandistic films.[5]

Hitler's government wanted to instruct the population in other ways too, such as through the Nazi propaganda effort that sought to promote the national community and nationalism, which was in line with another central goal of Nazi ideology, the resurrection of the fortunes of the German nation. Through the use of nationalistic slogans such as "One People! One Reich! One Führer!" educational and propaganda materials emphasized the greatness of Germany as well as the need for people to put the national interest before their own interests. The concept of the community before the individual (*Gemeinnutz vor Eigennutz*) was promoted to encourage a sense of nationalism and pride of homeland. Another illustration of this was the way in which the Nazi government linked home economy to the national economy, encouraging German housewives to think about their duties to the nation in their purchasing decisions. Historian Nancy Reagin has shown how the regime "persistently linked women's frugal use of available resources to the national interest," especially after the introduction of the Four Year Plan and during the war. A further instance of this was the Nazi concept of health as a responsibility to the nation (*Volk*), not just as the concern of the individual. The close relationship between "individual and collective physical health" was a central component of the state and society in the Third Reich. The national community was central to concepts of German society under National Socialism, and a vast array of educational and propaganda materials highlighted its importance to the German populace.[6]

The Nazi regime perpetuated a number of myths that were accepted by the German population for much of the Nazi period. Prominent examples include the following: the Hitler myth; the myth of a classless society, in which the working class could gain access to the privileges that previously had been restricted to the middle and upper classes, which persuaded the workers that the regime benefited them; and the myth of the omniscient and omnipresent Gestapo, which created fear among the German population and inhibited criticism of the regime or its policies. Furthermore, the Nazi propaganda machine was used to conceal the nationwide impact of Allied bombing, to persuade the Germans that they were winning the war, and to motivate them to continue fighting until the bitter end. Propaganda was far removed from reality by this time.[7]

Propaganda, in particular the use of radio, posters, films, magazines, and newspapers, as well as education at all levels—both formal and informal—contributed to the continued belief in such myths. Rallies, parades, posters, feature films, newsreels, the press, and the radio were all employed to generate consent for the regime and its aims. Nazi propaganda addressed itself to large masses of people in order to achieve uniformity of opinion and action and a reeducation of society based upon National Socialist principles. David Welch has shown that Nazi propaganda was "as much about confirming as about converting public opinion." In this sense, Nazi misinformation and disinformation perpetuated ignorance among the German population. Welch shows how Nazi propaganda reinforced "existing trends and beliefs, to sharpen and focus them" for the regime's own ends.[8]

The Hitler myth, as Ian Kershaw has argued, was "consciously devised as an integrating force by a regime acutely aware of the need to manufacture consensus." Hitler was portrayed as the great leader of Germany's destiny. Yet, this heroic leadership was "as much an image created by the masses as it was imposed on them." Pierre Ayçoberry states that "the construction of the Hitlerian myth resulted from a combination of autosuggestion, deliberate fabrication and a quasi-universal acceptance." After the Night of the Long Knives (June 30, 1934), for example, Hitler's reputation was elevated, despite the illegality of his actions. He was portrayed and regarded as the defender of the "little man" against the "big shots" and as the upholder of public morality. Curbing the excesses of the SA (*Sturmabteilungen*, or stormtroopers) appealed to the public's desire for law and order, while the murder of its openly homosexual leader, Ernst Röhm, accorded to "the healthy instincts of the people." In addition, the recovery of the nation's economy; the massive public works schemes, such as motorway construction; and the elimination of mass unemployment in the mid-1930s were regarded as the personal achievements of the *Führer*.[9]

Standing above and beyond the day-to-day realities of the regime, Hitler remained disassociated from unpopular decisions and from the avarice and hypocrisy of the party functionaries. Hitler was popular as a leader among all social groups and was personally exempted from criticisms of the regime. Any blame was directed at other Nazi leaders or officials. The Hitler myth enabled people to voice their quotidian grumbles and concerns and yet consent to the Nazi regime as a whole. This coexistence of complaint and compliance is significant to our understanding of the nature of popular opinion in the Third Reich. Much of the appeal of the Hitler myth was the yearning for security and leadership by a population in disarray and despair. While the regime continued to make achievements and restore order, with its visionary leader at its center, popular acceptance persisted. It appeared to offer the German population a solution to its social, economic, and political problems, and to this extent it succeeded in its capture of the popular mind and in its manufacture of consensus.[10]

Additionally, the Nazi government won people to the national cause by means of its foreign policy successes. A nation brought to its knees by the punitive Treaty of Versailles in 1919 could only rejoice in Hitler's wholesale revocation of its terms. The "bringing home" of the Saarland in 1935 and the march into the Rhineland on March 7, 1936, met with rapturous popular approval. These events signified success and recovery for the German nation under their great "leader." The Anschluss with Austria in 1938 was another massive triumph for the national community and its *Führer*. While Nazi foreign policy continued to succeed, the sense of national community was enhanced. Even a nation hesitant to go to war again in September 1939 accepted Hitler's decision and reaped the benefits during the initial blitzkrieg successes of World War II. Popular support reached its high point after the Nazi occupation of Paris in 1940. Hitler was regarded throughout this period as a great wartime leader. Kershaw argues that even on "the eve of the invasion of the Soviet Union . . . Hitler's popular standing was undiminished, and confidence in his leadership among the great majority of the population unbroken." It was only after the tide of the war turned against Germany, and in particular after the Battle of Stalingrad in January 1943, that both Hitler's infallibility and the strength of the nation began to be called into question. Yet this disillusionment was not translated into any determined resistance or revolt. There were a number of reasons for this: the Nazis' apparatus of terror remained intact until the end of the war; the burdens and strains of the war led to resignation rather than rebellion; and Nazism had created an "atomization of social relations" that stood in the way of a communal resistance effort. In addition, the manufacture of popular opinion through manipulation of both information and emotion contributed to this situation.[11]

Before the Battle of Stalingrad, direct, personal criticism of Hitler was extremely rare, but after Stalingrad the Hitler myth began to falter. The Allied bombing campaigns were another significant factor in the decline of the Hitler myth, particularly from 1943 on. The bombing caused considerable demoralization and anger directed against the Nazi leadership for failing to prevent it, despite continued Nazi propaganda directed at the home front. In the months following the July 20, 1944, bomb plot on Hitler's life, the *Führer* retreated from public life. In the last phase of the war, the Hitler myth collapsed entirely. Hence, significant as it was, this myth does not on its own explain the continuing acceptance of the Nazi regime, particularly after the tide of the war turned against Germany.[12]

As popular consensus broke down, the use of terror escalated. Within Himmler's vast SS (*Schutzstaffeln*) and SD (*Sicherheitsdienst*) police complex, the Gestapo, headed by the ambitious Heinrich Müller, became the central component in the system of terror and surveillance in the Nazi state. The Gestapo certainly instigated fear and terror among the German populace. Recent research has shown, however, that the popular image of the Gestapo as omniscient and omnipotent was a myth instigated by Gestapo leaders and perpetuated in the postwar period by historians who accepted the statements of Gestapo leaders at face value. Historians Klaus-Michael Mallmann and Gerhard Paul have shown that the Gestapo simply did not have the necessary manpower resources available to be omniscient and omnipotent. Its leaders carefully adopted the propaganda image of the ubiquitous Gestapo both to intimidate German society and to conceal its own deficiencies. But in reality the Gestapo never achieved these mythical levels of repression. Instead, it was an understaffed and overbureaucratized organization, incapable of comprehensive surveillance. Nor were members of the Gestapo ardent believers in Nazi ideology. Many career policemen from the Weimar era who stayed on and adjusted to the new regime made up its ranks. In the Gestapo, as Robert Gellately notes, they "played the part of loyal enforcers of the dictatorship's will." Most joined the Nazi Party sooner or later, but in 1939 only 3,000 of the Gestapo's approximately 20,000 employees held an SS rank.[13]

In reality, there were remarkably few Gestapo agents on the ground, and they relied on both amateur and professional helpers. For example, in 1937 the Dusseldorf Gestapo office, manned by only 126 agents, controlled a town of 500,000 inhabitants. The Essen Gestapo office had 43 agents for a town with 650,000 inhabitants. Gellately has shown the largely reactive nature of the Gestapo, describing its method as "a kind of auto-policing, or at least an auto-surveillance system." With the cooperation of neighbors, friends, acquaintances, and family members, the Gestapo could infiltrate even the private realm of the home to monitor compliance with the dictates of the regime. Without the help of informers, the Gestapo would have been virtually blind. Denunciations were key in the interactions between the police and the population. The Gestapo did not have adequate resources to generate its own cases but relied heavily upon the supply of information from outside. Following denunciations, the Gestapo relentlessly pursued and interrogated individuals, often using torture, in order to extort statements that led to arrests. But on its own, that is, without the help of informers, the Gestapo could not engage in comprehensive surveillance or perfect repression. Hence, a variety of myths played an important part in the Nazi disinformation and miseducation of the German populace. These myths were integral to the creation and perpetuation of ignorance in Nazi Germany.[14]

The Censorship of Art and Literature

Another key Nazi strategy that contributed to ignorance-making was the use of censorship. The Nazi regime banned many types of cultural expression, eclipsing opportunities for Germans to learn about and experience, for instance, jazz music and atonal music as well as Expressionist art. Art forms were censored, so that the population believed in the existence of only a certain type of German culture. In particular this section examines Nazi policies toward art and literature. These two aspects of cultural life presented the Nazi government significant opportunities to educate and instruct the German population. The Nazis used the arts to legitimize their rule by posing as the guardians of true German culture. They purged "alien" influences from German artistic and cultural life, claiming that racial degeneration was the main cause of aesthetic deterioration. Alan Steinweis has argued that "the cultural policies of the Nazi regime were inextricably intertwined with the policies of persecution and marginalization, driven by racist ideology, that were targeted at Jews, Roma and Sinti, homosexuals and other groups." Hence, cultural purification constituted a central component in the development of the national community throughout the duration of the Nazi dictatorship and in the broad educative process of the German populace. The Reich Chamber for the Visual Arts closely controlled the work of its 42,000 members. In addition, it strictly monitored membership, so that politically unreliable or "racially inferior" artists were ineligible and therefore not allowed to continue to practice their profession. Nazi officials removed prominent artists such as Paul Klee and Otto Dix from their positions. Many artists and painters whose work was disapproved of by the Nazi regime went into exile.[15]

The conveying of a particular type of art and culture played a critical part in Nazi educational imperatives. The desire of the Nazi regime to purify and cleanse society was reflected in its policy toward the arts, with its simultaneous encouragement of pure and wholesome contemporary German art and purging of decadent, "degenerate" or unwholesome art. The Nazi administration aimed to eliminate all forms of art that it regarded as alien or degenerate. It excoriated many of the artistic innovations that had flourished during the Weimar era, such as Dadaism and Expressionism. The modern section of the National Gallery in Berlin was closed down in 1934. In 1936–37 the Reich Chamber of Culture initiated and carried out a purge of German art in galleries and museums across the country. Adolf Ziegler, the president of the Reich Chamber for the Visual Arts, was assigned the task of selecting and collecting works of "degenerate" German art for the Degenerate Art Exhibition, which opened in Munich on July 19, 1937. Ziegler and his commission confiscated some 16,000 works of art by 1,400 artists. Among the most prominent of the artists whose works were seized were Max Beckmann, Ernst Barlach, Otto

Dix, Georg Grosz, Paul Klee, Oskar Kokoschka, and Emil Nolde. The commission selected works by 112 of these artists for the Degenerate Art Exhibition. In his opening speech Ziegler described the works that had been produced by modern artists as "monstrosities," "the crippled products of madness, insolence, lack of ability and degeneration." The paintings were purposely displayed in a poor manner, crammed into crowded galleries, many without frames, and accompanied by pejorative captions.[16]

The German Art Exhibition opened simultaneously in the House of German Art in Munich. This juxtaposition was designed to demonstrate the triumph of Nazi art over degenerate art and to (mis)inform the German populace about what were acceptable or suitable expressions of German art. At the opening of the House of German Art, Hitler described how the "Dadaist sensationalists, Cubist plasterers and Futurist canvas smearers" had damaged the achievements of the arts. He stated that "degenerate" artists created "deformed cripples and cretins, women who inspire nothing but disgust, human beings that are more animal than human. . . . But in the name of the German people I mean to forbid these pitiable unfortunates, who clearly suffer from visual disorders, from attempting to force the results of their defective vision onto their fellow human beings as reality, or indeed, from serving it up as 'art.'" Hitler spoke about laying "the foundations for a new and genuine German art." Robert Wistrich argues that National Socialist art, cleansed of abnormality, decomposition, and ugliness, aimed at the reconstruction of "a hierarchical, ordered society with a coherent vision of man, with a heroic ideal and the ability to totally reintegrate the individual into the community."[17]

The Nazi regime closely controlled German artistic life and the lessons derived from this life. It removed the influence of "undesirable" artists by imposing bans on the right to teach, to exhibit, and to paint. It supervised artists, museums, and academies, published magazines, organized exhibitions, undertook cultural exchanges with other nations, expropriated the artistic property of enemies, and engaged in large-scale plundering campaigns. Art was used instrumentally in the Third Reich to express the Nazi *Weltanschauung*, "a moral force permeating the whole of German society." It was, as Peter Adam argues, "considered one of the most important elements in building the new Reich and the new man." Historian Jonathan Petropoulos concurs that culture in the Third Reich was "a defining feature of the regime"; it was regarded as "an expression of ideological tenets, as a vehicle for creating a group ethos, and as a means of self-definition."[18]

Furthermore, National Socialism was keen to promote a particular type of German writing and to remove the influence of Jewish and Bolshevik writers from German literature. The Nazi regime approved literature that was pure and wholesome, a racially based literature that would serve to underpin its political and educative aims. In place of the "literature of the city," which had made a great impact during the Weimar era, it called for the "literature of the peasant." Nazism rejected works it considered to be decadent or degenerate, as well as the writings of so-called enemies of the regime and the German national community. The Reich Chamber of Literature excluded all Jewish writers from membership. Radical left-wing writers such as Carl von Ossietzky and Erich Mühsam were arrested within weeks of the Nazi takeover of the state. Ernst Toller, a German Jewish Socialist writer, lived in exile from 1933 to 1939, when he committed suicide. Thomas Mann, who had won the Nobel Prize in Literature in 1929, lived in exile in Switzerland from February 1933 on. Many other prominent writers, including Arnold Zweig, Alfred Döblin, Erich Maria Remarque, and Lion Feuchtwanger, went into exile. Heinrich Mann, author of *Der blaue Engel* (*The Blue Angel*), was forced to give up the presidency of the Prussian Academy of Poetry, and his position was filled by the writer Hanns Johst. The academy was purged of "racial aliens" and "degenerate" writers. Richard Grunberger estimates that twenty-five hundred writers left Germany during the Nazi era either voluntarily or under pressure from the Nazi state. Historian Richard Evans argues that by the end of 1933 "there was scarcely a writer of any talent or reputation left in Germany." It also meant that German literature became homogenized into a form that was acceptable to the Nazi government and provided the German populace with its view of what German literature should entail and encompass.[19]

Writers including Werner Beumelberg, Hans Grimm, Hans Friedrich Blunck, Agnes Miegel, Hermann Stehr, Erwin Guido Kolbenheyer, Emil Strauss, and Börries von Münchhausen filled the gap left by this great exodus of German literary talent. Beumelberg wrote stories about the war experience (*Fronterlebnis*), such as *Gruppe Bosemüller*. The *Fronterlebnis* genre became a major category of literature under National Socialism. Grimm's most famous work was his thirteen-hundred-page novel *Volk ohne Raume* (People without space), published in 1926, which underlined the Nazi argument for living space (*Lebensraum*). Kolbenheyer was another advocate of German power. A Sudeten German, Kolbenheyer was fiercely nationalistic and anti-Slav. He opportunistically concurred with Nazi ideology, hoping this would help him to gain recognition for his work. Blunck, president of the Reich Chamber of Literature, wrote nationalist sagas, such as *Urvätersaga* (Forefathers' saga), and advocated German racial supremacy. Miegel and Strauss wrote about the land and people.[20]

Blood and soil (*Blut und Boden*) novels were immensely popular during the Nazi era. They described a rural idyll that appealed to the traditional sentiments of their readers. Such literature engendered a sense of security in the German countryside, away from the evils associated with urbanization. This type of literature was not only regressive but, as James Ritchie points out, "also extremely aggressive." The peasant—like the soldier—was to defend himself and his land. For example, Hermann Löns's *Der Wehrwolf* (*The Warwolf*), first published in 1910, had a significant message in the Third Reich. The book described the actions taken by peasants against invading soldiers and looters in the Thirty Years' War. The novel showed that the task of the peasant was greater than simply to tend the land. It was to defend the land, with blood if necessary. Löns's novel suggested the exclusion of foreigners and demonstrated the importance of an undivided community working together industriously for the benefit of the whole. Hermann Burte, author of *Wiltfeber, der ewige Deutsche* (Wiltfeber, the eternal German), published in 1912, was another writer assured of success in Nazi Germany, as his work too was steeped in nationalist ideology and represented a particular, idealized type of German writing.[21]

The regime also promoted anti-Semitic works. Artur Dinter's *Die Sünde wider das Blut* (The sin against the blood; 1922) became one of the most acclaimed novels of the Nazi era. It told the story of Johanna, a German maiden seduced by a Jewish officer, who when she later married a German man, Hermann, gave birth to a "Jewish child." Dinter's novel explained how "racial miscegenation" ruined the German girl for good and concluded with a moral: "Now consider the damage which year in, year out, is inflicted upon the German race by Jewish youths who every year seduce thousands upon thousands of German maidens!" Hence, German literature played its part in the promotion of particular values and ideas. The Nazi government employed the censorship of art and literature to deny a broad cultural "public" education to the German population. It replicated this type of policy in the case of textbooks in schools.[22]

The Censorship of School Textbooks

Education policy was underpinned by the desire to disseminate Nazi ideology as widely as possible, and in this context other educational aims were subordinated. The curriculum was transformed in order to emphasize certain subject areas in which nation and race could be expounded and to decrease the significance of other subject areas. This effort by the Nazi administration to educate the German people was in line with its ideology and amounted to a direct form of miseducation. For example, in 1940 a Nazi directive for elementary education stated that "it is not the task of the elementary school to impart a multiplicity of knowledge for the personal use of the individual. It has to develop and harness all physical and mental powers of youth for the service of the people and the state. Therefore, the only subject that has any place in the school curriculum is that which is necessary to achieve this aim. All other subjects, springing from obsolete educational ideas, must be discarded." In secondary education, Nazi educationalists believed that "German, history, geography and biology require a deeper treatment." This skewed the balance of the curriculum at all levels. Control over the content of school textbooks created homogeneity and uniformity in what was taught and what was omitted from the curriculum.[23]

Bernhard Rust, the minister for education, believed that the purpose of school textbooks was to achieve the ideological education of young German people. Strict censorship was imposed upon textbook publishers. This censoring was the particular responsibility of Philipp Bouhler, the director of the National Socialist Party's Censorship Office, who worked in conjunction with Rust on this task. However, such was its extent and its complexity that for the first four years of Nazi rule, control was sporadic. Many textbooks of the Weimar era were reprinted with only slight amendments, such as the insertion of photographs of Hitler, swastika flags, and Nazi slogans. By the late 1930s, as new writing and illustrations became available in greater quantities, the older textbooks were removed from circulation and replaced with new, standardized ones that incorporated the central tenets of Nazi ideology. The Ministry of Education approved the authors and over time removed "blacklisted" works and replaced them with "whitelisted" works.[24]

The introduction and use of new school textbooks further assisted Nazi pedagogues in their aim of inculcating pupils with Nazi ideology. Bouhler examined all the textbooks in use for their ideological content. His work demonstrated the need for a process that entailed more than a simple weeding out and worked toward the creation of a uniform "Reich Reader" for the entire nation. The Ministry of Education began to actively implement this idea, removing old readers from the curriculum and replacing them with new ones. Ernst Krieck, a prominent Nazi educational theorist and professor at the University of Heidelberg, was involved in the educational theory behind the introduction of these readers. New editions were to include the themes of *Blut und Boden*; leadership; honor and loyalty; service and sacrifice; struggle and work. Between 1935 and 1940 Rust introduced new Reich readers for different age levels. Editors selected specific reading material for them, based on the key themes of *Blut und Boden* and the *Volk*. Texts had to conform to Nazi Party censorship requirements. They included extracts from German and Nordic folklore and sagas, with the selections tailored to the ideological values of National Socialism. Rust defined the purpose of the readers as to "serve the ideological education of young German people, so as to develop them into fit members of the national community—members who are ready to serve and to sacrifice." The Nazi regime utilized school textbooks and controlled classroom practice more widely to spread its *Weltanschauung*. Many subjects within the school curriculum were used to expound Nazi ideology, especially history and geography, arithmetic, biology and physics, as well as the German language itself.[25]

History lessons were used as an opportunity to demonstrate to pupils the greatness of Germany and to inculcate nationalism in children. History was interpreted as a struggle for existence between nations. It was further employed to highlight the leadership principle, emphasizing the role of Germany's "great leaders" and their "world-historical" achievements. In history textbooks such as *Volk und Führer: Deutsche Geschichte für Schulen* (Nation and leader: German history for schools; 1943), priority was given to periods when Germany had been a dominant power in Europe and especially to the triumph of Nazism. Defense history and frontier studies were added to the history curriculum. There was a concentrated effort to write a new history curriculum in the Third Reich. Publishers, teachers, professors, and school administrators became involved, sometimes as a way of achieving professional advancement. While the Hitler regime was not the first to use nationalistic ideas in history instruction, it was, according to Gregory Wegner, "the first and only regime to fully institutionalize a racist and anti-Semitic history curriculum."[26]

Geography teaching was expanded to include, and indeed even to justify, *Lebensraum* and racial expansion. Geography educators also exploited traditional and new forms of anti-Semitism in their publications. For example, the image of the eternally wandering Jew was a favorite theme. Walter Jantzen was one of the most prominent geography educators in the Third Reich. In his book *Die Geographie im Dienste der nationalpolitischen Erziehung* (Geography in the service of national political education; 1936) he integrated Nazi ideology into the geography curriculum, particularly racial concerns about Jews, blacks, and Gypsies but also issues of *Lebensraum*, *Blut und Boden*, and the decline in the birthrate since the end of World War I. In the geography textbook *Deutschland als Ganze* (Germany as a whole; 1938), Konrad Olbricht and Hermann Kärgel highlighted the ideological distinctions between National Socialism and "Jewish Bolshevism." Hence, geography was taught with a particular National Socialist slant, with specific kinds of occlusion and ignorance production, and the geography curriculum was designed to convey Nazi ideology to German pupils in this way.[27]

Arithmetic lessons dealt with "national political problems." Calculations of sums were presented in terms of bullet trajectories, aircraft, bombs, and so on. Numerical problems based on state expenditure on "hereditarily ill" and "inferior" people also exemplified the way in which Nazi ideology pervaded the curriculum. The Nazi regime used arithmetic exercises to convey its racial and political ideas. In a typical exercise, pupils were presented with this information: "Every day, the state spends RM. 6 on one cripple; RM. 4 1/4 on one mentally-ill person; RM. 5 1/2 on one deaf and dumb person; RM. 5 3/5 on one feeble-minded person; RM. 3 1/2 on one alcoholic; RM. 4 4/5 on one pupil in care; RM. 2 1/20 on one pupil at a special school; and RM. 9/20 on one pupil at an ordinary school." They were asked a series of questions relating to this data, such as "What total cost do one cripple and one feeble-minded person create, if one takes a lifespan of 45 years for each?" and "Calculate the expenditure of the state for one pupil in a special school and one pupil in an ordinary school over eight years, and state the amount of higher cost engendered by the special school pupil." These questions had clear implications for how "inferior" people were to be viewed by their compatriots, and children were exposed to such ideas at a particularly impressionable age.[28]

Biology teaching was of great significance in the Third Reich and was energetically promoted by the regime as a whole and the National Socialist Teachers' Association (NS-Lehrerbund) in particular. In no other subject area were the Nazi themes of *Blut und Boden*, race, and *Lebensraum* so directly linked to the subject matter. Biology had the task of instructing pupils about the essential nature of German *Lebensraum* for the German *Volk*. Laws of heredity, fertility, selection, and blood purity were central to the teaching of the subject. Paul Brohmer was one of the leading writers of a new biology curriculum under the National Socialist regime. In 1938, after the new curriculum was introduced, new textbooks appeared that took into account the changes. Brohmer utilized "race biology" as a means of encouraging German children to struggle to maintain racial purity. He underlined the dangers of "racial miscegenation" and justified the regime's racial, population, and eugenics policies in his writings. Biology lessons became vehicles for Nazi racial doctrine, emphasizing themes such as race, heredity, and the "selection of the fittest." Pupils were instructed in the classification of racial types and craniology. Films and slides were produced as teaching aids. Visual presentations were deemed to be particularly useful in showing the distinctions between examples of "racially pure" and "inferior" or "hereditarily ill" individuals. "Hereditary biology" formed a significant part of "biological thinking" under National Socialism. The realization of the hereditary health of the German people was to be "drummed into" children in school education so that it became "second nature to them." Lore of the family, race, genetics, eugenics, and population policy formed the core of hereditary biology. This subject purported to demonstrate that "racial mixing" and an increase in the number of the "hereditarily ill" damaged the integrity and value of the German population. The hereditary health of the nation was central to all work in this area. The promotion of "valuable" hereditary lines and the concurrent prevention of the reproduction of the "hereditarily ill" were emphasized. Pupils were given an understanding of the need for sterilization and of Nazi eugenic laws. They were shown the power of genetic transmission from one generation to the next. The preservation of the "Nordic character" and an understanding of racial differences were the central aims of lessons in hereditary biology.[29]

Nazi pedagogues, as well as Nazi leaders, regarded "racial miscegenation" and "bastardization" as serious threats to the German nation. In 1938 Alfred Vogel, a biology curriculum writer and primary school headmaster in Baden, produced a series of anti-Semitic teaching charts to be used as teaching aids for the new curriculum. These charts accompanied a teachers' book designed for the instruction of biology to primary school children. Vogel encouraged teachers to instruct children about the laws of nature and heredity, as well as racial consciousness and the "blood community" of the German nation. He drew parallels between cross breeding in plant biology and racial mixing in society. Vogel advocated a "race corner" on the school grounds where experiments could be carried out on plants and pupils could see the strength of the "pure bred" plant over the mixed bred one. Vogel examined "hidden" inherited tendencies in biology, claiming that it was not correct to judge a living thing by its outwardly visible characteristics. The implication was that heredity was the only important signifier of race, so that a Jew posed a danger to the German nation even if he did not look like a Jew or did not practice his religion, which was in line with Nazi racial anti-Semitic ideology.[30]

Topics for biology instruction included "the heredity of physical characteristics," "the heredity of mental and spiritual characteristics," "the heredity of frailties and illnesses," "the heredity of physical and spiritual characteristics of the German race," "the care of racial inheritance," "the law of selection," and "the Jews and the German people." Vogel advocated educating young Germans about "the racial value of our people and the tireless struggle over the preservation of our racial character" and about "the complete rejection of the Jews." His illustrated charts, including *The Racial Composition of the Jews* and *German Ways–Jewish Ways*, were used to show the perils presented to the German nation by the "parasitic," "wandering Jew." His charts stereotyped the Jews, both as stateless intruders and as financial and political dominators. He also linked Jews with Freemasons and suggested that they were engaged in a conspiracy for world domination. While biology teaching did not propose overtly the policy of the genocide of the Jews, it did provide a legitimization for this policy, as well as for the Nazi "euthanasia" campaign.[31]

Physics was used to teach pupils the achievements of German physicists and the contributions of German research to scientific knowledge in order to enhance national pride and underline the significance of the National Socialist "Aryan" worldview. The work of Jewish physicists was excluded in line with Nazi racial ideology. As the "racial heredity" of a person directly influenced his work, only the "Aryan" scientist could be seen to truly achieve and create. Most significantly, this meant that Albert Einstein's name was removed from the physics textbooks of the Nazi era. Einstein's theory of relativity was rejected on the grounds that it was "theoretical magic" and a "great worldwide Jewish bluff." By contrast, "Aryan" physicists such as Philipp Lenard and Johannes Stark flourished under National Socialism. Their "Aryan" physics aimed to preserve the national community and to express the life of the nation.

Stark, in particular, called for there to be "no Jewish propaganda" in German physics textbooks.[32]

Physics teaching under National Socialism had "to contribute its part to national political instruction and forming willpower." Furthermore, Nazi pedagogues argued, "knowledge of the natural conditions and requirements of the national community in the German living space, the ability and the will to do further work on the questions of physics research and technology in this connection, and also teaching to think realistically, which is so important for forming National Socialist willpower, can only be fostered organically on the basis of the pupils' horizon, which schools should help to expand gradually." Although topics such as classical mechanics and electromagnetism continued to be taught, much of the practical emphasis took the form of instruction in physics related to military topics for boys and domestic topics for girls. In particular, the significance of physics for warfare was emphasized, as physics teachers tried to justify their subject by relating it to military objectives. A new branch of physics teaching under National Socialism, "the physics of weapons," was designed to awaken the ability to bear arms. Military physics increased in significance from 1936 on. Pupils were instructed in orientation, measurement, communications, ballistics, and military engineering. A new handbook was provided for teachers on this subject. Physics under National Socialism, according to its author, Erich Günther, had the purpose "of awakening not only the ability to bear arms but the will to do so, and beyond that, to show the ways and technical means to carry out the decision to bear arms."[33]

Lessons in German were designed to foster a "consciousness of being German" and to encourage national pride and unity among pupils. Teachers of German language and literature were urged to emphasize the nation as "a community of blood," "a community of fate and struggle," "a community of work," and "a community of mind." Traditional German tales and sagas were supplemented with Nazi myths, war stories, and *Blut und Boden* literature. Indeed, as Christa Kamenetsky, author of *Children's Literature in Hitler's Germany*, has pointed out, many classics were adjusted to the requirements of the National Socialist state by means of "slanted abridgements or reinterpretations," while others that did not fit into the Nazi *Weltanschauung* were banned. National Socialist heroes such as Horst Wessel found their way into school textbooks. Political socialization in readers took the form of stories about "helping the *Führer*" and the national community, for example, by taking part in the regime's Winter Relief Agency (the Winterhilfswerk) and "one-pot dish" (*Ein-*

topf) campaigns. Such stories were designed to foster a sense of togetherness, unity, and belonging to the national community. Stories depicting "the enemies of Germany," in particular Bolsheviks, Slavs, and Jews, exemplified the more negative side of political socialization. Such negative stereotypes were used to underline the differences between "national comrades," who belonged to the national community, and "community aliens," who did not.[34]

In addition, a new subject, racial science (*Rassenkunde*), formed an integral part of the school curriculum under National Socialism. Its inclusion was formalized in a Nazi curriculum policy directive from the Ministry of Education on September 13, 1933. Racial science covered race and heredity, using racial science textbooks. "Aryan" and Jewish "racial types," often in the form of caricatures, were juxtaposed so that German children would identify with the former and reject the latter as "enemies." Anti-Semitic textbooks and teachers' manuals were produced by a number of writers, such as Fritz Fink, whose book *Die Judenfrage im Unterricht* (*The Jewish Question in Instruction*; 1937) integrated pictorial distinctions between Jews and "Aryans" for teachers' use. Fink called for anti-Semitism to pervade the entire curriculum as the most effective way of getting the message across in the classroom and at all age levels. Much use was also made in schools of *Der Stürmer*, as well as various publications under its aegis, such as Elvira Bauer's *Trau keinem Fuchs auf grüner Heid und keinem Jud auf seinem Eid* (*Trust No Fox on His Green Heath and No Jew upon His Oath*; 1936), which depicted vulgar stereotypes of Jews. Ernst Hieme r's *Der Giftpilz* (*The Poisonous Mushroom*; 1938) employed a whole array of anti-Semitic imagery, employing caricatures, graphic illustrations, and vivid descriptions of Jews as hideous, hook-nosed seducers of "Aryan" women, Christ-slayers, and usurers. Such subject matter brought anti-Semitism blatantly into the classroom. It was used to underline and provide legitimization for anti-Semitic ideology and policies and to misinform the German populace about the Jews.[35]

Mythmaking, Miseducation, and Disinformation during the Third Reich

This chapter advances our understanding of agnotology by examining miseducation, disinformation, and occlusion in a dictatorial regime. The Nazi government certainly intended to keep the German populace ignorant of the mass murder of the Jews and the Gypsies, as well as that of the asylum population, although all of these became open secrets eventually. However, by analyzing a variety of different aspects of agnotology under National Socialism, we can see the pervasiveness of the regime in everyday life and how these factors together add up to create a larger portrait of disinformation and miseducation in the Third Reich. The Nazi administration actively constructed and preserved ignorance among the German population by not only banning and censoring information but also replacing and obscuring knowledge. The Nazi government attempted to "reeducate" German "national comrades" by promoting its ideology through propaganda and the creation of a variety of myths, by banning and censoring entire areas of artistic life, and by controlling the content of school textbooks. Its policies in these areas were purposeful and pervasive. At times, however, the regime chose not to use blatant political propaganda to get its message across to the German population and could successfully manipulate the popular mind by using feature films or literary narratives that conveyed their purport without being overtly ideological. Words and symbols could be powerful in denoting meaning. Strong and emotive messages were conveyed. Propaganda on its own was not enough to persuade the populace—this was a brutal dictatorship that underpinned its message with fear and terror.

The speedy transformation and "coordination" of German society was achieved in part as a result of these policies, yet to a considerable extent the population assented to the regime and desired to accept what it conveyed. As we have seen, for instance, the Hitler myth was as much a self-imposed phenomenon as one imposed from above. The manipulation of the popular mind was subtle at times, but the control of information was critical for the Nazi regime at all times. The National Socialists were motivated by their ideological imperatives to keep the German population in ignorance. They were by no means alone in this. There are many historical and contemporaneous examples of information controlled by governments or organizations, of cultural and literary output limited and censored, as well as of regimes (not solely dictatorial ones) persuading their populations in different ways, keeping them in ignorance and subjecting them to political propaganda. Power and ideology allow regimes to get particular messages across to the people they rule, including messages that are distortions of truth. The Nazi state used its education policy to socially construct ignorance and develop normative socialization devices. The pervasion of the school curriculum with ideology was not exclusive to the Nazi administration, but the Third Reich clearly exemplified the ways in which governments can maneuver the popular mind and distort knowledge and truth through the tools of mythmaking and miseducation.

NOTES

1. Richard Evans, *The Coming of the Third Reich* (New York: Penguin, 2004), 196; Adolf Hitler, *Mein Kampf*, trans. Ralph Manheim, with an introduction by Donald C. Watt (London: Pimlico, 1992), 387; David Welch, *The Third Reich: Politics and Propaganda*, rev. ed. (London: Routledge, 1993), 24.

2. On the study of ignorance, see Robert N. Proctor and Londa Schiebinger, eds., *Agnotology: The Making and Unmaking of Ignorance* (Stanford, CA: Stanford University Press, 2008); and Proctor, "Agnotology: A Missing Term to Describe the Cultural Production of Ignorance (and Its Study)," in ibid., 9.

3. See Randall Bytwerk, *Julius Streicher: Nazi Editor of the Notorious Anti-Semitic Newspaper "Der Stürmer"* (New York: First Cooper Square, 2001); and Bytwerk, *Bending Spines: The Propagandas of Nazi Germany and the German Democratic Republic* (East Lansing: Michigan State University Press, 2004), 26.

4. Eric Rentschler, "The Legacy of Nazi Cinema: *Triumph of the Will* and *Jew Süss* Revisited," in *The Arts in Nazi Germany: Continuity, Conformity, Change*, ed. Jonathan Huener and Francis Nicosia (Oxford: Berghahn, 2006), 72, 74.

5. Michael Burleigh and Wolfgang Wippermann, *The Racial State: Germany, 1933–1945* (Cambridge: Cambridge University Press, 1991), 154, 155, 157.

6. On national renewal and rebirth, see Roger Griffin, *The Nature of Fascism* (London: Pinter, 1991), 98–99; Welch, *Third Reich*, 53; Nancy Reagin, "Comparing Apples and Oranges: Housewives and the Politics of Consumption in Interwar Germany," in *Getting and Spending: European and American Societies in the Twentieth Century*, ed. Susan Strasser, Charles McGovern, and Matthias Judt (Cambridge: Cambridge University Press, 1998), 241–61; Reagin, *Sweeping the German Nation: Domesticity and National Identity in Germany, 1870–1945* (Cambridge: Cambridge University Press, 2007), 145–46; and Geoffrey Cocks, *The State of Health: Illness in Nazi Germany* (Oxford: Oxford University Press, 2012), 46. On the national community, see Lisa Pine, *Hitler's "National Community": Society and Culture in Nazi Germany* (London: Hodder Arnold, 2007). See also Martina Steber and Bernhard Gotto, eds., *Visions of Community in Nazi Germany: Social Engineering and Private Lives* (Oxford: Oxford University Press, 2014).

7. On the disparity between Nazi propaganda and the true impact of the war, see Ian Kershaw, *The End: Germany, 1944–45* (London: Penguin, 2012).

8. Welch, *Third Reich*, 9, 5.

9. Ian Kershaw, *The "Hitler Myth": Image and Reality in the Third Reich* (Oxford: Oxford University Press, 1989), 3; Joseph Stern, *Hitler: The Führer and the People* (Berkeley: University of California Press, 1975), 111; Pierre Ayçoberry, *The Social History of the Third Reich, 1933–1945* (New York: New Press, 1999), 68; Kershaw, *"Hitler Myth,"* 91–92.

10. Detlev Peukert, *Inside Nazi Germany: Conformity, Opposition and Resistance in Everyday Life* (London: Batsford, 1987), 73.

11. Kershaw, *"Hitler Myth,"* 160; Peukert, *Inside Nazi Germany*, 63.

12. Kershaw, *"Hitler Myth,"* 193; Gerald Kirwan, "Allied Bombing and Nazi Domestic Propaganda," *European History Quarterly* 15 (1985): 341–62. See also Kirwan, "Waiting for Retaliation—A Study in Nazi Propaganda Behaviour and German Civilian Morale," *Journal of Contemporary History* 16 (1981): 565–83; and Kershaw, *"Hitler Myth,"* 219.

13. Klaus-Michael Mallmann and Gerhard Paul, "Omniscient, Omnipotent, Omnipresent? Gestapo, Society and Resistance," in *Nazism and German Society*,

ed. David Crew (London: Routledge, 1994), 166–96. See also Gerhard Paul and Klaus-Michael Mallmann, eds., *Die Gestapo: Mythos und Realität* (Darmstadt: Wissenschaftliche Buchgesellschaft, 1995); and Robert Gellately, *The Gestapo and German Society: Enforcing Racial Policy, 1933–1945* (Oxford: Clarendon, 1990), 254.

14. Ayçoberry, *Social History of the Third Reich*, 31; Gellately, *Gestapo and German Society*, 258, 75.

15. Alan Steinweis, "The Nazi Purge of German Artistic and Cultural Life," in *Social Outsiders in Nazi Germany*, ed. Robert Gellately and Nathan Stoltzfus (Princeton, NJ: Princeton University Press, 2001), 99.

16. Stephanie Barron, ed., *Degenerate Art: The Fate of the Avant-Garde in Nazi Germany* (Los Angeles: Los Angeles County Museum of Art, 1991); Peter-Klaus Schuster, ed., *Die "Kunststadt" München 1937: Nationalsozialismus und "entartete Kunst"* (Munich: Prestel-Verlag, 1988); Karl Arndt, *Nationalsozialismus und "Entartete Kunst": Die "Kunststadt" München 1937* (Munich: Prestel-Verlag, 1988); Berthold Hinz, *Art in the Third Reich* (New York: Pantheon Books, 1979), 39–40; Robert Wistrich, *Weekend in Munich: Art, Propaganda and Terror in the Third Reich* (London: Pavilion, 1995), 57.

17. Adolf Hitler, quoted in Hinz, *Art in the Third Reich*, 42; Wistrich, *Weekend in Munich*, 61.

18. Brandon Taylor and Wilfried van der Will, "Aesthetics and National Socialism," in *The Nazification of Art: Art, Design, Music, Architecture and Film in the Third Reich*, ed. Taylor and van der Will (Winchester, NH: Winchester Press, 1990), 4; Peter Adam, *Art of the Third Reich* (New York: H. N. Abrams, 1992), 9; Jonathan Petropoulos, "The Art World in Nazi Germany: Choices, Rationalization, and Justice," in Huener and Nicosia, *Arts in Nazi Germany*, 149. See also Petropoulos, *Art as Politics in the Third Reich* (Chapel Hill: University of North Carolina Press, 1996).

19. Richard Grunberger, *The 12-Year Reich: A Social History of the Third Reich, 1933–1945* (London: Weidenfeld & Nicolson, 1971), 431; Evans, *Coming of the Third Reich*, 412.

20. On Blunck, see W. Scott Hoerle, *Hans Friedrich Blunck: Poet and Nazi Collaborator, 1888–1961* (Oxford: Peter Lang, 2003).

21. James Ritchie, *German Literature under National Socialism* (London: C. Helm, 1983), 13.

22. Quoted in ibid., 20.

23. Richard Samuel and Richard Hinton Thomas, *Education and Society in Modern Germany* (London: Routledge & Kegan Paul, 1949), 83; Bundesarchiv Berlin R 4901/1 4620/1, "Betrifft: Neuordnung des höheren Schulwesens," 5.

24. See Lisa Pine, "The Dissemination of Nazi Ideology and Family Values through School Textbooks," *History of Education* 25 (1996): 91–109.

25. Christa Kamenetsky, *Children's Literature in Hitler's Germany: The Cultural Policy of National Socialism* (Athens: Ohio University Press, 1984), 187–88.

26. Gilmer Blackburn, *Education in the Third Reich: A Study of Race and History in Nazi Textbooks* (Albany: State University of New York Press, 1985); Hans J. Hahn, *Education and Society in Germany* (Oxford: Berg, 1998), 82; Gregory Wegner, *Anti-Semitism and Schooling under the Third Reich* (New York: Routledge, 2002), 126.

27. Wegner, *Anti-Semitism and Schooling*, 153.

28. Pine, " Dissemination of Nazi Ideology," 105.

29. See, e.g., Paul Brohmer, *Biologischer Unterricht und völkischer Erziehung* (Frankfurt am Main: Diesterweg, 1933); Änne Bäumer-Schleinkofer, *NS-Biologie und Schule*

(Frankfurt am Main: Verlag Peter Lang, 1990), 152–54; Bundesarchiv Berlin NS 12/41, "Erbbiologie in der Praxis des biologischen Unterrichts," 1, 2, 5.

30. Alfred Vogel, *Erblehre und Rassenkunde für die Grund- und Hauptschule* (Baden: Konkordia, 1937); Gregory Wegner, "Schooling for a New Mythos: Race, Anti-Semitism and the Curriculum Materials of a Nazi Race Educator," *Paedagogica Historica* 27 (1992): 197.

31. Wegner, "Schooling for a New Mythos," 198, 200.

32. Johannes Stark, "Zur Neuordnung des physikalischen Unterrichts," *Unterrichtsblätter für Mathematik und Naturwissenschaften* 45 (1939): 82.

33. O. Brandt, "Die neuen Lehrbücher," *Unterrichtsblätter für Mathematik und Naturwissenschaften* 46 (1940): 152; Erich Günther, ed., *Wehrphysik—Ein Handbuch für Lehrer* (Frankfurt am Main: Disterweg, 1936).

34. Kamenetsky, *Children's Literature in Hitler's Germany*, 149.

35. Wegner, *Anti-Semitism and Schooling*, 54–55.

12

USSR

E. THOMAS EWING

In 1938, O. Filippova had cause to celebrate. She was a successful history teacher in Moscow, and her star was on the rise. A regional education journal published an article by her about successes in the classroom and approaches to instruction. Filippova's reflection captured some of her ideas about preparing and delivering history lessons. When the class discussed the Bolshevik leaders of the revolution and civil war, students asked their teacher for specific quotations from Communist leaders so that they could "look them up in my mama's Lenin and Stalin." Filippova also described at length how she used visual aids such as maps, portraits, and other illustrations, as well as songs, literature, and memoirs, to enhance history lessons. In her classes, her students held discussions, prepared presentations, and even played games such as "What, where, and when?" to learn ancient history. After they played with these questions for a year, they introduced "why?" and "what was the significance?" to the game.[1]

Two years later, Filippova wrote another reflection, once again beginning with a nod to how she used quotations from Marxism-Leninism. But the bulk of the article dealt with more "practical," personal, and professional issues in the teaching of history. She urged fellow teachers to make use of primary documents, citing such examples as a speech by Pericles, excerpts from the writings of Herodotus, texts composed under the direction of Charlemagne, and literary works from premodern France, India, and China. From these primary sources, students were amused to discover the extent of illiteracy in the medieval era, as even famous leaders could not write their names. Filippova also continued to recommend the use of visual aids—portraits of leaders, diagrams of battles with pictures of weapons, and depictions of the daily life of a range of social groups—in lessons on the medieval period. Teachers were urged to stimulate their students' interest in out-of-class reading, suggesting that they bring in their own copies of Greek myths, Homeric epics, Arabian tales, and other collections of stories.[2]

Moving beyond the walls of the classroom, Filippova encouraged students to attend off-campus historical events and invited academics to come to her classes with the goal of stimulating interest in history. After a Professor Kun spoke about Greek and Japanese myths, students copied down the stories and then assembled them into a little book entitled *What Professor Kun Told Us*. This inspired the class to "self-publish" their own books about great explorers of the past, the Russian poet A. S. Pushkin, the proverbs of Arabian culture, and more. According to Filippova, these books were significant because they were produced by the children themselves, even if their appearance "is not all that pretty." The goal of these activities was to cultivate personal interest in history, which helped students learn the curriculum, while also developing "independent activity and creativity."[3]

Filippova's reflections illustrate how a Soviet teacher could achieve a seemingly idealized blend of political indoctrination, professional effectiveness, and personal engagement in her teaching. Content and method coexisted in a positive relationship, as the teacher encouraged students to learn for themselves and not just to meet standards or to conform. The key to her success was clearly the combination of knowledge, ability, and approach that made it possible to engage personally with her class and stimulate their desire to learn. Yet the celebration of Filippova's approach is not consistent with the general view of Stalinist history teaching, which emphasizes the ways that Soviet classrooms became instruments of the regime. It is precisely for this reason that these examples from Filippova's teaching help us to explore the social construction of ignorance in the Soviet context. Filippova was negotiating the difficult line between mastery of knowledge and strategic deployment of that knowledge in ways that satisfied both official expectations and her own perceptions of what worked in the classroom.

The creativity, sensitivity, and effectiveness of Filippova's teaching are certainly quite distant from the most familiar expressions of Stalinist political control, especially those having to do with the manipulation of historical knowledge. This control demanded that "enemies of the people" who had been physically eliminated through execution, imprisonment, or exile should also be eliminated in every way from public historical records by deleting names, manipulating pictures, and fabricating narratives. These practices also occurred in Soviet education, as documented in both archival records and the memories of former Soviet citizens.[4] According to one former teacher, after the purge of the Red Army commanders in late summer 1937, "students were told to paste black paper over their portraits in the text books. They couldn't cut out the portraits because on the reverse side of the photographs there would always be printed materials. Afterwards the books were revised."[5] Another former teacher recalled the consequences of showing unapproved material to students. In this account, the school's director, "a decent fellow . . . the one decent Communist whom I met," was subject to a trick orchestrated by another school official: "She obtained an old film which had Trotsky and Bukharin on it. She misplaced the film, so that the director did not have a chance to look it over and the film was shown. It caused a great deal of difficulty. The director was removed to another school, although the teacher stayed."[6]

Even when it came to the hanging of pictures, teachers had to take care not to trigger serious consequences. A history teacher named Fedorov discovered this source of potential vulnerability after being dismissed in 1936 for allegedly hanging a portrait of Stalin's foe Trotsky next to that of Lenin in a school office.[7] Not surprisingly, teachers feared these familiar forms of Stalinist control. They talked generally of the "many cases when teachers were sent to prison, because of what they had taught, to have taught something that had not been written from above." And they knew of extreme cases, such as when a teacher was punished for having a book by Bukharin, which was being used for toilet paper: "Somebody informed on her, and the NKVD came and took her away. We never heard anything about her again. We never heard if she had a trial, or went to prison, never heard a thing."[8]

These examples of repression and efforts to remove content, courses, faculty, and staff from schools align with standard accounts in Soviet historiography and offer obvious cases in the process of ignorance-making.[9] The imperative to teach new content that advanced a specific political message required the deliberate loss of knowledge. Teachers who disappeared because they transgressed political boundaries were certainly evidence of the process of forgetting. Repression and compulsory omissions provide rich, illustrative evidence of how the political context demanded that teachers and even students "forget" facts that were personally known, recorded in documents, or part of lesson content. They also demonstrate the consequence of "knowing" facts that were officially denied, suppressed, or forgotten. During this time of terror the punishment could be extreme, including dismissal, arrest, imprisonment, or execution.

This chapter, however, moves in a different direction. It casts new light on the question embedded in Stalinist history teaching, a question raised a decade later in George Orwell's fictional account of a dystopian totalitarian state: who controls the past?[10] Communist Party leaders, including Stalin, acted upon a belief that the past should be used as an instrument to shape the present and the future. The past did not have a truth or an autonomy independent of the needs of the party. Yet the examples and analysis presented in this chapter suggest that the complex relationship between ideology, agency, and structures shaped these efforts to control the past. Robert Proctor's assertions that ignorance is "an active production" that can be "actively engineered as part of a deliberate plan," Nancy Tuana's claim that ignorance is "a complex phenomenon which, like knowledge is interrelated with power," and David Magnus's statement that the "construction of ignorance" can be used to "bring about specific ends" are all suggestive interpretive positions that this chapter explores relative to the specific question how Stalinist history teachers struggled with, and survived along, the contested boundaries between ignorance and knowledge in the Soviet context.[11] This case study argues for the importance of selective ignorance as an analytical category for understanding classroom practices in the context of a school system characterized by rapid expansion and political repression. Teachers' classroom actions, their interactions with students, and the choice of pedagogical strategies required selective exercise of knowledge, as well as actual (or native) ignorance.

Based on extensive documentation of educational practices in the Stalinist history classroom, this chapter concludes that controlling the past involved the operation of agency, the capacities of actors, and the dynamics of social interaction as much as, and at times more than, demands and threats made by a dictatorial regime. Accounts by teachers and about them during this era indicate that the Soviet state was stymied in its efforts to control its past largely because of the "naïve" or native ignorance of history teachers. These accounts describe how the social construction of ignorance trumped the diffusion of knowledge and political indoctrination in Soviet schools of the mid- to late 1930s. Soviet history teachers in this era of repression drew upon multiple sources of knowledge, yet they had to engage that knowledge selectively, strategically, and at times silently. At the same time, they had to forget other kinds of knowledge in ways that were just as strategic. This case study of Stalinist history teaching provides new perspectives on agnotology by arguing that the value of studying ignorance lies in the recognition that contextual factors demand the display of ignorance to the extent that certain kinds of knowledge, if enacted or expressed inappropriately, produced extreme vulnerability.

Stalinism and History Teaching

In a ninth-grade class in the city of Elan', north of Stalingrad, none of the students in the spring of 1935 correctly answered the question "What were the goals of Tsarist Russia in the Imperial War?" Fewer than a tenth of them knew the correct answer to the question "What were the essential tasks that Lenin placed before the proletariat in the April theses?" Fewer than a third could explain why the Bolsheviks had dissolved the constituent assembly in January 1918. Yet more than a third correctly described Bismarck's policies toward the working class, and the same proportion knew which parties in the English Parliament were led by Gladstone and Disraeli. When asked about significant historical dates, half of the students gave correct answers to questions about the fall of Bismarck, the founding of the Communist International, the bourgeois transformation of Japan, the Boer War, and the Franco-Russian Alliance.[12] It would appear, judging by these results, that students were more successful in answering questions about European or world history than they were in answering questions about the recent Soviet past or even the history of the Communist Party.

In citing these results, the Stalingrad regional educational journal declared that students' achievement levels in classes on modern and Soviet history were "completely unsatisfactory." Poor performance on exams suggested a broader problem in history teaching. Students were being taught to memorize dates, but they were not being taught to understand the broader context of particular events. Moreover, students were learning about events in such a schematic manner that they lost all sense of the conditions that had provoked the events, the motivations of the participants, and their broader significance. Finally, these lessons were criticized for their lack of "party acumen" (*partiinaia zaostrennost'*). Students learned about historical figures as individuals, but without any attention to their class position. As a result, students came to believe, much as prerevolutionary textbooks had argued, that personality was the most important determinant of historical events, and they received little instruction in the heroic role of the proletariat in the development of Communism.[13]

But not all history teachers fell into these patterns. Teachers Penteleeva and Miroedova of Stalingrad School No. 1 were celebrated as models for other teachers to follow. They were praised for both their creative approach to teaching and their friendly collaboration with each other, which promoted mutual exchange of experiences. These teachers made lessons lively and interesting by painstakingly and conscientiously preparing materials and by carefully selecting the facts that students needed to learn. As a result, children acquired a "thirst for knowledge" and came "to know and to love the subject" of history. In striking contrast to the low scores elsewhere, the scores on final examinations taken in Stalingrad School No. 1 demonstrated that the students of Penteleeva and Miroedova had acquired "durable and consistent knowledge" of the materials covered in the course.[14]

The contrast between the "completely unsatisfactory" results in the Elan' ninth-grade class and the "durable and consistent knowledge" acquired by students of Penteleeva and Miroedova illustrates three seemingly contradictory aspects of history teaching: first, political indoctrination, or the transmission of specific ideological messages to students; second, professional effectiveness, or teachers' ability to comprehend and effectively communicate the content of lessons; and third, personal engagement, or the extent to which teachers persuaded students of the advantages and attractions of learning history. The requirements of political indoctrination in a single-party dictatorship specified the content as well as the purposes of historical materials. Professional effectiveness was measured by students' performance on examinations. Yet the example of Penteleeva and Miroedova suggests that personal engagement of teachers in the work of teaching was not only a requirement but also could in fact be the most important aspect of teaching history. Rather than demanding the kind of memorization that yielded "completely unsatisfactory" results, these model teachers demonstrated that getting students to personally engage with the material was the key to their better understanding. The balance between professional effectiveness, political indoctrination, and personal engagement posed a critical dilemma for Soviet history teachers in the 1930s, one resolved on occasion, as these model teachers had, by focusing on engagement in order to meet the other two demands.

The political significance of history teaching in Stalinist schools was determined by policies set at the highest levels of the political system. In the immediate aftermath of the Russian revolution of 1917 and the establishment of the world's first self-proclaimed socialist state, history was subordinated within the broader category of the social studies (*obshchestvovedenie*). With the shift in educational policy initiated by the Soviet government in 1931, however, the restoration of history to the elementary and secondary curricula was further evidence of the return to more traditional pedagogical approaches in the schools. This transition was confirmed on May 15, 1934, when the Central Committee of the Soviet Communist Party criticized the "unsatisfactory" state of history instruction in elementary and secondary schools as "schematic" and "abstract." Instead of using these methods, the Central Committee declared, history should be taught "in a lively and entertaining form that sets out the most important events and facts in chronological sequence and includes descriptions of historical figures." History lessons needed to follow chronological sequence, and students had to "memorize the most important historical events, historical figures, and chronological dates." Such an approach was necessary, according to the Central Committee, to provide students with the knowledge required for "correct analysis" of current events and for a "Marxist understanding of history." To pursue these goals, the Central Committee called for the publication of new textbooks devoted to a wide range of historical topics, including the ancient, medieval, and modern worlds, the Soviet period, and contemporary imperialism.[15]

As the Central Committee decree suggests, the political significance of history remained an absolute and unquestioned aspect of Stalinist schooling. While all lessons in all subjects were expected to provide political lessons to students, history was defined as one of the subjects deemed to have "exceptional political content."[16] Even before the Central Committee decrees, the main educational newspaper, *Za kommunisticheskoe prosveshchenie* (For Communist education), referred to history and social studies as "very important disciplines" for developing traits of creativity and initiative, providing the necessary knowledge and consciousness, and cultivating proper hatred of enemies and love of country.[17] A 1935 article by B. Levitan declared that "ideological consistency" and "a clear Party orientation" would ensure that every history lesson made students feel pride in their socialist homeland, their Leninist-Stalinist party, and their heroic past and present.[18] At this same time, an article by I. Klabunovskii in the journal *Bor'ba klassov* (Class struggle) called on teachers to fulfill "the responsible and honorable task of providing students with genuine historical knowledge and a genuinely Marxist understanding of history."[19] Two years later, the Stalingrad regional educational journal identified the "systematic and chronologically consistent exposition" of "concrete historical facts" as an "undeniable breakthrough" in the teaching of history.[20] More particularly, teachers who provided students with "a correct, ideologically consistent, and complete account" were praised in 1938 for ensuring that the necessary understanding of historical events was accompanied by "clear political lessons."[21] With the Central Committee decree setting the tone, public discussion after 1934 made political imperatives absolutely central to both the means and the end of the process of teaching history. To achieve these goals, however, the regime had to guide, or coerce, teachers into accepting the new definition of historical knowledge, while also selectively forgetting their previous understanding of the content, as well as the methods, of effective historical instruction.

Historians have long recognized the restoration of history instruction in elementary and secondary education as an important moment in the development of Soviet educational policy and more generally in the establishment of Stalinist political culture.[22] The restoration of history has been particularly central to interpretations of a "great retreat" in the second half of the 1930s.[23] Reforms in history teaching have also been interpreted in terms of how the content of the curriculum changed in response to new political imperatives defined by party leaders.[24] Yet these interpretations of Stalinist history teaching are, with a few exceptions, far more confident about the "success" of Central Committee reforms than were the teachers and certainly the educational administrators involved in making this shift in the post-1934 period.[25] Some, like Levitan, declared that despite the need for ideological consistency and party orientation, the teaching of history was often marked by "drift," as well as "schematicism and scholasticism," as teachers offered only "the most basic kinds of knowledge, formulae, and schema," while completely ignoring "clear examples of the heroic struggles of living people."[26] Others declared unequivocally that, in Kokovin's words, "there is no real grounds for discussing serious achievements in the teaching of history in all the mass schools."[27] A 1937 report on history instruction in rural schools in the Moscow region concluded that students had an "unsatisfactory" knowledge of history, and "the most important discipline in the school" was contributing little to "the development of a Marxist-Leninist worldview in students."[28] During this period, a report in the journal *Nachal'naia shkola* (Elementary school) declared that despite the 1934 Central Committee decree, the publication of new textbooks, and the greater availability of instructional materials, there were "many defects" in history teaching in elementary schools.[29]

Soviet educational publications from the 1930s reveal the complex process by which actors were called upon to carry out the policies imposed by the Soviet government. Central Committee resolutions defined what materials ought to be included and which perspectives had to be excluded, but these instructions "from above" could not determine how teachers actually taught in the classroom. Control of the past depended to a great extent on a form of "selective" and "strategic" forms of forgetting. History teachers contending with the actual demands of the classroom context, the extraordinary pressures of political repression, and the limits of their own training, expertise, and professionalism had to determine how to enact, and when to forget, their own definitions of historical knowledge, classroom management, and professional conduct.[30] In this respect, the question, who controls the past? as well as insights offered by agnotologists, calls attention to broader issues and tensions in the structure of Stalinist culture and politics, as well as to comparative issues in the work of history teachers and the process of learning about history.[31]

"The Teachers Themselves"

On January 10, 1934, the newspaper *Za kommunisticheskoe prosveshchenie* declared that "the main reason for the low quality of instruction in social studies and history . . . is the *teachers themselves, their general and political development, and their methods training.*"[32] Following the Central Committee resolutions that spring, educational publications continued to warn that many history teachers had a "very, very low" level of historical knowledge. *Bor'ba klassov* charged that deficiencies in training programs meant that the "average" history teacher threatened to convey a "distinct historical illiteracy" to students.[33] It also later argued that the main impediment to improving history instruction in schools was not the shortage of lesson materials or the absence of a good textbook but rather the poor training and inadequate knowledge of history teachers themselves, who were, "as a whole, poorly qualified people."[34] In early 1938 an inspector reported that "the majority of elementary school teachers have never studied history" and in fact cited teachers' own admissions that "they do not have the knowledge required by the history curriculum and textbook."[35] A report on history teaching in the Leningrad region that year began with the assertion that "teachers themselves have a poor knowledge of history," and, as a result, "many schools still cannot offer any kind of satisfactory instruction in history."[36] In a similar fashion, the "relatively poor quality" of history teaching in elementary schools was blamed on "the weak training" of most teachers.[37]

The limited evidence available regarding the academic qualifications of history teachers tends to confirm the claims that many history teachers lacked even a minimal level of education. At the start of the decade, one-quarter of social studies teachers in secondary schools had less than an elementary education, and fewer than one-fifth had any higher education.[38] In the city of Tula in early 1934 only two of more than thirty teachers of history had more than a secondary educations, but neither had studied history. In fact, none of the teachers in Tula had received any special training in the teaching of history. According to school inspector S. Dziubinskii, this widespread lack of specialized training meant that all teachers had "a weak command of history."[39]

The limits of teachers' knowledge were evident in surveys of teachers enrolled in training courses. A summer course held in Moscow in 1934 reportedly attracted some of the most qualified, experienced, and recognized history teachers in the region, yet 55 percent had five or fewer years of experience, and 63 percent did not have any higher education. Despite their professional standing, these teachers' limited knowledge was revealed by their performance on content-based examinations. Teachers were said to have "a schematic knowledge of social-economic formations" and forms of production and exploitation in medieval European history, and they were said to be "helpless" when it came to explaining specific historical events, locating sites on a map, or describing the significance of historical figures. More revealingly, some of these teachers openly stated that they were not interested in "personalities in history."[40] A 1935 report on retraining courses charged that teachers' "flimsy" knowledge of historical facts led to confusion about dates, transposition of locations, generalizations about materials, and incorrect conclusions.[41] These teachers' limited understanding of specific personalities was ridiculed in a cartoon with the headline "Nodding Acquaintance," which depicted a female teacher misidentifying a portrait of Danton as "the killer of Pushkin" (figure 12.1).

At the time of the Central Committee reforms, history teachers appear to have been less well educated than teachers of other subjects. A survey of more than six thousand teachers in the Urals region in 1933 found that 40 percent of history and social studies teachers had less than a complete secondary pedagogical education, compared with 10 percent of Russian language and physics teachers, 15 percent of math teachers, and 20 percent of geography teachers.[42]

Figure 12.1. "Nodding Acquaintance." Cartoon from the journal *Bor'ba klassov* (Class struggle). The caption reads, "Danton? Ah . . . he was the killer of Pushkin." Kan, "Za povyshenie kvalifikatsii," 57–59.

Two years later, however, a survey of more than one hundred thousand teachers in the Russian Soviet Federative Socialist Republic (RSFSR) indicated that the level of education of history teachers was comparable to, if not slightly higher than, those of teachers in other disciplines. Higher proportions of history teachers lacked secondary education when compared with literature, mathematics, and physics teachers, but history teachers were actually better educated on average when compared with all secondary teachers.[43] Similarly, reports from 1937 indicate that the training levels of secondary school history teachers were comparable to, if not higher than, those of teachers of other subjects.[44] In other words, history teachers as a whole were not necessarily less qualified than their colleagues in other fields, despite repeated complaints from educational publications.

The newspaper *Za kommunisticheskoe prosveshchenie* nevertheless declared that while the teaching of history and social studies was essential to the cultivation of proper political values in children, these subjects attracted low-quality teachers: "In the schools you often come across the completely incorrect and harmful belief that the teaching of social studies is an *uncomplicated* task that does not demand any special training or knowledge and that it is much simpler and easier than teaching some kind of math or physics."[45] This article seems to suggest that while teachers' level of formal education mattered, their approach to teaching was more important in determining their effectiveness. In particular, history teachers needed to recognize the particular demands of their lessons, which were different from, yet no less significant than, those of teachers of other disciplines. In this respect, even if history teachers were as well educated as their colleagues in other fields, they may have been perceived as less qualified because of the particular demands of their subject matter within the context of Stalinism.

Yet the prevailing impression of unqualified history teachers was repeatedly reinforced by the examples of poorly qualified teachers that filled the reports of inspectors and appeared in print. According to a report on schools near Kursk, for example, Komsomol history teacher I. A. Konorev did not know his subject material, skipped lessons, neglected students' notebooks, ignored directives from the school administration, and made up stories about other teachers.[46] A young Leningrad teacher named Podol'skii was accused of "permitting a series of political errors" in his teaching. But the real focus of criticism in the article was "the question-and-answer method" that Podol'skii used in the classroom, which meant, in practice, that "the teacher bombarded children with questions but showed little interest in their answers." Therefore, when a student asked for an explanation of an estate-based government, Podol'skii offered this "profound" explanation: "A government based on estates is a government that is made up of different estates." Instead of answering the question "Why is the lower house of parliament called lower?," Podol'skii answered, "No time to answer, let's keep going."[47]

In the North Caucasus, history teacher and party member Danil'iants had such a limited education that students often corrected his statements, which only served to undermine his authority in the classroom. In despair, parents told school officials: "Danil'iants is crippling our children because he offers them absolutely no knowledge."[48] While the explanation for the ineffectiveness of these teachers centered on their lack of training, the presentation of these "negative" examples also involved an assertion that both professional effectiveness and personal engagement mattered greatly. Lack of training may have caused Podol'skii to avoid answering students' questions, encouraged Konorev to ignore policies made by officials and neglect assignments turned in by students, and allowed Danil'iants to make misstatements in class, but such patterns may also be explained more directly by their lack of professional effectiveness in the classroom and their failure to personally engage with students. The Soviet aspirations to control the past coming "from above" were thwarted more by deficits in skill, commitment, and knowledge than by any determination to evade demands for political conformity. Rather than encouraging and rewarding the production and reproduction of officially sanctioned knowledge, the practices of these teachers demonstrate how history teaching involved both expressions of ignorance and assertions of multiple forms of knowledge.

The Practices of Teachers

The example of an outstanding teacher like O. Filippova, who shared her teaching practices with her peers, illustrates how Stalinist discourse of "ordinary celebrities" extended into the educational realm.[49] Yet a closer look at inspectors' reports suggests that such "outstanding" teachers were the exception rather than the norm. In Leningrad, a history teacher and school director named Stepanov made no effort to "enliven the dry materials in the textbook"; in one instance Stepanov did not even show up for a lesson but instead instructed students in a ninth-grade class to read the textbook aloud, work through the topics in order, and then complete a written assignment. Students were so indignant at these "original" methods that at the end of the lesson they turned in blank pieces of paper.[50] In the Stalingrad region, the majority of schools' history instruction consisted in nothing more than the teacher reading a "dry and uninteresting" summary of the textbook and having students take notes on this summary. At most, teachers followed up with questions about the summarized materials, but students were never encouraged to discuss the material. Teachers almost never used primary documents, slides, maps, excursions to historical sites, or any kind of fiction.[51]

Drawing from observations of history teachers enrolled in retraining courses, some inspectors concluded that inadequate knowledge of facts led

Figure 12.2. Cartoon from the Soviet newspaper *Za kommunisticheskoe prosveshchenie* (For Communist education). The caption reads, “A mysterious picture: is this a history lesson or a singing lesson?” *ZKP*, 20 September 1935, 1.

many history teachers to rely on “empty, contentless, and often mistaken schemas” intended to conceal their ignorance of personalities and events.[52] In the Moscow region, dictation and memorization were the main methods used in history instruction. In rural schools, teachers had a weak understanding of factual materials, frequently made mistakes during lessons, relied on examples from out-of-date textbooks, mixed together various examples “that miss the main point,” made spelling or grammatical errors, or “swamped” students with so many facts that the entire lesson was “incomprehensible.”[53] History teachers’ dependence on mindless repetition is nicely illustrated in figure 12.2, a cartoon published approximately one year after the Central Committee decree restoring history teaching to the primary and secondary curricula. In this cartoon, two rows of children recite disconnected sections of history lessons, including one snippet about the leader of a seventeenth-century Cossack revolt, Stenka Razin, under the direction of their teacher, who acts as a choir director. The caption asks rhetorically whether this “mysterious picture” is “a history lesson or a singing lesson?”

Writing about his experience teaching in Soviet schools during the war, émigré Abraham Kreusler offered this description of a teacher whose approach was strikingly similar to that of teachers criticized by Soviet inspectors, even if the political inflection was different. The teacher, Pastukhov, was described by Kreusler as “a very good man but extremely limited both in knowledge and abilities”:

> He had once been a herdsman and had made his way through school and college by dogged perseverance. Such people are the pride of the Communist Party and they are always whole-heartedly supported. He knew the *Short History of the Communist Party* backward and forward, but he was completely lost when the answers of the students were a little different, or when they asked him questions outside the scope of that book. Once he ran out of the class bewildered and confused, for the students had asked him who Byron was. Being utterly ignorant of the subject, he had to engage in elaborate preparations before he could again face his class. Of course his prestige among the students was the lowest in the school, but the principal was patient, encouraged him, and asked others to help him, for he was a self-made proletarian-intellectual.[54]

In the years following the Central Committee decree, school inspectors near Kalinin reported that instructional methods showed little improvement. The most common method of teaching history was for the teacher to read aloud from the textbook and then have students "retell" the same text.[55] Inspectors in Leningrad lamented that most history instruction "boils down" to inadequate summaries and endless repetition of textbook information, which in turn resulted in confusion of dates, distortion of facts, and incorrect interpretation of events. Because teachers themselves often had a limited understanding of history, inspectors concluded, they could not provide explanations that went beyond the textbook, and so they were afraid to ask questions of students. As a result, history instruction, one inspector observed, "turns into cramming." When students were tested, they were unable to provide "a connected narrative," displayed a poor understanding of chronology, failed to connect events and locations, and answered questions only after considerable prompting.[56]

Teachers who were either unable or unwilling to provide effective instruction embodied a model of ignorance production in schools through their documented inability to engage in required forms of political indoctrination. A 1938 inspection denounced a teacher's "monotonous and barely audible" description of the battle of Kulikovo Fields in the fourteenth century as an "insufficient—but accurate—presentation," because the teacher made no effort to use the example of the Russian struggle against the Mongols to develop "love for the socialist fatherland" and "hatred of enemies."[57] In Arkhangel'sk schools, teachers were criticized for missing the opportunity to use biographies of historical figures and the memory of past struggles to cultivate students' patriotism. In discussing the lives of Marx and Engels, for example, history teachers tended to limit themselves to a list of dates and to rhetorical statements about all they had done for the working people, without describing them "as living people, with remarkable aspects of their personalities that could be used to instruct our children." Even the revolutionary activism of these founding fathers of the socialist movement was described by most teachers in Arkhangel'sk "using general phrases, and without presenting [the] concrete facts" that might have generated an emotional response among students.[58] Whereas official Soviet pedagogy called for political instruction to be included in every lesson on every subject, history teachers tended to limit political instruction to certain key issues (such as battles with foreign enemies), rather than making political indoctrination a part of all their instructional activities.[59]

"Celebrity" teachers like Filippova, by contrast, stand out not so much for their demonstrated professional effectiveness or successful political indoctrination as for their remarkably high level of personal engagement with students. In the Chernigov region of the Ukraine, teacher and party member Kulik had the highest percentage of passing students (96%) in the district, even though he had no specialized training in history. The key to Kulik's effective instruction, according to a celebratory account, was his ability to get students involved in the process of learning:

> During the year, comrade Kulik taught students not only how to answer assigned questions but also how to describe concrete historical events independently and colorfully. Comrade Kulik devotes all of his spare time to lesson preparation. He searches constantly for additional literature and materials that will allow students to learn lessons in a more thorough, vivid, and lively fashion. With his active methods, the students of Model School No. 1 have been taught to know and love history. Thanks to the efforts of comrade Kulik in the school, well-defined political information is being presented in lively and colorful ways.[60]

Other "heroic" teachers were praised for similar kinds of innovative and engaged approaches to teaching. In the Black Sea region, history teacher Belov annually spent more than a month's salary to add to his personal library of more than two thousand books on history, as well as subscribing to important journals in the field.[61] In the Moscow region, P. I. Potemkin issued a regular newsletter on historical topics, M. A. Andreeva organized study circles for students interested in this subject, and Ogarev put on plays with historical themes.[62] Stalingrad teachers Panteleeva and Miroedova used an epidiascope for lessons about ancient Greece and Rome and medieval Europe. By "enlivening teaching and raising interest," these teachers made history the favorite subject of their students.[63]

These accounts of "outstanding" teachers were most often accompanied by admissions that such teachers were unusual. A report on Moscow schools in 1937 concluded that only "the most advanced" teachers planned lessons in such a way as to provide a clear explanation of facts, dates, and people from a Marxist-Leninist perspective, while "a significant minority" of teachers conducted lessons "without planning or any sense of organization."[64] Although educational officials tried repeatedly to demonstrate that political education did not have to make lessons less interesting, their own reports reveal that few teachers managed—or even attempted—to achieve the difficult synthesis of content and practice.

A closer look at inspectors' reports suggests that few history teachers achieved the desired synthesis of pedagogical engagement and political indoctrination. School inspector Dziubinskii declared that the "usual method" of teaching history was for the teacher to read aloud from the textbook, followed by an "exposition" of the lesson first by the teacher and then by students, an explanation of words that were not understood, and then further reading of the textbook by students. He clearly blamed teachers for shortcomings in instruction: "Unsure of their knowledge and only superficially familiar with the content of the lesson and the textbook, teachers often mix up historical facts and chronological dates, misuse names and titles, misplace the stress in old Russian words, and encounter difficulty explaining many concepts." As a result of their lack of training and generally low level of historical understanding, teachers often taught material that they had only just learned—and about which they may not have known any more than was in the textbook. In such cases, ignorance was "real" or native, rather than selective or strategic.[65]

Dziubinskii went beyond these complaints about the preparation of teachers to make a more general point about the extent to which history lessons deteriorated into a particularly limited exchange between teachers and students: "Generally, it appears that the 'question-answer method' of teaching history is nothing more than a 'catechism,' which repels any interest children might have in history. Class readings of the textbook turn a 'history lesson' into a 'lesson in explaining the assigned reading.' This would not be so bad if it were not the *only approach to teaching* and if the teacher were *more prepared for the lesson.*"[66] Dziubinskii elaborated on these criticisms in later published works. In a 1939 article he offered this assessment of the current state of historical instruction in schools:

> The 1937/1938 exams showed that students who knew historical facts best of all had teachers who paid attention not just to memorization but also to children's imagination and emotions. The greatest interest in history was shown by students whose teachers did not simply lay out the facts but offered lively and emotional accounts of heroic actions, the victories of the people over enemies, and the exploits of those who had given their lives to the people and the homeland. The challenge of teaching history in the elementary school is thus to provide students with a basic and accurate knowledge about the past and to develop pride in their own people, love and dedication to their Soviet government and its leaders, and a willingness to stand among the defenders of socialist society.[67]

In this article as well as in an earlier report, Dziubinskii described the remarkable efforts of E. E. Kozlova, a Leningrad elementary school teacher who put a tremendous amount of spirit into her history lessons. According to Dziubinskii, students were so emotionally engaged with these lessons, especially when discussing attacks on Russian lands by occupying armies and Russian victories over foreign invaders, that it seemed they were discussing contemporary events. Even though Kozlova's school had a large supply of maps and other visual aids, Kozlova seemed not to need these additional materials, because she made such effective use of historical narratives. Kozlova went beyond the materials in the required textbook and added her own examples drawn from the literature of the period under study (each of which was carefully spelled out in outlines to make it easy for students to understand and remember). In this respect, Kozlova provided the ideal synthesis of "historically accurate and politically sound" content with the living models and concrete examples introduced by the teacher. The key to Kozlova's effectiveness as a teacher, Dziubinskii concluded, was that her lessons were substantive, comprehensible, and engaging. Long after the lesson, students continued to talk with one another about what they had learned and how their opinions had been changed.[68] In other words, the ideal lesson, which provided students with clearly defined historical content and correct political views, was one in which the teacher engaged and even excited children.

These tensions embedded in aspirations to control the past were spelled out even more directly by school inspector Skaspina with regard to two Moscow teachers. The first teacher, Petukhov, was praised because he prepared a great deal for class (including using his own meager salary to buy books), his students were very good at using maps to describe historical events, they had excellent knowledge of chronology, and their exam results were better than those achieved by students of more experienced teachers. Petukhov's students were also praised because "when they talk about historical figures, it seems as if they are living people." When the lesson was about Cleopatra and Anthony, for example, the class "seemed like a theatrical production of *Arabian Nights*." Despite Petukhov's obvious abilities, however, inspector Skaspina found grounds for concern:

> What can you say about his teaching? This teacher has carried out a colossal amount of work. His students have excellent knowledge of dates and maps, and when they talk, their personalities come alive, but they come alive because of their dramatic exposition. But there is no class analysis in his teaching. When students talk about Tiberius or Caligula, they tell why they were killed and describe the content of their reforms—but they do not know for whom this was all necessary. None of this was made clear. The director said that he had fought with the teacher (Petukhov) all year because there was no social analysis in his teaching, but there was nothing that could be done. . . . When I talked with the teacher about this, he agreed with me, but he also said that he could not cover everything and that he covered everything needed to master the materials. As a result, children know history, but their knowledge is insufficient.

By contrast, Skaskina reported, the young teacher Kravtsova "had many defects in her work," including a lack of maps, posters, and other visual aids. Yet her approach was more correct: "She asks this kind of question about Tiberius and Caligula: 'Tell what the meaning is, for whom was this necessary?' Children get things mixed up, but they give answers. This young teacher is clearly providing children with the Marxist foundation without which all facts are worthless."[69] Despite her misgivings about the relative effectiveness of the two teachers, Skaskina clearly preferred a teacher who provided proper political indoctrination over a teacher who engaged students' interest and encouraged mastery of lesson content. Whereas the idealized synthesis described by Dziubinskii and others involved a harmonic balance of these three strategies for effective teaching, the actual practices of teachers suggest a more uneven, often precarious, and at times seemingly hopeless combination of professional, political, and personal elements.

"Students were left in a state of doubt—which choice was correct?"

Much was expected of history teachers in the political culture of Stalinism. According to Soviet publications of the era, they not only needed to master the material in the curriculum, including concrete facts and the chronology of events, as well as a Marxist understanding of the process of history, but they also had to "command the methods required for the skillful transfer of knowledge to students" and be able to make history "the most lively and fascinating subject in the school."[70] History teachers' most important task was "to develop interest in history among students" by stimulating a love for knowledge and the desire to continue learning through their own reading and activities outside of school.[71] In the words of one publication, "The greater the initiative, enterprise, and concern on the part of the teacher, the greater the results." Teachers who enhanced their lessons with maps, illustrations, and other materials were said to be most successful in cultivating students' love for their homeland and loyalty to the leaders and heroes of the socialist fatherland.[72] In this idealized synthesis, the best teachers combined personal engagement with students and professionally effective knowledge of methods and content in order to accomplish the designated goals of political indoctrination.

Yet this apparent dichotomy between memorization of dates and terms, on the one hand, and encouraging students to develop the ability to draw conclusions and make generalizations, on the other hand, underscores a deeper issue that agnotological insights help to explore. To what extent was the latter goal of developing skills in fact expected to lead to the same conclusions and generalizations for all students, and if it was, what was wrong with simply requiring students to memorize the "correct" conclusions and generalizations? Soviet educators demanded that history teachers follow the seemingly impossible path of teaching a monolithic yet mutable understanding of the past using creative and innovative approaches to get students interested and engaged in learning so that they all reached identical conclusions and acquired the exact same knowledge.

The examples cited here suggest that history teachers taught history in many different ways, but they were constrained by the regime's underlying assumption, enforced through coercive means, that the content of history itself was consistent and complete. Teachers were being judged on how well they conveyed this content to students, but not on their ability to construct this knowledge themselves. A few final examples can illustrate both the potential for contested meanings and the implications for allowing multiplicity to emerge in a monolithic context. In 1938 a teacher in a Leningrad region school was sharply criticized for a lesson on prehistoric peoples in which she failed to define terms used in the textbooks, introduced new terms without defining their meaning, and generally confused students about the important points in the lesson. According to the regional educational journal, the deficiency of the lesson was evident in the *different* answers that students gave when asked where information could be found about the lives of people in the past: "from history," "from excavations," or "from objects preserved to the present day."[73] In the Moscow region, teacher Zhil'tsova decided to teach about Kievan Rus' in a discussion format rather than in the more conventional lesson format involving a lecture and then questions of students. During this lesson, according to a published account, Zhil'tsova spent a considerable amount of time discussing the origins of the word *Rus'* and in fact provided three different explanations. Yet this method drew sharp criticism. In particular, the report charged, the strategy failed to meet the required outcome: "students were left in a state of doubt—which choice was correct?"[74]

As these final two examples suggest, teachers who used their own methods to encourage students to think for themselves in trying to understand the past were criticized for leaving their students "in a state of doubt." The implication of both articles was that students should end the lesson with a single answer to every question and that to end up with multiple answers was proof that students had been "confused" by their teacher's presentation of materials. Despite unceasing rhetoric about "stimulating" students' interest in acquiring knowledge, a powerful undercurrent in Stalinist discourse on history teaching remained the quest for the single correct answer. The question, who controls the past? had to be answered in such a way that the past had an absolute meaning transmitted directly to students. At bottom, the absence of knowledge, skill, and engagement on the part of the vast majority of Soviet history teachers perpetuated ignorance on the part of students in an era marked by great political intensity embedded in the Stalinist relationship between power and knowledge. Teachers were not permitted to use their power in the classroom to empower their students by offering them the opportunity to explore, acquire, and make use of knowledge, because such a path challenged the more fundamental—and "totalizing"—claims of Stalinist authorities to the exclusive control of the past.

NOTES

Earlier versions of this chapter were presented at the annual meeting of the Southern Conference on Slavic Studies (2001) and in the working papers series sponsored by the ASPECT program at Virginia Tech. For helpful comments, the author thanks Aaron Ansell, Francois Debrix, David Hicks, Larry Holmes, Amy Nelson, Gary Marker, Roy Robson, and A. J. Angulo. Financial support was provided by the University of Michigan, the Spencer Foundation, the International Research and Exchanges Board (IREX), the National Council for Eurasian and East European Research, and Virginia Tech.

1. O. Filippova, "Istoriia SSSR v 5-kh i 6-kh klassakh," *Istoricheskii zhurnal*, no. 12 (December 1938): 110–12.

2. O. Filippova, "Iz opyta prepodavaniia istorii v 29-i i 41-i shkolakh Frunzenskogo raiona Moskvy," ibid., no. 11 (November 1940): 112–14.

3. Ibid.

4. On the political repression of Stalinist teachers, see E. Thomas Ewing, *The Teachers of Stalinism: Policy, Practice, and Power in Soviet Schools of the 1930s* (New York: Peter Lang, 2002), 227–63. For the cleansing of purged leaders more generally in Soviet society, see David King, *The Commissar Vanishes: The Falsification of Photographs and Art in Stalin's Russia* (London: Canongate Books, 1997).

5. Harvard Interview Project on the Soviet Social System, Schedule A, No. 142, 7–8, Davis Center for Russian and Eurasian Studies Collection, H. C. Fung Library, Harvard.

6. Ibid., No. 91, 6–7.

7. Russian State Archive of Social and Political History (Rossiiskii Gosudarstvennyi Arkhiv Sotsial'no-Politicheskoi Istorii), formerly the Center for the Preservation of Documents of Youth Organizations (Tsentr Khraneniia Dokumentov Molodezhnykh Organizatsii) and originally designated the Central Archive of the All-Union Lenin Komsomol Youth Union (Komsomol) (Tsentral'nyi Arkhiv Vsesoiuznogo Leninskogo Kommunisticheskogo Soiuza Molodezhei), fond 1, opis' 3, delo 159, listy 75, 78, 82–83. In 1935 a party member and teacher named Leskov was charged with spreading "Trotskyist contraband" because a school poster included texts by Zinoviev and Kamenev, former Bolshevik leaders now in disgrace. *Za kommunisticheskoe prosveshchenie* (hereafter *ZKP*), 16 February 1935), 4. For other allegations of "Trotskyist contraband" in lessons, see ibid., 6 May 1935, 4.

8. Harvard Interview Project on the Soviet Social System, Schedule A, No. 111, 12. For more discussion of the experience and perspectives of this teacher, see E. Thomas Ewing, "If the Teacher was a Man: Masculinity and Power in Stalinist Schools," *Gender & History* 21 (2009): 107–29.

9. Robert N. Proctor and Londa Schiebinger, preface to *Agnotology: The Making and Unmaking of Ignorance* (Stanford, CA: Stanford University Press, 2008), vii.

10. For the most pronounced elaboration of this ambition, see the interrogation of the hero, Winston Smith, near the end of George Orwell's *1984* (New York: Signet, 1950), 204–5.

11. Robert N. Proctor, "Agnotology: A Missing Term to Describe the Cultural Production of Ignorance (and Its Study)," in Proctor and Schiebinger, *Agnotology*, 8–9;

Nancy Tuana, "Coming to Understand: Orgasm and the Epistemology of Ignorance," in ibid., 140; David Magnus, "Risk Management versus the Precautionary Principle: Agnotology as a Strategy in the Debate over Genetically Engineered Organisms," in ibid., 250.

12. A. I. Kokovin, "Osnovnye nedostatki v prepodavanii grazhdanskoi istorii v shkolakh Stalingradskogo kraia," *Pedagogicheskii zhurnal*, no. 1 (June 1935): 28–30.

13. Ibid., 30–32.

14. Ibid., 31–32.

15. "O prepodavanii grazhdanskoi istorii v shkolakh SSSR," in *Narodnoe obrazovanie v SSSR: Obshcheobrazovatel'naia shkola; Sbornik dokumentov 1917–1973 gg.* (Moscow: Pedagogika, 1973), 166. Whereas the "unified labor school" curriculum of the 1920s has replaced history with contemporary social studies (*obshchesvovedenie*), the curricular reforms of the 1930s restored history to the elementary and secondary school curricula. After 1934 Soviet history was taught in grades 4, 7, and 10; the ancient world was taught in grade 5, the Middle Ages in grades 6 and 7, and modern history in grades 8 and 9. Konstantin Shteppa, *Russian Historians and the Soviet State* (New Brunswick: Rutgers University Press, 1962), 136.

16. Two other subjects listed in this same category in a 1937 editorial were geography and natural sciences, the former because of the economic and military implications, the latter for the antireligious potential. "Ovladet' bolshevizmom!," *Nachal'naia shkola*, no. 10 (October 1937): 3.

17. "Obshchestvovedenie i istoriia—vazhneishie distsipliny," *ZKP*, 10 January 1934, 1.

18. B. Levitan, "Za bol'shevistskuiu bditel'nost' v vospitanii podrastaiushego pokoleniia," *Nachal'naia shkola*, no. 5 (May 1935): 2.

19. I. Klabunovskii, "Uchitel' istorii i povyshenie ego kvalifikatsii," *Bor'ba klassov*, no. 5 (May 1935): 60.

20. Kokovin, "Osnovnye nedostatki," 27.

21. E. M. Khazanova, "Prepodavanie istorii SSSR v 5–7 klassakh," *Istoricheskii zhurnal*, no. 2 (February 1938): 113.

22. For historian Klas-Goran Karlsson, the 1934 decree "reached far beyond the area of history teaching and became a turning point for the entire cultural life in Stalin's Soviet Union." Karlsson, "History Teaching in Twentieth-century Russia and the Soviet Union: Classicism and its Alternatives," in *School and Society in Tsarist and Soviet Russia*, ed. Ben Eklof (London: St. Martin's, 1993), 215. Soviet émigré and academic historian Konstantin Shteppa referred to the Central Committee decree as a document of "primary importance" that signaled important changes not only in the teaching of history "but in Soviet life in its entirety, in deep-seated habits and opinions," and hence "the beginning of a new era in the official world view." Shteppa, *Russian Historians*, 123, 127. Soviet historian Zakhar Ravkin cited the 1934 decrees on teaching history as an "essential" step toward the cultivation of the "Communist worldview" of young Soviet citizens. Ravkin declared that the policy changes and textbook revisions meant that "systematic, concrete, and content-rich courses" provided students with the knowledge of the "basic facts and events of history," without which a Marxist and Communist understanding of the world would be "unthinkable." Z. I. Ravkin, "Sovetskaia shkola v period bor'by za zavershenie stroitel'stva sotsialisticheskogo obshchestva i postepennogo perekhoda ot sotsializma k kommunizmu (1935–1941 gg.)," *Sovetskaia pedagogika*, no. 6 (June 1950): 107–8.

23. One of the first advocates of this interpretation, Russian émigré sociologist Nicholas Timasheff, argued that the return of history to the school curriculum

demonstrated that Soviet leaders had recognized that "the Great Experiment in education was a cultural catastrophe." More particularly, Timasheff asserted that Soviet history textbooks and teaching methods had been simplified, clarified, and systematized so that all teachers and students would understand history the same way. Timasheff, *The Great Retreat: The Growth and Decline of Communism in Russia* (New York: E. P. Dutton., 1946), 211–12, 221. Historian Sheila Fitzpatrick identified the 1934 decrees on history teaching as part of the broader effort to achieve the "restoration of order" in this postrevolutionary society. Echoing Timasheff's interpretation, Fitzpatrick also concluded that the heavy emphasis placed on specific details in textbooks and curricula was in response to the perceived need for a simplified approach that poorly trained teachers could easily use and upon which students could be easily examined. Fitzpatrick, *Education and Social Mobility in the Soviet Union, 1921–1934* (London: Oxford University Press, 1979), 211, 232–33.

24. According to the American sociologist Alex Inkeles, the rewriting of history to emphasize "the role of the individual as a historical force" and to glorify Russian national heroes was a strategy to make the "inculcation of patriotism" into "a prime responsibility of the school." More generally, Inkeles includes the restoration of history in the schools as an example of the Soviet regime's efforts to build the stable foundations for a modern industrialized society while also producing "disciplined, compliant, obedient individuals with respect for authority, who yet had a strong sense of individual responsibility and were active, goal-oriented, optimistic, and stable." Inkeles, *Social Change in Soviet Russia* (New York: Vintage, 1968), 15–16. Soviet émigré Dora Shturman argues that history instruction was intended to politicize students in a particular manner, by making them "efficient, diligent, and moderately active (on the school's orders)." Shturman, *The Soviet Secondary School* (New York: Routledge, 1979), 54. Historian Robert C. Tucker argues that the shift in approaches to history instruction was decreed directly by Stalin, who wanted a version of Russian history that glorified great leaders of the past as a means to justify his own emerging "cult of personality." Tucker, *Stalin in Power: Revolution from Above, 1929–1941* (New York: Norton, 1990), 279–82, 530–36. For scholars such as George M. Enteen and John Barber, interested in the development of the Soviet historical profession, the 1934 decrees marked an important shift away from more complex thematic and ideological approaches to the past and toward more conventional narrative, nationalist, and partisan approaches. Enteen, *The Soviet Scholar-Bureaucrat: M. N. Pokrovskii and the Society of Marxist Historians* (University Park: Pennsylvania State University Press, 1978), 189–90; Barber, *Soviet Historians in Crisis, 1928–1932* (London: Holmes & Meier, 1981), 139–40. More recently, David Brandenberger has examined archival documents regarding the formulation and revisions of party policies regarding history teaching and textbooks in order to argue that changes in historical understanding occupied a central place in the establishment of "Russocentric etatism" in the late 1930s. Brandenberger, *The "Short Course" to Modernity: Stalinist History Textbooks, Mass Culture, and the Formation of Popular Russian National Identity, 1934–1956* (PhD diss., Harvard University, 1999); Brandenberger, "Soviet Social Mentalite and Russocentrism on the Eve of War, 1936–1941," *Jahrbucher for Geschichte Osteuropas* 48, no. 3 (2000): 388–406; Brandenberger, *National Bolshevism: Stalinist Mass Culture and the Formation of Modern Russian National Identity, 1931–1956* (Cambridge: Cambridge University Press, 2002), 63–76.

25. A few scholars have recognized the complexities involved in the Soviet effort to change history teaching. Shteppa recognized the tensions in the work of teachers by noting the difficulty that teachers faced in "presenting historical facts graphically and

descriptively" in order to combine "vivid presentation and a patriotic education." More intriguingly, Shteppa identified the following as a "peculiar paradox": "on the one hand, the teaching of history was subjected to very severe regulation; and on the other, in view of the lack of classroom materials and established programs, the teacher was occasionally given an opportunity to exercise a great deal of initiative." Shteppa, *Russian Historians*, 134–36. Brandenberger uses unpublished reports as well as articles from the 1930s to describe how the Stalinist regime went about addressing, discussing, and then initiating reforms in the area of history instruction, with particular attention to the uneven pattern of implementation. Brandenberger, *"Short Course" to Modernity*, 45–58, 89–95; Brandenberger, *National Bolshevism*, 66–70. Karlsson's broad comparative study of "history teaching as a process" includes a brief mention of how history teachers "had difficulties adjusting to the new hermeneutic and idealistic Stalinist idea of history." Karlsson, "History Teaching," 216. Fitzpatrick makes a brief mention of the Central Committee's inability to understand that changing the teaching of history was more complicated than changing the teaching of the sciences, but she explains these differences in terms of the lack of acceptable textbooks and the poor qualifications of teachers, rather than the dilemmas of practices within the classroom. Fitzpatrick, *Education and Social Mobility*, 231. See also Larry E. Holmes's descriptions of history lessons in Model School No. 25, in which even excellent teachers encountered difficulties with the new curriculum and textbooks. Holmes, *Stalin's School: Moscow's Model School No. 25, 1931–1937* (Pittsburgh: University of Pittsburgh Press, 1999), 79–80, 96–97.

26. Levitan, "Za bol'shevistskuiu bditel'nost'," 2.

27. Kokovin, "Osnovnye nedostatki," 27–28.

28. E. Khazanova, "Itogi 1936–1937 uchebnogo goda po Moskovskoi oblasti," *Istoricheskii zhurnal*, no. 7 (July 1937): 110–11.

29. N. I. Ulanov, "O nedostatkakh v prepodovanii istorii v nachal'noi shkole," *Nachal'naia shkola*, nos. 2–3 (February–March 1940): 53–59.

30. Proctor, "Agnotology," 8–18.

31. Important issues relevant to this topic, but beyond the scope of this chapter, include the greater vulnerability of history teachers to political terror (discussed in chapter 7 of my *Teachers of Stalinism*), the contribution of history instruction to new constructions of Soviet nationalism, the problems teachers encountered in obtaining and using new textbooks, the place of Communist Party history in the secondary curriculum, and the potential advantages of a cross-cultural and comparative approach to these questions.

32. "Obshchestvovedenie i istoriia," 1. Emphasis in the original.

33. S. Kan, "Za povyshenie kvalifikatsii prepodavatelei istorii (iz opyta proverki ikh znanii)," *Bor'ba klassov*, nos. 5–6 (May 1934): 55.

34. Klabunovskii, "Uchitel' istorii," 60.

35. Scientific Archive of the Russian Academy of Education (Nauchnyi Arkhiv Rossiiskoi Akademii Obrazovaniia [hereafter NA RAO]), formerly Scientific Archive of the Academy of Pedagogical Sciences (Nauchnyi Arkhiv Akademii Pedagogicheskikh Naukov), f. 10, op. 1, d. 13, l. 139, and d. 21, l. 9.

36. "Prepodavanie istoriia," *V pomoshch' uchiteliu* (Leningrad), nos. 7–8 (August 1938): 25.

37. E. Khazanova and V. Klassen, "Prepodavanie istorii SSSR v nachal'nykh shkolakh," *Istoricheskii zhurnal*, no. 8 (August 1938): 97–98.

38. M. Pokrovskii, "Pedagogicheskie kadry," *Pravda*, 10 March 1930, 4.

39. S. Dziubinskii, "Istoriki ne znaiut istorii," *ZKP*, 26 April 1934, 3.

40. Kan, "Za povyshenie kvalifikatsii," 57–59.

41. Klabunovskii, "Uchitel' istorii," 63.

42. State Archive of the Russian Federation (Gosudarstvennyi Arkhiv Rossiiskoi Federatsii), formerly State Archive of the RSFSR (Tsentral'nyi Gosudarstvennyi Arkhiv RSFSR), and the Central State Archive of the October Revolution (Tsentral'nyi Gosudarstvennyi Arkhiv Oktiabr'skoi Revoliutsii i Sotsialisticheskogo Stroitel'stva SSSR), f. 2306, op. 70, d. 1895a, l. 3.

43. *Trud v SSSR* (Moscow: Gosizdat, 1935), 321.

44. State Archive of the Russian Federation, f. 2306, op. 69, d. 2405, l. 76.

45. "Obshchestvovedenie i istoriia," 1. Because of these views and the lack of control over teachers by local party and educational officials, this article warned, the teaching of social studies was often assigned to "class enemies, who use their positions to expose children to antiproletarian ideologies."

46. NA RAO, f. 17, op. 1, d. 18, l. 78.

47. D. V. Sadovskii and N. P. Irenev, "Ucheniki vozmushchalis ne zria," *ZKP*, 26 April 1934, 3.

48. Krapiva, "Uspekh dela reshaiut liudi," ibid., 22 June 1934, 3. The charges against Danil'iants were made during a "purge" and resulted in a recommendation that he be expelled from the party.

49. E. Thomas Ewing, "A Stalinist Celebrity Teacher: Gender and Professional Identities in the Soviet Union in the 1930s," *Journal of Women's History* 16 (Winter 2004): 92–118; Brandenberger, *National Bolshevism*, 64–65; Holmes, *Stalin's School*, 89, 96, 134.

50. Sadovskii and Irenev, "Ucheniki vozmushchalis ne zria," 3.

51. Kokovin, "Osnovnye nedostatki," 32. See similar descriptions of teachers' practices in Shteppa, *Russian Historians*, 134–35.

52. Klabunovskii, "Uchitel' istorii," 62.

53. Khazanova, "Itogi 1936–1937 uchebnogo goda," 106–9.

54. Abraham A. Kreusler, *A Teacher's Experiences in the Soviet Union* (Leiden: Brill, 1965), 138.

55. NA RAO, f. 10, op. 1, d. 13, l. 132. See also other criticisms of "dry and monotonous" lessons in NA RAO, f. 17, op. 1, d. 236, l. 15; N. P. Kuzin, N. M. Kolmakova, and Z. I. Ravkin, eds., *Ocherki istorii shkoly i pedagogicheskoi mysli narodov SSSR 1917–1941 gg.* (Moscow: Pedagogika, 1980), 373–74; and Brandenberger, *"Short Course" to Modernity*, 94–95, 97–98.

56. "Prepodavanie istoriia," 25–26.

57. Khazanova, "Prepodavanie istorii SSSR v 5–7 klassakh," 117.

58. Ulanov, "O nedostatkakh v prepodovanii istorii," 55–56. For similar complaints about the failure "to make use of clear examples of the heroic struggles of living people," see Levitan, "Za bol'shevistskuiu bditel'nost'," 2. For the importance attached to personalities, see Shteppa, *Russian Historians*, 127.

59. Khazanova and Klassen, "Prepodavanie istorii SSSR v nachal'nykh shkolakh," 97–98.

60. Vl. Pravdin, "Khoroshii kommunist, prekrasnyi pedagog," *ZKP*, 29 June 1934, 3. Kulik was not just an ordinary teacher, however, but also a member of the district party committee.

61. Klabunovskii, "Uchitel' istorii," 64.

62. Khazanova, "Itogi 1936–1937 uchebnogo goda," 112–13.

63. Kokovin, "Osnovnye nedostatki," 28.

64. Khazanova, "Itogi 1936–1937 uchebnogo goda," 106–7. See also examples and discussion in Brandenberger, *National Bolshevism*, 67–69.

65. Proctor, "Agnotology," 4–5.

66. NA RAO, f. 10, op. 1, d. 13, l. 139, and d. 21, l. 9, emphasis in the original.

67. Dziubinskii, "Vospitatel'naia rabota na urokakh istorii SSSR," 67–69. In a 1923 article, Dziubinskii charged that social studies teachers were merely reciting "naked historical facts" when they should have been teaching "a Marxist worldview." The article appeared in the journal *Na putiakh k novoi shkole*, nos. 7–8 (July–September 1923): 216–17, and is cited in Larry Holmes, *The Kremlin and the Schoolhouse: Reforming Education in Soviet Russia, 1917–1931* (Indianapolis: Indiana University Press, 1991), 36.

68. S. N. Dziubinskii, "Khoroshii urok istorii," *Nachal'naia shkola*, no. 3 (March 1938): 15–18; Dziubinskii, "Vospitatel'naia rabota na urokakh istorii SSSR," 67–69. Dziubinskii also provided extended descriptions of the methods of teaching history used by "celebrity" teacher O. F. Leonova, whose methods are discussed more fully in Ewing, "Stalinist Celebrity Teacher," 92–118. For further discussion of Dziubinskii's reports on history teaching, see Brandenberger, *"Short Course" to Modernity*, 116.

69. Central Municipal Archive for Moscow (Tsentral'nyi Munitsipal'nyi Arkhiv Moskvy), formerly the Central State Archive of the October Revolution and Socialist Development for Moscow city (Tsentral'nyi Gosudarstvennyi Arkhiv Oktiabr'skoi Revoliutsii i Sotsialisticheskogo Stroitel'stva goroda Moskvy), f. 528, op. 1, d. 337, ll. 37–39.

70. Klabunovskii, "Uchitel' istorii," 66–67.

71. A. Ioanisiani, "Prepodavanie istorii v shkolakh Kazakhstana," *Istoriia v shkole*, no. 2 (1936): 112.

72. "Prepodavanie istoriia," 25. In his study of history instruction in the prewar period, Shteppa concludes that the turning point in 1934 marked a shift from one pole to the other: "From the extreme of schematism and abstractness, the teaching of history moved to the mere memorizing of chronological dates, historical and geographical nomenclature, without promoting the development of skill in the drawing of conclusions or the making of generalizations that would have been necessary from the Party point of view." Despite his strikingly different political position, Shteppa's conclusions essentially confirmed the repeated complaints made by the Soviet educational establishment regarding the extent and particularly the nature of change in the teaching of history. Shteppa, *Russian Historians*, 142. See also the discussion and examples in Brandenberger, *National Bolshevism*, 68–69.

73. "Prepodavanie istoriia," 25.

74. Khazanova, "Itogi 1936–1937 uchebnogo goda," 112.

13

Israel

SOLI VERED AND DANIEL BAR-TAL

In 1955 Israel's minister of education Ben-Zion Dinur introduced the nation's first official history curriculum. The self-described aim of this new national curriculum was to bring the study of history into the service of glorifying Israel. To that end, ministry officials selectively chose events in the nation's past that would help build patriotic identity and reinforce myths about heroic national outlook. As Dinur described it, "Anything learned in school should help inculcate students with these values. . . . Nothing within the curriculum and life in school should even slightly contradict these goals." Dinur was not alone. Michael Ziv, head of the ministry's High School Education Department during the 1950s and later the history national inspector, viewed the teaching of history as a political task aimed at instilling "specific values, to guide the students towards a particular point of view, to encourage them to adopt defined attitudes sanctioned by society." This meant that in selecting historical material to be taught, heads of the educational system were expected to ensure that it would help the teacher "to instill in the student a love and respect for our most important and cherished values, to encourage the young to identify wholeheartedly with the goals that the society is struggling to achieve, and to inspire them to defend society's existence and actively contribute to its development and nurturing." This nationalistic approach to the study of history was particularly salient until the 1970s, when scholars, educators, and community leaders raised concerns about how the national curriculum complicated the teacher's ability to adhere to "historical truth" and foster critical approach in educational settings.[1]

Israel's use of history for political ends is nothing new. Nation-states have long used the past to inculcate patriotism, allegiance, and assimilation. This is a well-worn area of inquiry among historians of education. The case of history instruction in Israel from the 1950s through the 1970s, however, offers an opportunity to explore connections between recent insights in agnotology and recent contributions to the sociopsychological literature on collective memory in the context of intractable conflict. This chapter seeks to highlight these connections by presenting modes and practices of miseducation aimed at the social construction of ignorance through the use of conflict-supportive collective narratives. These modes and practices are examined within the context of the Israeli-Palestinian conflict. It also aims to highlight processes and contents common in school systems of societies engaged in protracted and violent conflicts. In this illustration, we focus on the Israeli Jewish case[2] as it emerged during the climax of the Israeli–Arab/Palestinian conflict in the period from the 1950s through the 1970s.[3]

What follows is a discussion of the themes of conflict-supportive narratives as they were reflected in Israel's national master narrative (often called the Zionist narrative), which was dominant from the late 1940s to the 1970s. The analysis then moves to a description of the methods used at the climax of the conflict in their construction by the Israeli Ministry of Education, which contributed to social ignorance among generations of students about Israeli-Palestinian history. These themes and methods are then analyzed through one particular event in the Israeli-Palestinian conflict—the 1948 Palestinian exodus—as it was presented in schools. Finally, the implications of miseducation used in the construction of conflict-supportive narratives are discussed.

While the present chapter focuses on Israeli narratives, its underlying premise is that similar conflict-supportive narratives and similar modes of miseducation can also be found among Palestinians—as mirror images of Israeli narratives—as well as in other societies involved in intractable conflicts.[4]

The Zionist Narrative as a Conflict-Supportive Narrative

Israeli society has been involved in conflict with the Arab states and the Palestinians for more than a century. Since the Israeli-Arab War of 1948, the Israeli–Arab/Palestinian conflict has become the primary issue in Israeli Jewish existence, ideology, and identity. From the war until the mid-1970s, life for Israelis and Palestinians took on all the characteristics of an intractable conflict. After the peace treaty with Egypt in 1979, tensions decreased, suggesting a move away from violence—a tendency that was accelerated by the Oslo Accords with the Palestine Liberation Organization in the 1990s and the peace treaty with Jordan in 1994. Nevertheless, with the failure of the Camp David summit in 2000 and the eruption of the second *intifada*, the conflict reescalated. Ever since, peace negotiations between Israelis and Palestinians have all ended in failure. The conflict, however, has not returned to its midcentury levels.[5]

After the conflict lost some of the characteristics of intractability in the 1970s, the educational system in Israel went through major changes, as indicated by various scholars. Especially after the mid-1990s, learning materials began to present a more complex historical narrative of the conflict and a spirit permitting more openness and self-criticism. The materials recognized Palestinian identity and aspirations and even mentioned historical misdeeds carried out by Jews against Palestinians. Nevertheless, in spite of positive changes that have taken place, the school system of early twenty-first century Israel continues to carry and reflect core themes of the Zionist ethos and of the conflict-supportive narratives, although without the extreme manifestations of four decades ago. Given the ongoing violence, it is not surprising that formal institutions still seek to maintain the dominance of conflict-supportive narratives.[6] Taking this context into account, the period from the late 1940s through the 1970s marked the height of the conflict. The culture and ethos of conflict dominated Israeli belief structures and social order. This dominancy was well reflected in the educational system's maintenance and regulation of conflict-supportive narratives.[7]

Conflict-supportive narratives include at least eight common themes. These themes appear in both the collective memory and the ethos of conflict and can be found in almost every society involved in intractable conflict. All of these themes appeared prominently in the Zionist narrative, which dominated Israeli society for about three decades in the second half of the twentieth century, during the climax of the Israeli–Arab/Palestinian conflict. First, conflict-supportive narratives justify the involvement of the society in the conflict and the course of its development. They outline the reasons for the supreme importance of the conflictive in-group's goals, stressing that failure to achieve them may threaten the very existence of the group. At the same time, they disregard and dismiss the goals of the rival, describing them as unjustified and unreasonable.

The Zionist narrative followed this model, telling the story of the return of the Jews to the Land of Israel after two thousand years of exile to establish a state in their historical homeland, while offering historical, theological, national, existential, political, societal, and cultural justifications for exclusive Jewish rights over the land. Among these arguments could be found references to the facts that the Jewish nation was formed in the land of ancient Israel; that during their exile Jews maintained close spiritual as well as physical relations with the Land of Israel, continuously aspiring to return to it; and that the repeated experience of anti-Semitism in exile was a constant reminder of the need to secure the existence of the Jewish people in their old homeland. In the 1950s and 1960s the Israeli narrative did not recognize the existence of a Palestinian people. It referred to the Palestinians as "Arabs," thereby denying the existence of a people and their claims to their own state.[8]

Second, conflict-supportive narratives emphasize dangers inherent to the conflict situation: threats to the physical existence of the in-group, to the society's cherished values, to the shared identity, to territory, and so on. In addition, they elaborate the sources of these threats and the conditions needed in order to overcome them. These may pertain to geopolitical, economic, diplomatic, educational, or societal domains; for instance, they may refer to the geopolitical borders that ensure security, the crucial importance of military strength and military industries for maintenance of security, or security-related legislature.

Achieving a sense of security was one of Zionists' basic reasons for returning to Israel and establishing a Jewish state. The Israeli narrative also delineated a legacy of wars and battles and glorified heroism, while referring to Jewish history, historical fighters, and their acts of heroism, all presented as models for identification and admiration. As to the Israeli-Palestinian conflict, it was presented as inseparable from the larger Israeli-Arab conflict. By doing so, the narrative told a story of a nation continuously under existential threat, fighting against numerous enemies united in their effort to destroy it. Moreover, present perceived threats were often viewed as the continuation of past experiences, and especially of the climax of Jews' persecution—the Holocaust—an attempt to exterminate the Jewish people by the Nazi regime during World War II.[9]

Third, the narratives delegitimize the opponent. Delegitimization is defined as a "categorization of a group, or groups, into extremely negative social categories that exclude it, or them, from the sphere of human groups that act within the limits of acceptable norms and/or values, since these groups are viewed as violating basic human norms or values and therefore deserve maltreatment." In essence, delegitimization denies the adversary's humanity and serves as a psychological permit to harm the rival group. In addition, conflict-supportive narratives focus on the rival's violence, atrocities, irrationality, intransigence, and lack of concern for human life. The Israeli narrative accordingly referred to Arabs at the height of the conflict collectively, without differentiation, as a violent and hateful people aiming to exterminate the Jews in Israel and refusing to seek a peaceful resolution to the conflict.[10]

Fourth, in contrast to the description of the opponent, these narratives present a positive image of the in-group. They depict the favored society using glorifying terms, describe events that demonstrate its positive characteristics, and especially emphasize the humane and moral aspects of the society. Also prevalent among positive collective self-image narratives are images that tell a story of suffering followed by successful redemption and of the emergence of power and strength out of weakness and persecution. The Zionist narrative, as it appeared at midcentury, was a typical example of such a story, highlighting the miraculous national rebirth of the Jewish people from the abyss of the Holocaust. It also emphasized the moral nature of Israeli society and its army as manifested in specific stories of specific wars and events.[11]

Fifth, conflict-supportive narratives present the in-group as the sole victim of the conflict, suffering from the opponent. This view is formed over the long period of violence as a result of the society's hardship and losses. The content of the Israelis' narrative about their collective self-victimhood was based on Jewish history and the traditional Jewish ethos, whereby Jews were seen as victims in a hostile world. Arabs were viewed as the sole aggressor in the conflict, forcing violent confrontations upon the Israeli nation, which sought only to defend itself.[12]

Sixth, the narratives celebrate patriotism, which is needed in order to mobilize citizens for the purpose of achieving group goals, especially for violent confrontations with the rival that involve sacrificing lives for the good of the state. The Israeli narrative praised citizens' loyalty and sacrifices for the country and glorified patriotic models from the past and current historical periods in the Israeli-Palestinian conflict.[13]

Seventh, conflict-supportive narratives refer to the importance of maintaining unity within society by ignoring internal discord and disagreement in the face of external threats. The sense of commonality and solidarity strengthens society from within and allows energy to be directed toward coping with the enemy. The Israeli narrative stressed the common heritage and fate of the Jews in Israel, minimized disputes within Israeli society, and emphasized its social solidarity as the key to its strength. It promoted social consensus regarding the Israeli-Arab conflict, in particular about the justness of Israel's goals and the means of ensuring security.[14]

Finally, the eighth theme of conflict-supportive narratives describes the in-group society as peace loving. The Israeli narrative accordingly presented peace as the "ultimate desideratum" of the Jewish people. It portrayed Israelis as peace-loving people forced by circumstances to engage in violent conflict. In particular, the narrative presented Israeli Jews as ready to negotiate and achieve peace, whereas the Arabs and Palestinians, rejecting any peaceful resolution of the conflict, were viewed as the sole obstacle to progress.[15]

These eight themes of conflict-supportive narratives can also be found in other societies undergoing intractable conflict—societies such as the Kurds and the Turks, Catholics and Protestants in Northern Ireland, the Hutus and the Tutsis in Rwanda, and Greek and Turkish Cypriots. Some of these themes (e.g., a positive collective self-image) exist as well in societies that are not engaged in intractable conflict. Yet the combination of these eight themes is unique to societies involved in this pattern of conflict, as their content is dominated by the conflict experience.[16]

As such, conflict-supportive narratives play a major role in helping society members—on both the individual and the collective level—to adapt to the harsh, stressful, and demanding conditions of the intractable violence. The narratives fulfill the epistemic function of illuminating and interpreting the situation, which is characterized by uncertainty, stress, and unpredictability. Conflict-supportive narratives also serve to justify the negative acts of the in-group toward the enemy, including violence against human beings and destruction of property. Providing justifications for such acts, these narratives allow society members to maintain a positive personal self-image and a positive collective social identity with a sense of worthiness, esteem, and integrity, which every group strives to preserve. Likewise, they differentiate between the in-group and the rival and provide a sense of superiority. By attuning society members to information that signals potential harm and violent confrontations, conflict-supportive narratives enhance individuals' psychological preparedness for negative experiences. They function as a motivating force that encourages mobilization and a willingness to make sacrifices on behalf of the group. As a result, conflict-supportive narratives serve the important function of representing the society positively and as the victim before the international community.[17]

Given these important functions that conflict-supportive narratives fulfill, it's not surprising that societies engaged in intractable conflict make significant efforts to uphold these narratives and maintain their dominance. Under these circumstances, historical accuracy does not figure as an overriding imperative in their formulation. Rather, these societies tend to propagate one-sided depictions of reality and history.[18]

Various means and practices are used by leaders, scholars, and officials in the construction of narratives, including leaders, scholars, officials, channels of communication, and informal and formal institutions. Among these agents, the educational system serves as one of the most important means by which societies construct and reinforce their national identity and collective narratives. Israeli formal education partook in the co-construction of the Zionist narrative as directed by the Ministry of Education's efforts to establish conflict-supportive themes and as evidenced by educational policies and available learning materials during the climax of the Israeli–Arab/Palestinian conflict.[19]

Educational Construction of Conflict-Supportive Narratives: The Israeli Case

During the climax of the Israeli-Arab intractable conflict in the 1950s through the 1970s, the educational system was perceived as an institution whose role was to hand down to the younger generation the Zionist narrative and its values in order to unite the nation, to form a clear collective identity in the face of the external Arab threat, and to instill in the students a desire to safeguard the newly established state. The Ministry of Education's overall approach to the teaching of history, as well as the content of learning materials and textbooks of the time, reflected the desire to ensure that the public discourse corresponded with the themes of the conflict-supportive narrative. This attitude was expressed in the writings of Naftali Zon, superintendent of high school education in the ministry from 1958 to 1972. Zon believed that history teachers' obligation was "to select those facts indicating the existence of the values of sacrifice and heroism." Stressing the difference between educators and historians, he noted that the task of the teacher was to "encourage true motivation in the student for identifying with his people and acting for them." The intent, then, was to keep the public ignorant of information that contradicted its themes, such as evidence that might undermine the justness of the Israeli goals in the conflict or the positive collective self-image of the Jewish people.[20]

A representative history curriculum published in 1961 presented ideological goals similar to those of previous and later programs that served the national narrative reflecting Zionist values. Learning materials and textbooks used in schools until the 1970s included all the themes of a typical conflict-supportive narrative. They presented justifications for the Jewish people's claims on the land and for the Zionists' position in the conflict; and they focused on inculcating Zionist values and transmitting national myths, while presenting a simplistic, one-sided picture of central historical events. At the same time, the educational materials excluded the Palestinian narrative, rejected the Palestinian national movement, refused to recognize a Palestinian entity, and presented the most negative stereotypical depiction of Arabs. The Jews, on the other hand, were presented as victims, with an overwhelming emphasis on Jewish suffering through the centuries. In addition, textbooks used in schools at the time stressed the cultural, moral, and social supremacy of the Jewish people and also glorified those who had sacrificed their lives for the sake of Israel and their commitment to Zionism.[21]

Presenting this black-and-white description of Israeli-Palestinian history required the deliberate construction of some degree of societal and historical ignorance among students. Indeed, engaged in the task of serving the Zionist narrative, the Ministry of Education used various methods of miseducation in order to establish the conflict-supportive themes presented in schools. These methods are described in the following analysis, which includes examples that illustrate how they have been used to construct the Zionist master narrative.

1. *Reliance on supportive sources.* To construct the conflict-supportive narrative, learning materials were based on sources that provided information consistent with the narrative's themes, while sources that contradicted these themes were disregarded or minimized. Therefore, in the fifties and sixties the Ministry of Education approached only scholars, teachers, journalists, and others who they assumed would write their publications in line with the Zionist narrative and support the eight themes described above.[22]

Similarly, textbooks used only those sources (e.g., documents, testimonies, archival materials, and historical accounts) that supported the Zionist narrative. For example, in conveying the story of the first *aliyah* (the first modern widespread Zionist immigration to Ottoman Palestine, which began in 1882), textbooks depicted the Land of Israel as an uninhabited, desolate land. To substantiate the "wilderness" theme of the Zionist narrative, the educational materials used photographs showing the land as empty and underdeveloped, while in contrast, other photos showed flourishing Jewish communities in the new settlements built from the 1880s on. With this orientation, students were led to believe that the immigrants of the first *aliyah* found a parched, derelict country sparsely inhabited by a backward Arab population. The Zionist myth that the land was empty served to reinforce the Jewish claim to the Land of Israel and to negate that of the Arabs.[23]

2. *Magnification of supportive themes.* In support of the Zionist narrative, events that strengthened the narrative and were in line with its conflict-supportive themes received special emphasis in educational material. The Ministry of Education used this method to justify Israel's national goals, to delegitimize the Palestinians, and to present the Israeli society in a positive light and as the victim of the conflict. A notable example of magnifying themes of the conflict-supportive narrative was the deliberate homogenization of the Arabs/Palestinians and the presentation of them in negative terms—"all of them are bad"—and as united with the intention to carry the conflict.[24]

Another theme that was emphasized was the allegedly close relationships between Arabs, Nazis, and other fascist regimes in Europe during the 1930s. As one textbook noted, "Inflammatory Italian and German political propaganda, which aimed at harnessing the Arab movement to the chariot of their own political interests, fell on the fertile ground of religious and national fanaticism." While in reality the ties between the mufti and the Italians or Germans were limited, without major mobilization on their behalf, the magnification of the connection between Arabs and fascist regimes, which were considered symbols of evil, was one of the salient ways to delegitimize Arabs.[25]

Ministry officials tended to focus on and magnify particular events that provided support for the negative characterization of the Arabs/Palestinians and the positive image of Jews/Israelis. Therefore, during the 1950s and the 1960s learning materials and school textbooks magnified acts of violence carried out by Arabs since the very beginning of the conflict. For example, they repeated the mythical story of the dramatic efforts by Joseph Trumpeldor and his friends to defend the Upper Galilee settlement of Tel-Hai in 1920. The story emphasized the bravery of the Jews against the treacherous Arabs who had deceitfully attacked them and describe their deaths as heroic. Some scholars note that the Tel-Hai myth was important to the Zionist narrative because of the negative image it conveyed of the Arabs as an insidious rival and the message to trust only Jews and to defend settlements even at the cost of one's life.[26]

Another example is the emphasis of the numerical superiority of the Arab armies over the Jews in the 1948 war. The emphasis glorified Israeli victories in this war—of the few and the weak over the many and the strong—as a central theme in the Zionist narrative, while in fact the Jewish side had the upper hand in terms of trained soldiers and military equipment, as well as planning, organization, and operation of equipment in the decisive part of the war.[27]

In addition to history textbooks, other educational means in schools have magnified the themes of the Zionist narrative. Avner Ben-Amos and Ilana Beth-El, for example, show how memorial ceremonies in schools expressed the basic concept underlying the national narrative—the country's movement from loss and bereavement to heroism and national rebirth—by stressing the symbolic meaning of the calendar sequence from the Day of Holocaust Remembrance, to Memorial Day, to Independence Day, and finally to Jerusalem Day (an Israeli national holiday commemorating the reunification of Jerusalem and the establishment of Israeli control over the Old City in June 1967).[28]

3. *Marginalization of contradictory information*. Historical content that contradicted themes of the Zionist narrative was minimized or hidden. This method was employed to lessen students' exposure to information that impinged badly on the justness of Israel's goals or its positive self-image. One way this was accomplished was to devote history curricula to the history of the Jewish people rather than to that of the Land of Israel. Learning materials focused on those periods in which the Jews lived in their homeland, as well as on the story of the Jewish people in exile, and left little room to describe other peoples' rule over the Land of Israel. As such, they gave the impression that these periods had been marginal and that various occupiers (including Arabs) had been unnatural to the land and that their occupations had been only temporary.[29]

A further example of marginalizing contradictory information was reflected in the ministry's reference to the 1936–39 events (what the Palestinians termed the "Arab Revolt"), which constituted the most significant anti-Jewish confrontation during the Mandate period. Portraying these events as simply another round in recurrent waves of violent acts initiated by inflamed Arab mobs, the learning materials disregarded their national and organized character, which helped undermine a Palestinian national movement.[30]

In other cases, in order to avoid undermining the moral image of the Israelis, killings of Palestinian civilians by Jewish forces were referred to as marginal acts made by deviants who did not represent the Jewish collective. Such was the case with the massacre in the Palestinian village of Deir Yassin in April 1948, which has been minimized by emphasizing that it was done by members of the *Etzel* (Irgun Tzva'i Leumi, or National Military Organization) and the *Lehi* (Lohamei Herut Yisrael, or Freedom Fighters of Israel), two right-wing dissident Jewish groups that splintered from the *Hagana* (the military organization of the Jewish community in Israel during the British Mandate).[31]

4. *Skewing of information processing.* Learning materials and textbooks used in schools until the 1970s provided inferences, evaluations, causal explanations, and interpretations of events and processes in a manner that supported Zionist narrative themes. This method was used especially with regard to ambiguous information and knowledge that could have been interpreted in more than one way. Such information and knowledge, however, was used to interpret, organize, and frame unambiguous information about the conflict in a way that sustained conflict-supportive beliefs. The 1956 war in Sinai, for instance, was presented as a preemptive attack on Israel—an inevitable war being the outcome of ongoing terrorist (*fedayyin*) border raids, Egypt's blockade of the Suez Canal, the Egyptian military's threatening development, and Nasser's alliance with the Soviet Union, which allegedly targeted Israel. History textbooks portrayed these events as serious attempts to form a ring of strangulation around Israel by land and sea. It presented the *fedayyin*'s activity as a preparatory action intended to lay the groundwork for a larger Arab operation. In order to reinforce the myth that the Sinai War was a war of "no alternative," these interpretations and inferences disregarded Israel's aggressive policy against Arab infiltration and also concealed its collusion with Britain and France prior to the war.[32]

5. *Omission of contradictory contents.* The Ministry of Education made sure to omit content about events, processes, or individuals that had an evidential basis but contradicted themes of the Zionist narrative supporting the conflict. This method, sometimes called "silence" or "collective amnesia," was mainly used to suppress evidence that contradicted the justness of Israel's goals in conflict and undermined its moral image. Until the 1970s, learning materials made no mention of misdeeds committed by Jews against the Palestinians during the *Yishuv* period (Jewish settlement in Palestine before the establishment

of the state of Israel) or the expulsion of Palestinians during the 1948 war and the destruction of hundreds of their villages. They also omitted the brutal killings of Palestinian civilians by the Israel Defense Force during the 1950s. Two notable examples involve massacres. In the first, which occurred at Qibya in the West Bank, in October 1953, Israeli special units killed approximately seventy civilian residents. In the second, at the Israeli Arab village of Kfar Qasim, in October 1956, Israel's Border Police killed forty-eight Palestinian civilians.[33]

6. *Use of framing language.* Language framing is based on the assumption that language influences how a society perceives reality and triggers existing emotions, memory, cognition, and motivation in a way that fosters continuation of the conflict. This means that the terms, concepts, and wordings that were used in textbooks and learning materials to frame the story of the Zionist narrative imprinted its understanding and impact on students. During the 1950s and 1960s, for example, the Ministry of Education avoided using the term *Palestine* in schools. Instead, the biblical term *Eretz Israel* (Land of Israel), or simply *Israel,* was used in textbooks to ensure that no negative connotation would disrupt the Jewish claim to the land or obscure the area's "true" name. Consistently, the Arab population of Palestine was referred to as "Arabs of Eretz Israel," who were part of the Arab nation, rather than Palestinians. By using these terms, learning materials blurred the distinction between the two national movements and denied Palestinian nationalism.[34]

Likewise, Zionist language framing involves the words used to describe the violent encounters between Arabs and Jews that took place during the Mandate period. The various waves of violence were referred to as *events*—a vague descriptor that stripped these episodes of national meaning. Other terms used to describe these encounters, such as *pogroms*, *riots*, *bloody disturbances*, and the like, were not only devoid of any national connotation but also associated with Jewish victimhood history. These terms established links between these events and other anti-Semitic attacks against the Jews throughout history. The links implied similarity between Arabs and other oppressors who had attacked and massacred Jews in other countries.[35]

In summary, at least six specific methods of constructing societal ignorance were used by the Ministry of Education to ensure the consistency of the Zionist narrative and its themes. To maintain Israel's hegemony and block dissemination of counternarratives, strict monitoring of school textbooks during the 1950s and 1960s ensured high levels of school conformity. Ministry-approved textbooks pervaded the Israeli educational system and presented the national narrative. These policies contributed to students' misperceptions and distorted knowledge regarding Israeli-Palestinian history; in particular, it prevented instilling knowledge and information that would contradict major themes central to the Zionist narrative. The educational policy of the era succeeded in using history to perpetuate conflict-supportive views and ideas.[36]

While there are many instances of the Ministry of Education's use of the methods described above, the best example for illustrative purposes has to do with the Palestinian exodus during the Arab-Israeli War of 1948 and the creation of the Palestinian refugee problem. This also happens to be one of the most controversial issues in the historiography of the Israeli-Palestinian conflict.

Miseducation in Action: The 1948 Palestinian Exodus in Israeli Curricula and School Textbooks

The focal event of the Israeli-Palestinian conflict is the displacement of from 600,000 to 760,000 Palestinians during the 1948 war and the destruction of hundreds of Palestinian villages. This major historical episode, coupled with the question of Israel's responsibility in the creation of the Palestinian refugee problem, carries great political, psychological, and social importance for both parties to this day. The memory of 1948—as passed down from generation to generation—turned the "Palestinian issue" into the central point of contention between Israel and its neighbors.[37]

The Zionist narrative blames the exodus exclusively on the Arabs and denies Israeli responsibility. The Palestinians, it argues, fled willingly, encouraged by their leadership and the Arab states to leave their localities until the end of fighting or, in some cases, because of their fear of the Jews and despite Israeli attempts to persuade them to stay. The narrative does not indicate, and even denies, acts of expulsion by the Jewish or Israeli forces.[38]

Curricula and textbooks written at the climax of the conflict, until around the 1970s, were produced in the context of high intractability and threat of conflict, alongside an intensive international Arab campaign against Israel. In this period, as noted earlier, there was a consensus in Israel among education professionals that schools should serve a national effort to mobilize Israeli youth and protect Israel's international image. Indeed, learning materials used in schools in the 1950s and 1960s presented the Zionist narrative exclusively. They introduced the traditional Israeli position on the Palestinian exodus during the 1948 war, portraying Israel as having had no part in the creation of the Palestinian refugee problem. According to one typical description from a history textbook of that time, "The Arabs fled the country a few weeks prior to the end of the mandate. A mass and panic-stricken flight began. The spirit of the Arab population was broken and they were in a state of utter terror. Destructive and malicious [Arab] propaganda only added fuel to troubled water. The Arabs were deluded into thinking that they would soon return victorious to the country, expel the Jews and seize their goods as spoils of war."[39]

In accordance with the Zionist historical narrative, the "Arabs of Eretz Israel" (not the "Palestinians") were never expelled or forced to leave but rather just "left," "preferred to leave," "fled," "ran away," "evacuated," "escaped," or "abandoned" their homes after seeing their towns and villages fall into Jewish hands in response to the fighting (e.g., in the case of Deir Yassin) or after receiving instructions from their leaders. The educational materials failed to mention cases of evacuation of Arabs by Jewish forces and skipped over the Hagana's "Plan D" (*Tokhnit dalet*), which included the controversial guidelines of destroying Arab villages and expelling the population outside the borders of Israel in cases of resistance.[40]

Instead, to support the notion that the Arabs had not been expelled by Israeli forces, some textbooks presented testimonies of Arabs condemning the Arab states or acknowledging their responsibility for their fleeing. Many others asserted that the Jews had tried in vain to persuade the Palestinians to stay in their localities, as noted also in the following common Zionist narrative: "The Arabs of Eretz Israel declared their obstinate objection to the establishment of a Jewish state. . . . In the dispute about the future of the refugees the Arabs assert that they were forcibly and atrociously expelled from their homes, but this is not the truth. Mainly, in localities which were seized—including in a big city like Haifa—the Jews asked their [Arab] neighbors to stay, but the latter preferred exit to exile, instead of accepting a Jewish rule."[41]

The majority of textbooks did not refer to the number of refugees involved. The few texts that did so gave spuriously low figures. They also denied Israeli responsibility for the refugee problem and condemned Arab states for using the refugees as a bargaining chip to exert pressure on world opinion. Moreover, the texts blamed Arabs for deliberately exacerbating the refugee problem. Arabs were cast as preventing effective humanitarian aid from reaching refugees and denying them official representation in order to turn the human crisis into a weapon against Israel.[42]

Learning materials also presented various historical interpretations colored by Israel's position on the nonreturn of refugees who supported ongoing war efforts. Natanel Lorch, a leading Israeli scholar of the 1948 war, argued that once the refugees returned, they would have only one aim: "to destroy the state of Israel from within. Acting as a fifth column, they will work to undermine the state's security and facilitate the task of the foreign troops who will arrive to annihilate Israel." Furthermore, Lorch suggested that the exodus of the Palestinians, corresponding to the exodus of more than five hundred thousand Jews who had fled the Arab states and been absorbed into the state Israel, was actually nothing more than a reasonable population exchange. As for Palestinian nationhood, the argument continued, Jordan's annexation of the West Bank and Egypt's control over the Gaza Strip had eroded claims about the sovereignty and autonomy of the Palestinian people.[43]

It appears that the national historical narrative regarding the Palestinian exodus and the creation of the refugee problem, as constructed in the educational system until the 1970s, had several social and political goals: to highlight the moral superiority of the Jewish people and of the state of Israel, to reinforce the right of the Jews to the land, and to exempt Israel from responsibility for solving the problem. To achieve these goals and reinforce the themes of the Zionist narrative of the conflict, the Ministry of Education used all the methods of miseducation described earlier: learning materials relied on sources that were consistent with the Israeli reconstruction of events and tragedies; magnified supportive themes of the Israeli narrative, while marginalizing or omitting content considered contradictory to the justness of Israel's goals or its positive image; presented skewed interpretations that sustained themes of the narrative; and used specific terms and concepts that framed the Zionist story of the 1948 war.[44]

From Ignorance and Conflict to the Possibility of Peace

The present chapter describes how ignorance can be produced and regulated in societies engulfed in conflict through the use of formal and informal institutions, channels of communication, and cultural and educational systems. As in the case of Israel, each of these potential avenues of ignorance production and regulation performs important functions in maintenance of conflict-supportive narratives. As a result, many learners and citizens receive systematic indoctrination and exposure to selective, biased, and distorted knowledge of the conflict, the rival, the in-group, and other related subjects.[45]

Educational systems that socialize the young and follow the mandates of the establishment serve a particularly critical role in the social construction of memory, especially during national crises. This chapter situates the construction of Israel's historical conflict-supportive narrative in the nation's schools at the height of the Israeli–Arab/Palestinian conflict.

During that period, Israel's goals for education were to nurture the collective national identity based on the dominant Zionist narrative and to encourage loyalty to and love of the homeland. Accordingly, the educational materials were designed to impart Zionist messages and values and to build the sociopsychological infrastructure that encouraged students to carry on the missions of defending the young state and establishing the Jewish settlement.

As described, various methods of miseducation have been used by the Ministry of Education to construct the historical account and to establish the conflict-supportive themes of the Zionist narrative. The ministry strategically identified certain content as suitable for learning, structured links between different historical periods, and framed historical and political issues to suit the national narrative. The result of these educational policies, practiced mainly during the 1950s and 1960s, was that children in their formative stage of development gained a partial, one-sided knowledge of Israeli-Palestinian history in schools.[46]

These midcentury children became the voters, politicians, and educators of early twenty-first century Israel. They were exposed to an educational system patterned after the Zionist narrative that shaped and imprinted their knowledge about the conflict. They received lessons about the goals of the conflict and the existential threat to the Jewish people of failure. Some of these goals were even viewed as reflecting sacred values. Currently, the majority of Jews in Israel find it difficult to entertain ideas about compromising sovereignty over Jerusalem, which to them is considered "the rock of Jewish existence." Also, for the large portion of them, ceding the Occupied Territories—which they consider to be their ancient homeland—is unthinkable because of the land's sacred status. The midcentury Israeli system of education helped fashion these beliefs that delegitimize Arabs and linger to this day. Israelis learned to view Arabs as violent, untrustworthy people who do not value human life and aspire to exterminate the state of Israel. At the same time, Israeli Jews were taught to consider themselves moral and humane, as well as the sole victims of the conflict.[47]

Despite changes that have taken place in contemporary Israeli education, ongoing public controversy remains over the "space of collective memory" and what should be taught in schools in the interest of the nation. While Israel has begun to turn away from the canonical Zionist narrative, many observers strongly criticize the teaching of new, balanced, and complex narratives. Critics perceive the new narratives as threatening to the national ethos and to the nation's ability to face the challenges posed by conflict.[48]

We recognize that other agents and sources of information, such as family and the media, also play a major role in shaping the way individuals in a society view their nation's past. Individuals can and often do change their beliefs and attitudes as they age. However, schools have a number of characteristics that make them especially meaningful agents in this process. First, they socialize students politically at a very early age, when the identity of children as well as their social and political attitudes, beliefs, and values are formulated and easily influenced by formal agents of the educational system. This influence in the formative years is of special significance in view of the fact that many parents do not consider young children ready for political socialization and therefore do not discuss political issues with them. Furthermore, because of compulsory-attendance laws, schools reach nearly an entire segment of a society (i.e., the entire young generation). As such, schools are often the only social institutions that can formally, intentionally, and extensively achieve the mission of providing knowledge, as they have the authority, the legitimacy, the resources, the methods, and the conditions to do so. Knowledge coming from books and teachers carries official status and is often viewed by children as trustworthy. Finally, children are required to learn the messages and information transmitted in schools. Even if details taught in school are forgotten, their general impressions are coded, stored, and even internalized. This is particularly the case when details are repeated and reinforced by other sources and institutions. Indeed, this happens when political leaders and formal institutions actively propagate the conflict-supportive narratives.[49]

For some, schools lay the foundation for their adult worldview about the Israeli-Palestinian conflict. For others, social and cultural forces reshape and change entrenched attitudes and views. Still, it is important to acknowledge that the knowledge and impressions learned in childhood remained stored in their minds. In fact, longstanding research demonstrates that political knowledge acquired early is not wiped out but is activated unconsciously and automatically even among people who acquire alternative beliefs over time.[50]

In this context, we need to know that embarking on the road of peace-building requires changing the conflict-supportive narratives, with their basic goals, premises, assumptions, and aspirations. Such transformation demands changing fundamental views about the conflict, the goals of the group, the rival and rival relationships, the image of self, the accepted assumptions about the past, and so on. The new repertoire should include ideas about the need to resolve the conflict peacefully, to personalize and legitimize the opponent, to change views of the conflict as being unsolvable, to decrease ambitions that fuel conflict, to develop peaceful relationships with the rival, and to build trust between parties. It should promote beliefs that peace agreements can be successfully implemented, as well as recognition of the need to construct a new climate that promotes peace-building. These ideas have to be adopted by society members, who must mobilize other society members to advance the peace process.[51]

This is a major challenge, because societies facing persistent conflict have to contend with populations who firmly believe that conflict-supportive narratives are accurate and trustworthy. The dominance of conflict-supportive narratives, therefore, has significant consequences for the future of any society that maintains them. They play a central role in perpetuating the conflict, feeding its continuation, and preventing its peaceful resolution.

The case of the Israeli-Palestinian conflict demonstrates that societies involved in harsh and violent conditions face significant barriers to the development of accurate historical self-assessment. Education, and historical instruction in particular, as this chapter argues, is central to societies facing intractable conflict and aspiring to peace.

NOTES

1. Arie Kizel, *Subservient History: A Critical Analysis of History Curricula and Textbooks in Israel, 1948–2006* [in Hebrew] (Tel Aviv: Mofet, 2008); history curricula for 1954–55, quoted in ibid., 26; Yehoshua Mathias, "The Nationalization of Education: History in State Education," in *History, Identity, and Memory: Images of the Past in Israeli Education*, ed. Avner Ben-Amos [in Hebrew] (Tel Aviv: Ramot, 2002), 15–65; Michael Vizinger (Ziv), "Improving the History Curriculum," in *Educating the Citizen*, ed. Haim Yehuda Ruth (Jerusalem: Magnes, 1950), 116; Michael Ziv, *Teaching History in the Classroom: Methods and Trends* (Tel Aviv: Urim, 1957), 16. For more on Israeli educational system's nationalistic attitude toward teaching history, see Elie Podeh, *The Arab-Israeli Conflict in Israeli History Textbooks, 1948–2000* (Westport, CT: Bergin & Gavey, 2002), 21–26.

2. Unless otherwise specified, the term *Israelis* will refer to Israeli Jews.

3. See David Coulby, "European Curricula, Xenophobia and Warfare," *Comparative Education* 33 (1997): 29–41; Robert D. Hess and Judith V. Torney, *The Development of Political Attitudes in Children* (Chicago: Aldine, 1967); Klaus Schleicher, ed., *Nationalism in Education* (Frankfurt am Main: Peter Lang, 1993); Judith Torney-Purta, John Schwille, and Jo-Ann Amadeo, eds., *Civic Education across Countries: Twenty-four National Case Studies from the IEA Civic Education Project* (Amsterdam: IEA, 1999). Daniel Bar-Tal counts several distinguishable features of intractable conflicts: they are fought over contradictory goals that are viewed as being existential; they are violent; they are perceived by the involved parties as being of a zero-sum nature and irresolvable; they deeply involve the engaged societies, which in turn invest resources in their continuation; and they last at least a generation. See Bar-Tal, "Societal Beliefs in Times of Intractable Conflict: The Israeli Case," *International Journal of Conflict Management* 9 (1998): 22–50; and Bar-Tal, *Intractable Conflicts: Socio-Psychological Foundations and Dynamics* (Cambridge: Cambridge University Press, 2013). For a similar characterization, see Louis Kriesberg, "Intractable Conflict," *Peace Review* 5 (1993): 417–21; and Kriesberg, "Nature, Dynamics, and Phases of Intractability," in *Grasping the Nettle: Analyzing Cases of Intractable Conflict*, ed. Chester A. Crocker, Fen O. Hampson, and Pamela Aall (Washington, DC: United States Institute of Peace Press, 2005), 65–97. In order to meet the challenges of living under these harsh conditions, societies develop a repertoire of functional beliefs, attitudes, emotions, values, motivations, norms, and practices, all of which gradually crystallize into a well-organized system of societal shared beliefs, attitudes, and emotions that penetrates institutions and communication channels of the society. See Daniel Bar-Tal, *Living with the Conflict: Socio-Psychological Analysis of the Israeli-Jewish Society* [in Hebrew] (Jerusalem: Carmel, 2007). Of special importance in this system are conflict-supportive collective narratives that provide explanation, rationalization, and justification for the eruption of the conflict and its continuation. In order to fulfill these roles, these narratives tend to be simplistic and partial in their depiction of reality and history so as to portray the society itself positively and its rivals negatively. In support of their interests, societies engaged in intractable conflicts strive to maintain the hegemony of their official memory of conflict among in-group members and put particular emphasis on imparting their conflict-supportive narratives to their young members through education.

4. Bar-Tal, *Intractable Conflicts*. See also, e.g., Sami Adwan, Daniel Bar-Tal, and Bruce Wexler, "Portrayal of the Other in Palestinian and Israel Schoolbook: A Comparative Study," *Political Psychology (2014): doi: 10.1111/pops.12227*; Ruth Firer and Sami Adwan, *The Israeli-Palestinian Conflict in History and Civics Textbooks of Both Nations* (Hannover, Germany: Georg Eckert Institute for International Textbook Research, 2004); and Robert I. Rotberg, ed., *Israeli and Palestinian Narratives of Conflict: History's Double Helix* (Bloomington: Indiana University Press, 2006).

5. Bar-Tal, *Living with the Conflict*. For a historical overview of the Israeli-Palestinian conflict, see Benny Morris, *Righteous Victims: A History of the Zionist-Arab Conflict, 1881–2001* (New York: Vintage Books, 2001); and Mark Tessler, *A History of the Israeli-Palestinian Conflict*, rev. ed. (Bloomington: Indiana University Press, 2009).

6. See Kizel, *Subservient History*; Ruth Firer, *The Agents of Zionist Education* [in Hebrew] (Tel Aviv: Hakibutz Hameuchad, Sifryat Hapoalim, 1985); Mathias, "Nationalization of Education"; Podeh, *Arab-Israeli Conflict*; Adwan, Bar-Tal, and Wexler, "Portrayal of the Other"; Firer and Adwan, *Israeli-Palestinian Conflict*; Eyal Nave and Esther Yogev, *Histories: Towards a Dialogue with Yesterday*, ed. Dafna Danon [in Hebrew Edition] (Tel Aviv: Bavel, 2002); Esther Yogev, "A Crossroads: History Textbooks and

Curricula in Israel," *Journal of Peace Education* 7 (2010): 1–14; and Neta Oren, Rafi Nets-Zehngut, and Daniel Bar-Tal, "Construction of the Israeli-Jewish Conflict Supportive Narrative and the Struggle over Its Dominance: A Case Study," *Political Psychology* 36 (2015): 215–30.

7. Bar-Tal, *Living with the Conflict.*

8. Shlomo Avineri, *The Making of Modern Zionism: The Intellectual Origins of the Jewish State* (London: Weidenfeld & Nicolson, 1981); Gideon Shimoni, *The Zionist Ideology* (Hanover, NH: University Press of New England, 1995); Neta Oren, "The Israeli Ethos of Conflict 1967–2005" (working Paper #27, Institute for Conflict Analysis and Resolution, George Mason University, Fairfax, VA, 2009).

9. Charles S. Liebman and Eliezer Don-Yehiya, *Civil Religion in Israel: Traditional Judaism and Political Culture in the Jewish State* (Berkeley: University of California Press, 1983). See, Daniel Bar-Tal, Dan Jacobson, and Aharon Klieman, eds., *Security Concerns: Insights from the Israeli Experience* (Stamford, CT: JAI, 1998).

10. Daniel Bar-Tal and Phillip L. Hammack, "Conflict, Delegitimization, and Violence," in *The Oxford Handbook of Intergroup Conflict*, ed. Linda R. Tropp (New York: Oxford University Press, 2012), 30; Daniel Bar-Tal and Yona Teichman, *Stereotypes and Prejudice in Conflict: Representation of Arabs in Israeli Jewish Society* (New York: Cambridge University Press, 2005); Robert R. Holt and Brett Silverstein, "On the Psychology of Enemy Images: Introduction and Overview," *Journal of Social Issues* 45 (1989): 1–11; Herbert C. Kelman, "The Role of National Identity in Conflict Resolution," in *Social Identity, Intergroup Conflict, and Conflict Reduction*, ed. Richard D. Ashmore, Lee Jussim, and David Wilder (New York: Oxford University Press, 2001), 187–212; Susan Opotow, "Moral Exclusion and Injustice: An Introduction," *Journal of Social Issue* 46 (1990): 1–20.

11. See, e.g., Roy F. Baumeister and Stephen Hastings, "Distortions of Collective Memory: How Groups Flatter and Deceive Themselves," in *Collective Memory of Political Events*, ed. James W. Pennebaker, Dario Paez, and Bernard Rim (Mahwah, NJ: Erlbaum, 1997), 277–93; Gerald N. Sande, George R. Goethals, Lisa Ferrari, and Leila T. Worth, "Value-Guided Attributions: Maintaining the Moral Self-Image and the Diabolical Enemy Image," *Journal of Social Issues* 45 (1989): 91–118; Ralph K. White, *Nobody Wanted War: Misperception in Vietnam and Other Wars* (Garden City, NY: Anchor Books, 1970); and Phillip L. Hammack, *Narrative and the Politics of Identity: The Cultural Psychology of Israeli and Palestinian Youth* (New York: Oxford University Press, 2011).

12. Daniel Bar-Tal, Lily Chernyak-Hai, Noa Schori, and Ayelet Gundar, "A Sense of Self-Perceived Collective Victimhood in Intractable Conflicts," *International Review of the Red Cross* 91 (2009): 229–77; J. E. Mack, "The Psychodynamics of Victimization among National Groups in Conflict," in *The Psychodynamics of International Relationships*, ed. Vamik D. Volkan, Demetrios A. Julius, and Joseph V. Montville (Lexington, MA: Lexington Books, 1990), 119–29; Johanna Ray Vollhardt, "Collective Victimization," in Tropp, *Handbook of Intergroup Conflict*, 136–57; Daniel Bar-Tal and Dikla Antebi, "Siege Mentality in Israel," *International Journal of Intercultural Relations* 16 (1992): 251–75.

13. Yael Zerubavel, *Recovered Roots: Collective Memory and the Making of Israeli National Tradition* (Chicago: University of Chicago Press, 1995); Avner Ben-Amos and Daniel Bar-Tal, eds., *Patriotism: Homeland Love* [in Hebrew] (Tel Aviv: Hakibbutz Hameuhad, 2004).

14. Pnina Lahav, "The Press and National Security," in *National Security and Democracy*, ed. Avner Yaniv (Boulder, CO: Lynne Rienner, 1993), 173–95; Moshe Negbi, *Paper Tiger: The Struggle for Press Freedom in Israel* [in Hebrew] (Tel Aviv: Sifriyat Hapoalim, 1985).

15. Bar-Tal and Teichman, *Stereotypes and Prejudice.*

16. For examples from around the world, see David Bruce MacDonald, *Balkan Holocausts? Serbian and Croatian Victim-Centered Propaganda and the War in Yugoslavia* (New York: Manchester University Press, 2002); Nikki R. Slocum-Bradley, "Discursive Production of Conflict in Rwanda," in *Global Conflict Resolution through Positioning Analysis*, ed. Fathali M. Moghaddam, Rom Harré, and Naomi Lee (New York: Springer, 2008), 207–26; Yiannis Papadakis, "Greek Cypriot Narratives of History and Collective Identity: Nationalism as a Contested Process," *American Ethnologist* 25 (1998): 149–65; and Papadakis, *History Education in Divided Cyprus: A Comparison of Greek Cypriot and Turkish Cypriot Schoolbooks on the "History of Cyprus,"* PRIO Report 2/2008 (Oslo: International Peace Research Institute, 2008).

17. Daniel Bar-Tal, Keren Sharvit, Eran Halperin, and Anat Zafran, "Ethos of Conflict: The Concept and Its Measurement," *Peace and Conflict: Journal of Peace Psychology* 18 (2012): 40–61; John W. Burton, ed., *Conflict: Human Needs Theory* (New York: St. Martin's, 1990). For examples of how conflict-supportive narratives fulfill these various functions, see David E. Apter, ed., *Legitimization of Violence* (New York: New York University Press, 1997); John T. Jost and Brenda Major, eds., *The Psychology of Legitimacy: Emerging Perspectives on Ideology, Justice, and Intergroup Relations* (New York: Cambridge University Press, 2001); Michael A. Hogg and Dominic Abrams, "Social Motivation, Self-Esteem and Social Identity," in *Social Identity Theory: Constructive and Critical Advances*, ed. Dominic Abrams and Michael A. Hogg (New York: Springer-Verlag, 1990), 28–47; Henry Tajfel and J. C. Turner, "The Social Identity Theory of Intergroup Relations," in *Psychology of Intergroup Relations*, ed. William G. Austin and Stephen Worchel (Chicago: Nelson-Hall, 1986), 7–24; Jim Sidanius and Felicia Pratto, *Social Dominance: An Intergroup Theory of Social Hierarchy and Oppression* (New York: Cambridge University Press, 1999); Nave Zohar, "The Image of the Arab as Presented in Textbook Readers" [in Hebrew] (master's thesis, Tel Aviv University, 1972); Aaron Antonovsky, *Unraveling the Mystery of Health: How People Manage Stress and Stay Well* (San Francisco: Jossey-Bass, 1987); Richard S. Lazarus and Susan Folkman, *Stress, Appraisal, and Coping* (New York: Springer, 1984); Daniel Bar-Tal and Ervin Staub, *Patriotism in the Lives of Individuals and Nations* (Chicago: Nelson-Hall, 1997); Eric Langenbacher, "Collective Memory as a Factor in Political Culture and International Relations," in *Power and the Past: Collective Memory and International Relations*, ed. Eric Langenbacher and Yossi Shain (Washington, DC: Georgetown University Press, 2010), 13–49.

18. Yehudith Auerbach, "National Narratives in a Conflict of Identity," in *Barriers to Peace in the Israeli-Palestinian Conflict*, ed. Yaacov Bar-Siman-Tov (Jerusalem: Jerusalem Institute for Israel Studies, 2010), 99–134; Baumeister and Hastings, "Distortions of Collective Memory"; Barbara Tint, "History, Memory, and Intractable Conflict," *Conflict Resolution Quarterly* 27 (2010): 239–56.

19. Daniel Bar-Tal, Neta Oren, and Rafi Nets-Zehngut, "Sociopsychological Analysis of Conflict-Supporting Narratives: A General Framework," *Journal of Peace Research* 51 (2014): 662–75; George H. Richardson, *The Death of the Good Canadian: Teachers, National Identities, and the Social Studies Curriculum* (New York: Peter Lang, 2002); Schleicher, *Nationalism in Education.*

20. Firer, *Agents of Zionist Education*; Kizel, *Subservient History*; Yogev, "Crossroads"; Zon's papers 6.29/21, no date given, as quoted in Podeh, *Arab-Israeli Conflict*, 23. For a similar view, see Ziv, *Teaching History in the Classroom*, 13–14.

21. The following comments made by Minister of Education Dinur in 1953 demonstrate just how dominant these themes of the conflict-supportive narrative were

in Israeli society and in the educational system in particular: "The position of our country must form the underlying premise of the civil education system. . . . We cannot stop preparing for the war of destruction planned against us. We are surrounded by enemies whom we fought during the War of Independence and who have yet to reconcile themselves to our existence. Meanwhile, they have grown even stronger. . . . They announced to the world their intention to throw us into the sea and erase our name from the map. . . . Consequently, the principle of defense, ensuring that we are able to withstand our enemies, is our primary civil duty." Quoted in Podeh, *Arab-Israeli Conflict*, 31–32. Understandably, as Podeh mentions, any other approach to teaching the history of the Israeli-Arab conflict was unthinkable at the time. See Bar-Tal, "Societal Beliefs"; Firer, *Agents of Zionist Education*; and Podeh, *Arab-Israeli Conflict*. For additional examples see, Yoram Bar-Gal, *Moledet and Geography in One Hundred Years of Zionist Education* [in Hebrew] (Tel Aviv: Am Oved, 1993); Y. Bezalel, "Changes in Zionist Values as Reflected in Elementary School Literature and Language Readers from the Late Fifties to the Mid-Eighties" [in Hebrew] (master's thesis, Ben Gurion University of the Negev, 1989); Ohad David, *A Nation's Statues: Jewish-Israeli Identity in Readers of Twentieth Century* [in Hebrew] (Tel Aviv: Dor Le'Dor, no. 41, 2012); and Zohar, "Image of the Arab."

22. Yehoshua Mathias, "The Thorny Way to Recognition: Palestinians and Arabs in Israeli Curriculum," in *Contested Past, Disputed Present: Curricula and Teaching in Israeli and Palestinian Schools*, ed. Falk Pingel (Hannover, Germany: Verlag Hahnsche Buchhandlung, 2003), 29–58; Podeh, *Arab-Israeli Conflict*.

23. Podeh, *Arab-Israeli Conflict*, 80.

24. Bar-Tal and Teichman, *Stereotypes and Prejudice*; Firer, *Agents of Zionist Education*; Podeh, *Arab-Israeli Conflict*; Zohar, "Image of the Arab."

25. Shlomo Horowitz, *Chronicles of Our Times, 1918–1950*, rev. ed. (Tel Aviv: Chechik, 1953), quoted in Podeh, *Arab-Israeli Conflict*, 98–99.

26. Podeh, *Arab-Israeli Conflict*, 95.

27. Ibid., 102–3.

28. Avner Ben-Amos and Ilana Beth-El, "Commemoration and National Identity: Memorial Ceremonies in Israeli Schools," in *Between "I" and "We": The Construction of Identities and Israeli Identity*, ed. Aazmi Bishara [in Hebrew] (Jerusalem: Van Leer, Hakibbutz Hameuchad, 1999), 129–51.

29. Firer, *Agents of Zionist Education*, 127–28.

30. Podeh, *Arab-Israeli Conflict*, 97–98.

31. Ibid., 106.

32. Ibid., 110–12.

33. Baumeister and Hastings, "Distortions of Collective Memory"; Nazan Maksudyan, "Walls of Silence: Translating the Armenian Genocide into Turkish and Self-Censorship," *Critique* 37 (2009): 635–49; Tint, "History, Memory, and Intractable Conflict"; Jay Winter, "Thinking about Silence," in *Shadows of War: A Social History of Silence in the Twentieth Century*, ed. Efrat Ben-Ze'ev, Ruth Ginio, and Jay Winter (Cambridge: Cambridge University Press, 2010), 3–31; Rafi Nets-Zehngut, "Israeli Approved Textbooks and the 1948 Palestinian Exodus," *Israel Studies* 18 (2013): 41–68; Podeh, *Arab-Israeli Conflict*.

34. Ageneza Bozic-Roberson, "Words before War: Milosevic's Use of Mass Media and Rhetoric to Provoke Ethnopolitical Conflict in Former Yugoslavia," *East European Quarterly* 38 (2004): 395–408; Winter, "Thinking about Silence." For the Israeli case, see Nadir Tsur, "Vocabulary and the Discourse on the 1967 Territories," in *The Impacts*

of Lasting Occupation: Lessons from Israeli Society, ed. Daniel Bar-Tal and Izhak Schnell (New York: Oxford University Press, 2013), 471–506; and Podeh, *Arab-Israeli Conflict*, 92.

35. Podeh, *Arab-Israeli Conflict*, 94.

36. Firer, *Agents of Zionist Education*; Podeh, *Arab-Israeli Conflict*.

37. Some have used the term *chosen trauma* to describe the Palestinian displacement, referring to Vamik D. Volka's definition in *Bloodlines: From Ethnic Pride to Ethnic Terrorism* (New York: Farrar, Straus & Giroux, 1997). On the centrality of the 1948 Palestinian exodus and the Palestinian refugee problem to the Israeli-Palestinian conflict, see, e.g., Benny Morris, *1948 and After: Israel and the Palestinians* (Oxford: Oxford University Press, 1994); and Nets-Zehngut, "Israeli Approved Textbooks."

38. Bar-Tal, *Living with the Conflict*; Rafi Nets-Zehngut, "Origins of the Palestinian Refugee Problem: Changes in the Historical Memory of Israelis/Jews, 1949–2004," *Journal of Peace Research* 48 (2011): 235–48; Nets-Zehngut, "Israeli Approved Textbooks."

39. Nets-Zehngut, "Israeli Approved Textbooks"; Yitzhak Spivak and Moshe Avidor, *The Jewish Nation in its Homeland and in Diaspora: From the Independence of the United States until the Establishment of the State of Israel* (Tel Aviv: Yavne, 1960), 217–19, quoted in Podeh, *Arab-Israeli Conflict*, 105.

40. Firer, *Agents of Zionist Education*, 133; Nets-Zehngut, "Israeli Approved Textbooks." It is noteworthy, as Podeh mentions in *The Arab-Israeli Conflict*, 110, that these omissions were not the result of educational scholars' and textbooks authors' ignorance, because information about the expulsion of Arabs by Jewish forces was known at least in part at the time.

41. Firer, *Agents of Zionist Education*, 133; Podeh, *Arab-Israeli Conflict*, 107; Michael Ziv, *The State of Israel and the Israeli Diaspora in our Generation* (Haifa: Yuval, 1964), 47–48, quoted in Nets-Zehngut, "Israeli Approved Textbooks," 50.

42. Elie Podeh, "History and Memory in the Educational System: The Israeli-Arab Conflict in History Textbooks in Israel, 1948–2000," in Avner Ben-Amos, *History, Identity, and Memory*, 87; Podeh, *Arab-Israeli Conflict*, 107.

43. Natanel Lorch, "Israel in the International Arena," in *The State of Israel and the Modern-Day Jewish Communities*, ed. Michael Ziv, Lessons for High Schools, rev. ed. (Haifa: Yuval, 1966), 50–51, quoted in Podeh, *Arab-Israeli Conflict*, 106–7; Podeh, *Arab-Israeli Conflict*, 107.

44. Firer, *Agents of Zionist Education*; Podeh, *Arab-Israeli Conflict*.

45. It is important to recognize that ignorance in this case, as Proctor distinguishes in his agnotology categories, is deliberately constructed. Holding back certain types of information while ignoring possible undesired perspectives allows societies involved in intractable conflict to construct the historical stories they chose to tell about themselves. These narratives serve different functions, such as providing a meaningful picture of the conflict situation, justifying the behavior of the society, facilitating mobilization for participation in the conflict, enabling maintenance of a positive social identity and self-collective image, and so on.

46. In this respect, agnotology as a framework helps to illuminate an important area in the study of Israeli history instruction, that is, those facts and events that for decades have been excluded or occluded, resulting in a public construction of ignorance in these matters.

47. On the concept of sacred values in this context, see Philip E. Tetlock, "Thinking the Unthinkable: Sacred Values and Taboo Cognitions," *Trends in Cognitive Science* 7 (2003): 320–24. Daniel Bar-Tal, Eran Halperin, and Neta Oren elaborate on these goals

and values in their "Socio-Psychological Barriers to Peace Making: The Case of the Israeli Jewish Society," *Social Issues and Policy Review* 4 (2010): 63–109.

48. Podeh, *Arab-Israeli Conflict*; Yogev, "Crossroads."

49. John C. Coleman, *The School Years: Current Issues in the Socialization of Young People* (London: Routledge, 1979); Jan W. van Deth, Simone Abdenschön, and Meike Vollmar, "Children and Politics: An Empirical Reassessment of Early Political Socialization," *Political Psychology* 32 (2011): 147–74.

50. Patricia G. Devine, "Stereotypes and Prejudice: Their Automatic and Controlled Components," *Journal of Personality and Social Psychology* 56 (1989): 5–18; John F. Dovidio, Nancy Evans, and Richard B Tyler, "Racial Stereotypes: The Content of their Cognitive Representations," *Journal of Experimental Social Psychology* 22 (1986): 22–37.

51. Michelle I. Gawerc, "Peace-Building: Theoretical and Concrete Perspectives," *Peace and Change* 31 (2006): 435–78; Bar-Tal, *Intractable Conflicts*.

14

China

DONGPING HAN AND STEPHEN SAMUEL SMITH

On several occasions in the mid-1960s a former middle school teacher and primary school principal in China took his nephew aside to share some poignant ideas about education. This occurred in the years leading up to and during China's Cultural Revolution, one of world history's most dramatic revolutionary struggles. The former educator asked, "All this attending classes and discussing things, what is the use of it?" He told his nephew to instead "go study reality," especially the reality for which "very many members of our family have given up their lives, killed by the Kuomintang and the American imperialists. You grew up eating honey, and thus far you have never known suffering. . . . What can you learn merely by relying on listening to lectures? The most important thing is to go and learn from practice."[1]

According to the uncle, this nephew had much more to gain from visiting the rural areas, in particular, where he could learn from peasants. The former teacher and administrator had been "deeply struck by how many things they knew." But first and foremost, the nephew needed to understand that "ideology comes first," before "real knowledge," and that class struggle was the most important subject of all. "Only when you have considered such a course of political training," the uncle emphasized, "can I consider you a graduate." The nephew eventually became a close confidant of the uncle, and the uncle, Mao Zedong, took those ideas with him into the pages of history as the leading figure of the Cultural Revolution.

It's difficult to overstate the significance of education to the Cultural Revolution, and Mao's conversations with his nephew provide some insight into what he had in mind. For Mao, the educational system had reached a significant limit. It was perpetuating inequality—rather than reducing it—by providing greater educational opportunities for those students whose families were better educated and had more of what is generally called cultural capital.[2] By so doing, it obscured mass suffering and poverty, masked class struggle, and contributed to an understanding of the past, both national and personal, that prevented the forming of identity and, more importantly, an understanding of ideology. What Mao wanted was a system of education that would alleviate inequality and reveal the political realities of the mid-twentieth century through the lens of class struggle and the ideologies relevant to understanding the relationships between wealth, poverty, and power. Based on Mao's views of education during the Cultural Revolution, China underwent one of the largest educational reform efforts ever attempted. Western observers and scholars immediately commented on this reform and the broader revolutionary context as unparalleled in history and on the difficulty of assessing its motives and potential outcomes.

In light of the significance of education to Mao, the Cultural Revolution, and how Westerners have learned and responded to it, the present chapter offers a brief exploration of this provocative and contested episode in global history. It does so in relation to agnotology and education, but with the goal of raising specific research questions rather than offering a formal case study. The relevant historical, political, and educational literature is vast, and the questions it raises are contentious. But recent work in the social construction of ignorance promises to help untangle distortions in knowledge and understanding about the Cultural Revolution as they occurred in China and abroad. This is particularly important when considering the way historians often define education as encompassing all of the ways a culture transmits itself across generations. Here the cultural transmission process entered a period of revolution and became enmeshed in patterns of ignorance-making.

What follows is a series of questions, contextualized and problematized, for future historians, Sinologists, and educational researchers to use to begin this process. The chapter offers a call for further research on how the production, maintenance, and regulation of ignorance might intersect with this rich, complex history. A formal historical case study is beyond the scope of this chapter. Rather, the goal is to highlight possible avenues of research on the social construction of ignorance in relation to the Cultural Revolution from three main perspectives: (1) education in China during the Cultural Revolution; (2) education in China after the Cultural Revolution about the revolution itself; and (3) education, popular works, and research about the Cultural Revolution in Western contexts during and after the revolution.

Education in China during the Cultural Revolution

One of the earliest events of the Cultural Revolution occurred in May 1966, when students and faculty at Beijing University posted on university walls a *dazibao* (big-character poster) condemning the university president for suppressing discussion of a literary and political controversy.[3] The poster summoned all "revolutionary intellectuals" to "go into battle."[4] Although punished by local party leaders, these revolutionaries subsequently received a ringing endorsement from Mao, the chairman of the central committee of the Chinese Communist Party (CCP) and the person who had led the successful battle to unify China and free it from foreign domination. That endorsement and other kinds of support from him quickly led to the organization throughout China of student groups, generally called the Red Guards, who launched attacks, both political and physical, against educational administrators, teachers, local party leaders, and, very frequently, other Red Guard groups.

From this perspective, the first salvo of the Cultural Revolution was fired over both agnotology and education. The university administration, in this case, wanted to suppress discussion and the exchange of what it considered to be disruptive ideas. Students and faculty wanted to fight for the right to learn and interact with suppressed content. As such, the Cultural Revolution was born in part out of a struggle between supporters and opponents of the decision to maintain and manage the production of ignorance over dissenting views and opinions.

This leads us to the first question: *To what extent did this agnotological suppression of literary and political controversy in Chinese education provide a significant impulse for the launching of the Cultural Revolution?*

For many Western observers, Chinese higher education mattered little, if at all. They tended to view events as spiraling out of control because of a power struggle within the top echelons of the Chinese Communist Party. In the mid-1960s Mao was immensely popular among the Chinese people as the person who had led the Chinese Revolution. But within the leading bodies of the party he thought he was treated as a dead ancestor, someone admired but ignored when it came to important decisions.[5] Underlying the power struggle were sharp and crucial differences over China's political and economic trajectory. In Mao's view, China was in danger of veering off the socialist road because of growing inequality, the use of material rather than ideological incentives, the increased bureaucratization of China's major institutions, the loss of revolutionary spirit, and the incipient development of a new capitalist class. To be sure, very little capital—as Marxists had historically understood the term—remained in private hands in China by the mid-1960s. But in Mao's view, leaders of a bureaucratized communist party posed an ongoing danger of becoming a capitalist class. In other words, while they lacked capital in the traditional Marxist understanding of the term, they had a large stock of what might be called political capital, which they could use to further concentrate power in pursing their own interests rather than those of the people. The struggle to maintain and develop a socialist worldview and socialist ideology at the highest levels was as crucial, if not more so, as the struggle to maintain it throughout society.

For other observers, however, particularly those closest to the revolution, the Beijing University controversy was a catalyzing event and an important expression of broader cultural and political transmission processes. In explaining this view to a foreign delegation, Mao distinguished between the main task of the Cultural Revolution and its goal. The task was to struggle against power holders who took the capitalist road, but that was "by no means the goal. The goal is to solve the problem of world outlook. . . . The masses must educate themselves and liberate themselves . . . because world outlook cannot be imposed on them. . . . If the world outlook is not transformed, then although there are 2,000 power holders taking the capitalist road in this Great Cultural Revolution, there may be 4,000 next time."[6]

Mao's emphasis on using education at all levels to transform world outlook—by which he meant the replacement of capitalist and precapitalist ideologies with socialist ones—explains why the upheaval was called a *cultural* revolution. Given the importance of an educational system in transmitting, developing, and inculcating cultural understandings and ideology, it is hardly surprising that China's educational system would be the arena in which the Cultural Revolution initially unfolded and in which many key battles were fought. To revolutionaries, the masses who had the most to gain from educating and liberating themselves were students. In this light, Mao's support for the Beijing students who criticized the university president and Mao's subsequent encouragement and embrace of the Red Guards merit closer attention and further inquiry.

Consider, for instance, the way students' criticism of the president of Beijing University opened the door to challenging the construction of ignorance resulting from the heavy hand of local authorities and lower-level party officials. As a student at Tsinghua University, China's leading technical school, remarked many years later in explaining why he had supported the Cultural Revolution, "Before the Cultural Revolution, everything came down from above, one level at a time. You had to listen to those right above you. Then all of the sudden, Mao went around the hierarchy and told the masses that the people between him and them had problems; that they should not listen to them. 'Think for yourselves.' That was the first time we had room to think for ourselves. That's why we supported the Cultural Revolution."[7]

The revolt against ignorance-making spread from universities to nonacademic life. Two Americans teaching in China recalled how, in the early days of the Cultural Revolution, Beijing had become

> a political festival of the masses. . . . And it was not only the students who participated in this orgy of writing and reading. . . . The population of a country which only twenty years before had been 80 percent illiterate conducting a national debate through the written word. . . . The formidable organization of the Chinese Communist Party . . . had been suddenly overturned and replaced by a communications and organizational network which embraced millions of ordinary citizens in a decision-making apparatus of their own. In the evenings, thousands of mass meetings occurred simultaneously throughout the capital. There the latest political developments were discussed, analyzed, and acted upon.[8]

These accounts indicate that students, higher education, and disputes over agnotological processes had much to do with the mobilizing of various segments of Chinese society. The widespread appeal generated enthusiasm for removing obstacles to free inquiry and launched a struggle that had global implications. But the concern over prerevolutionary repression and ignorance-making as catalysts for change prompts a second question: *What role did agnotological processes play in the Cultural Revolution?* In other words, did one system of agnotology and education replace another?

Numerous accounts describe how the Cultural Revolution mobilized the Chinese to struggle against the four "olds": old ideas, old customs, old culture, and old habits.[9] This led to the ravaging of museums, the destruction of books and important cultural artifacts, and the frequently violent persecution of intellectuals, the very segment of society whose presumed mission was to challenge the manufacture and perpetuation of ignorance. All of this constitutes not only what agnotologists call the social construction of ignorance, but also, in many cases, its strategic and coerced construction.

Ignorance-making during the Cultural Revolution also took the form of adulation and veneration of Mao as an uncontestable figure. There are parallels between the intellectual stultification imposed by the prerevolutionary bureaucracy and an equally great stultification imposed by Mao's power; the frequent reduction of his complex political thinking to a series of quotations in the widely distributed Red Book; and the widespread invocation of him as the "Great Helmsman," "the Bright Sun Shining in Our Hearts," and so forth. As one scholar has recently noted, as the Cultural Revolution unfolded, Mao's power was such that "he could change the course of events by uttering a few words."[10]

Agnotological processes were also manifest in the ambiguous nature of a key term in Cultural Revolution discourse: *class*. As might be expected from Mao's admonishments to his nephew, one of the main slogans of the Cultural Revolution was "Never Forget Class Struggle." If Mao's early works, such as his 1926 article "Analysis of the Classes in Chinese Society," are any guide, in his view it was of prime importance to distinguish sharply between the relatively few enemies of the revolution and its many actual and potential supporters. The article begins, "Who are our enemies? Who are our friends? This is a question of the first importance for the revolution."[11] It goes on to provide a detailed analysis of classes in Chinese society based on the largely objective and traditional Marxist criteria of their relation to the means of production. Drawing on that analysis, the essay specifies the extent to which each class is an agent of the revolution (e.g., poor peasants) or an obstacle to it (e.g., landlords). Upon Mao's subsequently assuming leadership of the Chinese Communist Party, that analysis proved crucial to the victory of the Chinese Revolution in 1949.[12]

Maoist guidelines for the Cultural Revolution—appearing in a document generally called the "Sixteen Points"—also ask, "Who are our enemies? Who are our friends? This is a question of the first importance for the revolution."[13] However, delineating class lines in a socialist society is more difficult than doing so for prerevolutionary China, and the Sixteen Points lacked a class analysis as detailed or as grounded in a largely objective relation to the means of production as that contained in Mao's 1926 article.[14] That absence made it easy for contending factions in the Cultural Revolution to view one another as enemies, which contributed to violence and ignorance on behalf of a struggle that at its inception challenged a previous form of violence and ignorance.

Recent scholarship and memoirs, however, have begun to identify the standard account of the Cultural Revolution as itself a product of what could be called passive agnotological processes.[15] This has occurred as scholarly attention has shined light on the themes of violence and ignorance, while leaving in the shadows counternarratives that complicate the dominant interpretation. Some new lines of thinking suggest that too much power and credence have been attributed to Mao himself. Others have written that personal experiences with the Cultural Revolution were neither violent nor actively ignorance-producing.

If, for instance, Mao could "change the course of events" with just a few words, then scholars have the opportunity to test the efficacy of calls to nonviolence that came from the leader and his circle. The Sixteen Points emphasized that, generally speaking, "the method to be used in debates is to present the facts, reason things out, and persuade through reasoning. . . . When there is a debate, it should be conducted by reasoning, not by coercion or force." Those guidelines notwithstanding, there was widespread violence as the Cultural Revolution unfolded in China's educational institutions. As such, the relationship between Mao and his followers was likely more tenuous than imagined, and calls to action that came from above took on a life of their own once they reached those struggling on the ground level. In short, the Cultural Revolution was interpreted in varied ways depending on specific contexts and individuals across China's vast landmass.[16]

This ambiguity takes on a special salience when considering how the Cultural Revolution unfolded in many of China's schools and universities. In the ideological contest that developed, what is often called "bloodline theory" allowed the student offspring of party bureaucrats to claim that they were more revolutionary than other student groups. This claim helped them mobilize support to protect their parents from other student groups that, in accord with Mao's view, saw these bureaucrats as targets of the Cultural Revolution. Moreover, while all groups in the Cultural Revolution engaged in violence, it was initially largely instigated, as Maurice Meisner notes, "by Party-organized 'rebels' intent on protecting Party bureaucrats from Maoist assaults."[17]

Education in China after the Cultural Revolution about the Cultural Revolution

After the defeat of the Cultural Revolution and Mao's death, the CCP and the Chinese state, initially under the leadership of Hua Guofeng and then under that of Deng Xiaoping, developed an officially sanctioned narrative that was backed by the arrest of four leaders of the Cultural Revolution, including Mao's wife, as well the purging and/or arrest of many of their supporters. The narrative's goal was to manage the interpretation and reception of Mao's legacy, depicting the former leader as someone who deserved credit and adulation for leading the Chinese Revolution but also immense blame for what was viewed as the ten-year national holocaust and national disaster of the Cultural Revolution. Some saw this move as personal. At Mao's direction, Deng had been removed from leadership positions on two occasions and therefore had a motive for overturning Mao's legacy. Others saw it as purely a matter of accurately documenting the past, including all the revolution's abuses, excesses, and downsides. Whatever the case, the official narrative appeared in several iterations leading up to the "Resolution on Certain Questions in the History of our Party since the Founding of the People's Republic of China," which was adopted by the Sixth Plenary Session of the Eleventh Central Committee of the Communist Party of China on June 27, 1981. The resolution repudiated the Cultural Revolution, saying that it had been responsible for the most severe setbacks and losses China had suffered since the founding of the People's Republic.

Official narratives were not the only means of communicating this interpretation of Mao's legacy to the Chinese public. Significant material relating to the Cultural Revolution was destroyed or otherwise made inaccessible. In 1978 the Chinese government required all college students to take a standard political studies course throughout their four-year program of study that involved condemning the Cultural Revolution. Formal political study was also a part of the required curriculum throughout the middle school and high school years. There was a standard textbook, the most prominent part of which was the Chinese Communist Party Central Committee's Resolution about the Cultural Revolution. This curriculum prepared students for national entrance examinations and graduate entrance examinations that tested student knowledge of the official historical narrative. Middle school, high school, undergraduate, and graduate programs were official sites used to socialize the public to accept the CCP's assessment of the Cultural Revolution.[18]

Similar attempts to manage Mao's legacy and to suppress understanding of the Cultural Revolution have continued. In 2006, the fortieth anniversary of the start of the Cultural Revolution saw no official comment, and the media was told not to mention it. Organizers of a conference in New York to commemorate the anniversary invited approximately a dozen Chinese scholars to attend, but none were allowed to leave China.[19] In 2011 the Chinese authorities broke up a rally commemorating the thirty-fifth anniversary of Mao's death at which participants gave speeches based on the Red Book.[20]

This leads us to a third question raised by this chapter: *To what extent did official narratives serve as an effective educational and agnotological tool in the hands of post-Cultural Revolution Chinese governments?*

Evidence suggests that the Chinese government facilitated the work of scholars, authors, and public officials who wrote and presented a one-sided, unbalanced picture of the Cultural Revolution.[21] But the agnotological question presented here has to do with how effective these portraits of the past were with the general Chinese public, in which parts, and at which times. Just as the Cultural Revolution played out differently in various places in China, so too, almost certainly, did the extent to which the official narrative took hold vary. In the absence of valid public opinion data, no definitive answer is possible, but several observations can be offered.

The first is that the official narrative resonated with the lived experience of a significant number of people, especially urban Chinese, as evidenced by the proliferation of what is usually called the "scar literature." This was a genre of firsthand, often lurid accounts of how persecution, brutality, and violence had devastated and victimized the authors and their family members, neighbors, colleagues, and many others.

Second, that experience was not universal, even in urban areas, as indicated by competing accounts like *Some of Us: Chinese Women Growing Up in the Mao Era*, a 2001 collection of nine memoirs. As the word *some* in the title indicates, the book makes little effort to generalize from the nine women's experience other than to indicate how different their experiences were from the conventional wisdom about the experiences of educated urban youth. One of the editors recalls reading the accounts of the Cultural Revolution that came out after Mao's death: "Everyone who was talking, including the once victimizing Red Guards, was a victim scarred by the Maoist dictatorship. But I could not think of any example in my life to present myself as a victim or a victimizer. I did not know how to feel about my many happy memories and cherished experiences of a time that most vocal people now call the dark age."[22] Included in these memories and experiences were times spent on state farms, and for many of the book's contributors, being sent to the countryside was an experience "especially valuable to their growth as individuals."[23]

Third, hard as it is to know the extent to which the official narrative took hold in urban areas, it is even harder to know how it took hold in rural areas, where more than 80 percent of China's population lived. But in the case of education, the main concern of this volume, there is considerable evidence that the lived experience in China's rural areas was different from that portrayed in the officially sanctioned narrative. Among such evidence is Dongping Han's study of Jimo, a rural county in Shandong Province. Han grew up there during the Cultural Revolution, and after coming to the United States for graduate study, he returned to Jimo to conduct research for his doctoral dissertation. His research strongly indicates that the Cultural Revolution resulted in a huge increase in educational opportunities in this village, with almost all students attending school in the early grades and about two-thirds attending upper middle school (roughly equivalent to a US high school).[24]

Similar results are reported by Mobo Gao, another Western-based scholar who grew up in a rural village in Jiangxi Province during the Cultural Revolution and subsequently studied its effects on his native village. As a teenager during the Cultural Revolution, he was put under house arrest and had to undergo two weeks of nightly struggle sessions because of charges of preserving feudalist genealogical records and reading feudalist books.[25] He thus understands why people victimized by the Cultural Revolution would condemn it. But as Gao puts it, "To participate in the official project of reducing everything to the label of 'ten years of calamities' is another matter."[26] His research on the village in which he was raised reveals that the Cultural Revolution resulted in a great boost in literacy that was facilitated by the egalitarian ethos of the Cultural Revolution.[27]

In his detailed study of education in Hebei Province's Lai Shui County, sociologist Joel Andreas reported results that generally parallel Han's and Gao's. Among other things, in the decade of the Cultural Revolution that county took "major steps toward universalizing 9-year education," with the number of children attending regular middle school increasing more than elevenfold.[28]

The locales studied by Han, Gao, and Andreas are in three different provinces, suggesting that increases in schooling were widespread. Evidence that this was indeed the case nationally can be found in data showing that between 1965 and 1976, roughly the bookends of the Cultural Revolution, the number of secondary school students increased by approximately 500 percent, and there was a similarly large increase in the number of secondary schools.[29] Accompanying this expansion in educational opportunities was a major curriculum shift. Previously, secondary education in rural areas had focused on what was often known as the "little pagoda system," a system aimed at preparing students to pass the examinations that would determine access to higher education and a ticket out of the village. But the Cultural Revolution stimulated significant changes that resulted in practical rural-oriented studies being included as a critical component of rural education rather than as a marginal, inferior program of studies for those lacking ambition to continue on to urban areas.[30]

To be sure, in many cases educational quality may have suffered because of a lack of adequate resources (e.g., appropriately trained teachers) for the many new schools. In an effort to restore quality, in the first decade of the Deng era many middle schools were closed so that resources could be concentrated in a few key schools aimed at preparing students to pass the examinations that were a gateway to higher education. Based on his study of Lai Shui County, Andreas noted that it could justifiably be argued that as a result of the Cultural Revolution, "the quality of little pagoda schools declined greatly." But, he added,

> did children who had the opportunity to attend newly built village middle schools, most of whom would not otherwise have gone to middle school, experience a decline in education quality? Rapid expansion during the Cultural Revolution certainly did severely stretch resources, especially in terms of trained teachers. The post-1977 decision to retrench in order to shore up quality, however, only makes sense in terms of the key-point logic. From the point of view of the majority of village children, who failed to test into the middle schools that survived retrenchment, it is hard to conclude that the quality of education improved. If the key-point logic had not prevailed, strategies to improve quality might have focused on gradually increasing the level of teacher training, rather than shutting down schools.[31]

As emphasized previously, given the absence of valid public opinion data, it is difficult to address the key agnotological question of the extent to which the official narrative of the Cultural Revolution, portrayed as an unmitigated disaster, took hold among the population. But with respect to education, especially in rural areas, there is strong evidence that the lived experience did not align with government-sanctioned views of the revolution.

Research, Higher Education, and the Cultural Revolution in the West

Western appraisals of the Cultural Revolution, historical and political, have depended heavily on sources from and scholarly observers of China as events unfolded from the mid-1960s on.[32] These sources have leaned overwhelmingly toward the Chinese government's official narrative of the revolution. This tendency presents three possible conclusions. First, it might indicate that the official narrative is correct. The revolution, accordingly, was an unmitigated disaster of epic proportions that yielded little more than political, economic, and cultural destruction. Second, it might suggest that the postrevolutionary Chinese government was effective in managing global perceptions of the revolution. This would mean that the Cultural Revolution was not an unmitigated disaster and that the Chinese government was effective in suppressing alternate perspectives. And third, it might be that it was both a disaster and not a disaster, depending on time, place, and condition. Correspondence between the official narrative and nonofficial sources might indicate other possibilities, such as the destruction of alternate accounts, the desire of Western authorities to diminish revolutionary tendencies in their own countries by taking a dim view of positive appraisals of the Cultural Revolution, and so on.

Current scholarship sides with the first of the three conclusions. In the recent volume *The Chinese Cultural Revolution as History* the book's three editors—all of whom hold prestigious academic positions in the United States—state that there are only a "few dissident academic voices who still defend aspects of the Cultural Revolution."[33] The implication is that contemporary scholarship rarely doubts that the Cultural Revolution was an unmitigated disaster. Given, as noted above, that important aspects of the lived experience of many Chinese, especially in China's rural areas, likely conflict with that view, the very fact that there is such a uniformly dominant view among Western scholars raises historical, educational, and agnotological questions that have to do with potential political factors influencing historical research on the Cultural Revolution, as well as the basic validity and reliability of sources and the handling of counterinterpretive accounts.

Such considerations raise a fourth and final question: *To what extent did Western interpretations of the Cultural Revolution get caught in the political, educational, and agnotological crossfire between right and left in Western societies?*

Political factors influencing research on China appeared as soon as the Cultural Revolution broke out.[34] In the late sixties and early seventies, the revolution resonated with and helped fuel the leftist challenges that rocked politics throughout the Western world. As Sinologist Arif Dirlik wrote, "Radicals, including a new generation of radical China scholars . . . saw in the Cultural Revolution a Third World renewal of the promise of revolutionary socialism that had been betrayed in the Soviet Union."[35] The flip side of radicals' embrace of the Cultural Revolution in the sixties is how the view of it as a disaster resonates with the turn since the seventies toward market economies and market societies. The preference for market-based solutions to social, political, and economic problems, as some scholars suggest, was reflected in and fueled by the Chinese government's official repudiation of the Cultural Revolution. This can be seen in the Deng administration's implementation of policies that have resulted in the world's most populous nation becoming a capitalist country in all but name and with immense disparities in wealth, power, privilege, and opportunity. These were the developments, of course, that Mao sought to prevent by initiating the Cultural Revolution. Given this turn of events, why, we ask with Dirlik, is the Cultural Revolution largely viewed "as an historically meaningless and deviant undertaking that had little purpose beyond satisfying the craving for power or the perverse ideological attachments of its leaders and proponents?"[36]

This question acquires additional weight if the concern is not Mao's fear of a capitalist restoration but the less controversial concern of economic growth. Contrary to widely held views, China's economy did not stagnate during the Cultural Revolution. Rather, annual GDP growth averaged about 6 percent during that decade, certainly a lower figure than in subsequent years but a respectable figure and one that far exceeded that of India (whose GDP growth was about 3 percent), the other Asian giant with which China is often compared.[37] Moreover, as political scientist Richard Curt Kraus has pointed out, the Cultural Revolution's success in increasing access to education—along with increasing access to health care and increasing the integration of women into workplaces—improved the quality of China's workforce, an improvement that would benefit any economic development strategy, even one as diametrically opposed to the Cultural Revolution policies as Deng's.[38]

Kraus supplies one part of the answer to this question when he notes that as the West wrestles with China's rise, it boosts the West's claim to moral superiority for there to be a "typical and mistaken Cultural Revolution narrative: the economy was a shambles, education destroyed, but Deng Xiaoping rescued China by copying *our* free market. Kicking Mao's corpse strengthens the position of globalism, proves that socialism does not work, and reveals China as unstable and dangerous."[39] To call attention to this consequence of the conventional Cultural Revolution narrative is most emphatically *not* to say that this consequence motivated that narrative's scholarly development. It is, however, to call attention to the extent to which agnotological processes, whether passive or active, are likely to have come into play in the access granted by Chinese authorities to sources, the production of research in terms of publishing decisions, successful grant applications and funding, and tenure and promotion decisions affecting the direction of Cultural Revolution scholarship. Thomas Kuhn's well-worn *Structure of Scientific Revolutions* still speaks to this kind of process whereby entrenched scholarly opinion fends off upending interpretations by marginalizing, obscuring, or ignoring competing and unsettling research agendas.[40] The pattern may have been followed in this history of higher education, and it deserves greater clarity, attention, and detail in terms of the explicit preset views on the Cultural Revolution of gatekeepers—those in authority to promote or suppress scholarly views—in universities across the West.

Implications

This chapter examines the role education and agnotology have played in the reception of the Chinese Cultural Revolution in China and abroad. It explores questions critical to understanding how the social construction of ignorance might indeed function as a prime catalyst in history rather than as a side-effect or consequence. Researchers have much to gain by considering these contextualized questions as starting points for looking at education and ignorance-making as agents of historical development and as confounding variables in historical understanding.

But this reconsideration of the Cultural Revolution through a history of education and agnotological lens has consequences beyond simply the advancement of knowledge. It also has consequences for how that knowledge can influence contemporary teaching of the revolution. An excellent example of how agnotological processes affect and are reproduced in the teaching of the

Cultural Revolution is provided by an article in *Social Education*, a journal of the National Council for Social Studies.[41] Drawing on several books from that council's list of "Notable Social Studies Trade Books for Young People," the article discusses ways of teaching the Cultural Revolution. Its summary of the Cultural Revolution reflects the standard narrative of the event as an unmitigated disaster. But even more interesting from an agnotological perspective is its sample lesson plan, which draws from a chapter in *Red Scarf Girl*, a memoir of Jiang Ji-li, who was a twelve-year-old schoolgirl in 1966 and came from a family with considerable cultural capital (her father was an actor who knew calculus and could explain Zeno's paradox and infinite series). Titled "Writing Da-Zi-Bao," the chapter recounts the posting of big-character posters in schools, including one made about her. In many ways the lesson plan encourages critical thinking, such as by posing a series of questions aimed at fostering empathetic understanding, suggesting that "readers might empathize with her [Jiang's] profound sense of shame" at being the target of a dazibao. In more telling ways, however, the article also facilitates the construction of ignorance. Nowhere, for example, does the lesson plan suggest that the reader might empathize with one of Jiang's schoolmates, Yin Lan-lan, who had posted a dazibao that read, "Being from a working-class family, I have a lot more housework than students from rich families. So I have difficulty passing exams. I was forced to repeat grades three times. And I was not allowed to be a Young Pioneer or to participate in the school choir."[42]

Nor does the lesson plan suggest that readers reflect on Jiang's reaction to Yin's dazibao, "'Yin Lan-lan a victim?' . . . She rarely spoke up in class. When she was asked to answer a question, she would just stand there without saying a word. She was not very bright." And it does not encourage readers to reflect on how, as the dazibaos continued to be posted, "Yin Lan-ln had changed a great deal. No longer hesitant and clumsy, she had become vocal, aggressive, and confident. She stood up straight and threw out her chest, whereas before she had always slouched."[43] Indeed, despite Yin's figuring prominently in the chapter, nowhere does the lesson plan mention her, even though her experience highlights some of the key goals of the Cultural Revolution: to alleviate educational inequalities, to increase educational opportunities for those otherwise lacking them, and to help students from families (such as Jiang's) with more cultural capital and higher levels of educational attainment better understand the situation, qualities, and worth of students from working-class families (such as Yin's) that lacked such levels of education and cultural capital.

Included in the Red Book, which was widely distributed during the Cultural Revolution, are excerpts from a paragraph Mao had written titled "Where Do Correct Ideas Come From?"[44] Whatever the merits of Mao's answer to that question, the paragraph overlooks an equally important question: where do *in*correct ideas come from? That question is at the heart of much agnotological inquiry. What this chapter argues and the *Social Education* lesson plan exemplifies is that agnotological processes have played a critical role in the teaching, learning, and study of the Chinese Cultural Revolution. Fresh interpretations of this critical period of China's past are long overdue. Substantive and methodological opportunities await historians attuned to the possibilities of education and ignorance as prime agents of historical development and understanding. Such scholars will also be in a promising position to bring about a historically grounded, agnotologically inspired perspective on what is (mis)understood about the Cultural Revolution and what is taught (or not taught) about it.

NOTES

Stephen Samuel Smith would like to thank David Pugh for helpful discussions on Maoist China. Of course, the authors are solely responsible for any errors of fact or interpretation.

1. The conversations were publicized during the Cultural Revolution. See Mao Zedong, "Talks with Mao Yüan-hsin," accessed 28 December 2014, https://www.marxists.org/reference/archive/mao/selected-works/volume-9/mswv9_23.htm and http://www.marxists.org/reference/archive/mao/selected-works/volume-9/mswv9_51.htm.

2. For an insightful discussion of cultural and political capital during the Cultural Revolution, see Joel Andreas, *Rise of the Red Engineers: The Cultural Revolution and the Origins of China's New Class* (Stanford, CA: Stanford University Press, 2009).

3. The controversy involved an early-1960s play called *Hai Rui Dismissed from Office*. Hai Rui was a Ming Dynasty official who criticized the emperor and consequently lost his position. Maoists viewed the play as an allegorical account of Mao's effecting the dismissal several years earlier of China's popular minister of defense, who had criticized important Maoist policy initiatives, as well as bad-mouthed them to Soviet leader Nikita Khrushchev at a time when Sino-Soviet relations were deteriorating rapidly. At Mao's instigation, a critique of the play was published that triggered an increasingly ferocious controversy about the content of the play, the playwright's intentions, and the extent to which Maoists' interpretation of the play was correct. For a brief but effective summary of the controversy and its antecedents, see Maurice Meisner, *Mao's China and After: A History of the People's Republic*, 3rd ed. (New York: Free Press, 1999), 230–33, 312–15.

4. As quoted in Meisner, *Mao's China and After*, 316. Dazibaos were widely used in the Cultural Revolution. Indeed, the Sixteen Points (see below) notes that "through the media of big-character posters and great debates" revolutionary young people "argue

things out, expose and criticize thoroughly, and launch resolute attacks on the open and hidden representatives of the bourgeoisie."

5. As noted in Erza F. Vogel, *Deng Xiaoping and the Transformation of China* (Cambridge, MA: Harvard University Press, 2011), 42.

6. Mao Zedong, "A Talk by Chairman Mao with a Foreign (Albanian) Military Delegation," in *People's China: Social Experimentation, Politics, Entry onto the World Scene, 1966–1972*, ed. David Milton, Nancy Milton, and Franz Schurmann (New York: Vintage Books, 1974), 261. Text from *Survey of Chinese Mainland Press*, no. 4200 (18 June 1968).

7. As quoted in Andreas, *Rise of the Red Engineers*, 110.

8. David Milton and Nancy Milton, *The Wind Will Not Subside: Years in Revolutionary China—1964–1969* (New York: Pantheon, 1976), 147–48. The Miltons taught at the Beijing First Foreign Languages Institute.

9. Meisner, *Mao's China and After*, 322.

10. Andreas, *Rise of the Red Engineers*, 110. These parallels became especially pronounced after the early years of the Cultural Revolution, when much of the CCP apparatus was reconstituted and the military was used to restore order, which it frequently did very violently.

11. Mao Zedong, "Analysis of the Classes in Chinese Society," in *Selected Readings from the Works of Mao Tse-tung* (Beijing: Foreign Language Press, 1967), 11–19, https://www.marxists.org/ . . . 1/mswv1_1.htm. As indicated by the 1967 publication date, these selected readings appeared during the Cultural Revolution. It is indicative of the importance that Mao attached to this article and the issue of distinguishing friends from enemies that the article is the first of the almost forty selections in the volume and that its first two sentences are, "Who are our enemies? Who are our friends?"

12. For further discussion on the issue of class and the Cultural Revolution, see Yiching Wu, *The Cultural Revolution at the Margins: Chinese Socialism in Crisis* (Cambridge, MA: Harvard University Press, 2014).

13. "Decision of the Central Committee of the Chinese Communist Party Concerning the Great Proletarian Cultural Revolution," *Peking Review* 9 (12 August 1966): 6–11, https://www.marxists.org/subject/china/peking-review/1966/PR1966-33g.htm.

14. Marxists have long debated the criteria for defining classes in a socialist society. In Maoists' view, the leaders of a bureaucratized communist party in a socialist country can become the equivalent of a capitalist class in a capitalist country.

15. Among these memoirs and scholarship are the works of Mobo Gao and of Xueping Zhong, Wang Zheng, and Bai Di discussed below.

16. It bears noting that while Maoist guidelines generally called for debate, reason, and persuasion rather than coercion in waging the Cultural Revolution, they also stated that the objective of the Cultural Revolution was to "overthrow those persons in authority who are taking the capitalist road" and that because the Cultural Revolution was a revolution, it was bound to meet resistance. The guidelines also contained a reminder that "revolution cannot be so very refined, so gentle, so temperate, kind, courteous, and magnanimous," an excerpt from Mao's definition of revolution (contained in the Red Book) as an "act of violence by which one class overthrows another." The point here centers on the agnotological consequences of the mixed messages coming from revolutionary leaders and the issue of which messages were privileged over others. The Sixteen Points omitted the part of the definition about revolution being an act of violence. But the document's obvious reference to Mao's definition of revolution, combined with the fact that Mao was calling for a cultural revolution, undoubtedly

facilitated some of the violence of the Cultural Revolution. In this respect, the agnotological consequences of these mixed messages resemble those stemming from the absence of a more precise definition of class in the Sixteen Points.

17. Meisner, *Mao's China and After*, 316.

18. On the destruction of material in the library of Fudan University—located in Shanghai, a city that figured prominently in the Cultural Revolution—see Jing Dalu, "Fudan de wenge ziliao dao nali qu le—Cao Chong Qing Banglian fangtan ji [Where are Fudan's CR materials? An interview with Cao Chong and Qing Banglian]," *Shilin* [Historical review], suppl. 1 (2006): 108, as quoted in Hongsheng Jiang, "The Paris Commune in Shanghai: The Masses, the State, and Dynamics of 'Continuous Revolution'" (PhD diss., Duke University, 2010), 18; for additional material on the suppression of information about the Cultural Revolution, see ibid., 14–19. For the curricula about the Cultural Revolution, see Wang Jiping, ed., *Zhongguo Jinxiandaishi Gangyao* [Contemporary Chinese history] (Xiangtan: Xiangtan University Press, 2010), 226–48; Institute of Chinese Communist Party History, ed., *Zhongguo Gongchandang Lishi* [Chinese Communist Party history], vol. 2 (Beijing: Chinese Communist Party History Press, 2011), 1053–69; and The Editorial Group, ed., *Zhongguo Jinxiandaishi Gangyao* [Contemporary Chinese history] (Beijing: Chinese Higher Education Press, 2013), 267–86.

19. Michel Bonnin, "The Threatened History and Collective Memory of the Cultural Revolution's Lost Generation," *China Perspectives* (2007), chinaperspectives.revues.org/pdf/2573.

20. Ségolène Malterre, "Modern-day Maoists Worry Chinese Authorities," France 24 International News, 23 September 2011, http://observers.france24.com/content/20110923-china-modern-day-maoists-worry-authorities-commemoration-unrest-taiyuan.

21. Mobo Gao, *The Battle for China's Past: Mao and the Cultural Revolution* (Ann Arbor, MI: Pluto, 2008).

22. Xueping Zhong, Wang Zheng, and Bai Di, introduction to *Some of Us: Chinese Women Growing Up in the Mao Era*, ed. Xueping Zhong, Wang Zheng, and Bai Di (New Brunswick, NJ: Rutgers University Press, 2001), xx.

23. Ibid., xxxi.

24. Dongping Han, *The Unknown Cultural Revolution: Life and Change in a Chinese Village* (New York: Monthly Review Press, 2008).

25. Mobo C. F. Gao, *Gao Village: Rural Life in Modern China* (Honolulu: University of Hawaii Press, 2007), 157.

26. Gao, *Battle for China's Past*, 16.

27. Gao, *Gao Village*, 159.

28. Joel Andreas, "Leveling the Little Pagoda: The Impact of College Examinations, and their Elimination, on Rural Education in China," *Comparative Education Review* 48 (2004): 16.

29. Calculations based on data in Suzanne Pepper, *Radicalism and Education Reform in Twentieth-Century China: The Search for an Ideal Development Model* (New York: Cambridge University Press, 1996), 416, 425.

30. Andreas, "Leveling the Little Pagoda," 42.

31. Ibid., 45.

32. Joseph W. Esherick, Paul G. Pickowicz, and Andrew G. Walder, "The Cultural Revolution as History: An Introduction," in *The Chinese Cultural Revolution as History*, ed. Esherick, Pickowicz, and Walder (Stanford, CA: Stanford University Press, 2006), 27.

33. Ibid., 27.

34. Arif Dirlik, "The Politics of the Cultural Revolution in Historical Perspective," in *The Chinese Cultural Revolution Considered: Beyond Purge and Holocaust*, ed. Kam-yee Law (New York: Palgrave, 2003).

35. Ibid., 160.

36. Ibid., 158. On Deng's reforms and the rise of neoliberalism, see David Harvey, *A Brief History of Neoliberalism* (New York: Oxford University Press, 2005).

37. Richard Curt Kraus, *The Cultural Revolution: A Very Short Introduction* (New York: Oxford University Press, 2012), 66.

38. Ibid., 80.

39. Ibid., 117.

40. Thomas S. Kuhn, *The Structure of Scientific Revolutions*, 3rd ed. (Chicago: University of Chicago Press, 1996).

41. Lindsey Cafarella and Chara Haeussler Bohan, "The Top Five Narratives for Teaching about China's Cultural Revolution," *Social Education* 76 (2012): 121–24, scholarworks.gsu.edu/cgi/viewcontent.cgi?article=1022&context.

42. Ji-li Jiang, *Red Scarf Girl: A Memoir of the Cultural Revolution*, rev. paperback ed. (New York: HarperCollins, 2008), 41.

43. Ibid., 45.

44. Mao Zedong, *Quotations from Chairman Mao Tse-tung*, 2nd ed. (Beijing: Foreign Language Press, 1966), 206, http://www.marx2mao.com/PDFs/QCM66.pdf.

Reflections

A. J. ANGULO

Tobacco, asbestos, climate change, orgasms, acid rain, genetic engineering, and military secrets—all of these and more have recently caught the attention of those seeking to understand how ignorance is produced, diffused, and regulated. At the start of the twenty-first century, the quest to study the contours of ignorance spread like wildfire from philosophy and history to political economy and anthropology. In this excitement to explore intellectual blind spots, many sped past one of the most promising sites for investigating the social construction of ignorance: education.[1]

It's an unusual claim, since education offers the best hope for combating ignorance. No developed society today denies the potential of education in advancing quality of life, social cohesion, intellectual exchange, and political values. Political leaders of any stripe would face career suicide to suggest otherwise. For all of its flaws when implemented, education will continue to serve as a bedrock in the fight against ignorance for the foreseeable future.

But as the chapters in this book illustrate, education is more than schools, colleges, and universities. Education broadly construed involves all the processes a society uses to transmit its culture from one generation to the next. This is hardly a new insight. Rather, as suggested in this volume, it has shaped the standard approach to exploring the educational past since the mid-twentieth century. What distinguishes these chapters from standard accounts is their use of agnotological perspectives. They demonstrate how these relatively new ways of studying the past can reinvigorate well-established lines of research and light up new ones in the study of education. The chapters also show how the fruits of these case studies can circle back and do the same for agnotological methods.[2]

We can see how agnotology generates new research possibilities when it's used to examine the intersection of law, education, and ignorance production. The opening chapters tell us about how slaveholders, radio broadcasters, politicians, religious figures, and corporate executives all shared the common goal of suppressing forms of knowledge considered threatening or disruptive to their interests. These groups took an active role in thwarting the diffusion of knowledge through legal and regulatory means. In doing so, they sidestepped questions of validity and reliability and skirted problems having to do with accuracy, public health, and human rights in order to protect narrow interests or socially constructed norms. The middle chapters turn from courthouses and regulatory agencies to some of the most cherished assumptions—myths really—in American history. These essays paint new, compelling portraits aimed at exploding some enduring myths. They focus on poverty and its relation to progressive education, Asian Americans and the model minority mythology, and approaches to history itself as informed by religious and cultural predispositions. Exploding myths in general is dangerous business. The task involves unsettling the mind and the self in uncomfortable ways. But these historical case studies ease us into an understanding of the way the mythmaking process works, evolves, adapts, and takes surprising detours. The final chapters explore the agnotological interplay between education, the law, and mythmaking at national and global levels. There, themes of ignorance-producing nationalism—in the United States, Germany, the Soviet Union, Israel, and China—converge everywhere, from the classroom to the global exchange of ideas.

Education history, without question, stands to benefit in countless ways from having agnotology among its methods, approaches, and frameworks—and these chapters lay the foundation for future exploration from this perspective. The field is ripe for new ideas and imperatives, and agnotology hints at the possibility of an epistemic shift in the way historians approach education. Additional study of ignorance-making along the lines established in this book has the potential to secure miseducation's place among the primary forces influencing constancy and change in global history.

The benefits, however, flow in both directions. These chapters not only show the impact of agnotology on historical studies of education but also raise, directly and indirectly, new perspectives and problems for agnotologists. Consider the issue of literacy. Throughout the volume, literacy stands at the center of debates over the known and unknown. Literacy—the basic, scientific, and legal kind—is only the starting point. These contributors bring the agnotology discussion to the doorstep of other literacies, such as those having to do with myths, media, history, politics, survival, hermeneutics, nationalism, imperialism, and globalization. In one sense or another all understanding boils down to language, and the process of developing command over that language is literacy. The project of agnotology, in this light, has much to gain from directing attention to subtleties and nuances influencing the expansion and contraction of life's literacies. These chapters demonstrate that approaching agnotology in this way is likely to stimulate new and productive lines of research on ignorance.[3]

Collectively, chapter contributors also wrestled with some fundamental problems of interest to agnotologists. In the process of reframing their areas of expertise in agnotological terms, they encountered methodological problems having to do with defining the boundary between knowledge and ignorance. *Ignorance*, it turns out, has a great deal in common with other conceptually pesky terms, like *art*, *love*, and *freedom*. As the US Supreme Court once claimed about pornography, we know it when we see it. But once we do, it's not so easy to put conceptual fences around it or make a case for what the parameters are that will satisfy most of us. It's in defining ignorance that matters become more challenging, and in some cases inscrutable. In other words, what can we mean by and say about ignorance and its social construction?[4]

This volume traces basic historical explanations and responses to this question in the context of education. In doing so, it reveals a history littered with what the authors present as winners and losers. When legalizing ignorance, governments have passed laws and established regulations that worked against the expansion of literacies critical to understanding the self, the body, and the natural order. In this light, the legalization of ignorance presents a clear picture of who won and who lost along the way. The same could be said of ignorance and socially constructed myths. These were stories told with enough frequency that they took on a reality of their own. They were stories that focused attention on some and obscured, neglected, excluded, omitted, or ignored others. Myths of this kind competed directly with other ways of knowing or experiencing the world. When laws and myths converged in countries around the world and across the twentieth century, ignorance became a tool for the powerful. Ignorance production in these contexts became a national and global asset traded by power brokers who saw great value in controlling the past and present as they planned for the future. Ignorance served specific nation-building and globalizing needs within the United States, Germany, the Soviet

Union, Israel, and China. These cases indicate that the social construction of ignorance has always played an important role in the creation of national identity and that all countries at one level or another engage in the practice.

Along the way, however, chapter contributors made fundamental decisions about what counts as knowledge and what counts as ignorance. Those decisions were informed by the view that all knowledge is political and that, by implication, so too is the act of identifying, categorizing, and defining ignorance. These are inescapable realities. All of us make judgments about these matters, using criteria and frameworks that agree with our beliefs and understandings. Decisions to expand or limit knowledge are acts of power between those who wield it (e.g., states, vested interests, religious groups) and those who don't (e.g., enslaved people, the general public, children). The interplay between knowledge and power, put simply, means that all present knowledge production has triumphed over previous forms of knowledge that have ended up—ignored and forgotten—in history's wastebin of ideas.[5]

These chapters implicitly or explicitly make judgment calls about ignorance and knowledge. Some appeal to human rights to draw the boundary line and highlight instances of ignorance-making—as in cases having to do with slaves, sexual orientation, imperialism, the characterization of Asian Americans, and the representation of Jews in Nazi Germany. Others turn to science, empiricism, and nonreligious appeals to knowledge—as in the chapters on sex education, evolution, the environment, and the mainstream presentation of history. Still others appeal to inclusivity and accuracy—as in the cases having to do with poverty, education historians, and entire nations such as the Soviet Union, Israel, and China. Whatever the criteria, these authors use their expertise and best judgment in the study of ignorance as situated within educational contexts to distinguish between knowledge and non-knowledge. Their goal is to move beyond a discussion of good guys versus bad guys and toward enlightening readers about the process by which ignorance occurs and, when it does, how it transforms that absence into a presence.

In example after example, this volume illustrates how ignorance and its social production through naïveté, passivity, and active agency have been central to the human experience. This will come as no surprise to cognitive researchers, who tell us that the human brain is hard-wired to forget and ignore. We aren't supposed to remember all those faces we see in the subway or in a crowded venue. It would be traumatic if we did retain all that information. As such, humans are ignorance-producing beings. That's one of our survival mechanisms, researchers explain, born out of evolutionary processes, and perhaps one that helps keep us sane. All of this means that we might want to consider changing what we call our species, exchanging *Homo sapiens* for *Homo agnotologicus*. Ignorance production is as much a trademark of human nature as knowledge or wisdom production.[6]

Looking ahead, research on agnotology—in terms of substance and methods—will likely keep scholars busy thinking about this vital part of the human experience for many years to come. Each of these chapters points out the need for further substantive analysis. They should serve as calls for research on how the social construction of ignorance operates within these specific contexts, although there are countless other contexts that stand beyond the scope of this volume. The intersection of gender, education, and agnotology, for instance, demands its own separate study. Individuals like Edward Clarke, a late nineteenth-century Harvard professor of medicine, believed women damaged their private parts if they took up college-level studies. He wasn't alone and their ideas beg for agnotological analysis. Commercials in educational settings raise significant questions about law, propaganda, and ignorance that also demand agnotological scrutiny. What the public "learns" from advertising and media more broadly, about meaning, purpose, desire, and values, is an area that could benefit from a social construction of ignorance perspective. The field with the lowest hanging fruit, however, has to do with food and pharmaceuticals. Scholars have only begun to offer an agnotological analysis of what children and adults know and ignore about what they ingest three times per day or what's prescribed to them by their physicians. Investigations into the kind of information that reaches children in schools or adults through mass education campaigns could sustain several active programs of research.[7]

A dizzying array of additional substantive possibilities await those with "sharp eyes and ears ready to hear," as American Indian scholars have long argued. Extending, elaborating, and complicating the case studies included in this volume is a project worthy of a generation of scholars. Comparing how nation-states and regions manage the processes of ignorance transmission from one age to the next has the potential to energize at least another generation or two. Given current problems of global proportions—such as religious extremism; public health concerns like Ebola, which know no borders; ecological degradations that fill the headlines almost every day; and intractable conflicts around the world—the call for understanding the historical and educational origins of ignorance production is urgent, immediate, and real.[8]

The most urgent of all substantive issues, arguably, is the dramatic and increasing rate of economic inequality, which points to very troubling paths ahead for 99 percent of the people on this planet. Thomas Piketty's *Capital in the Twenty-First Century* attempted to warn us about projected levels of economic inequality and their unsustainability. He's argued, broadly, that Cold War economics and its legacy account for many of the problems we face today. Currently, the top 0.1 percent of the US population owns nearly the same amount of wealth as the bottom 90 percent. Raising the issue at all, however, has been an uphill battle. Piketty's *Capital* stood on the receiving end of active agnotological strategies the moment the book came off the press. Many in the financial media immediately cast doubt on his findings, data, and conclusions in an effort to manage the book's reception and manufacture debate over his multidecade research program. The identity of those who stand to gain from these agnotological distortions, namely, the beneficiaries of extreme inequality, is clear and predictable. For those on the receiving end of such distortions, agnotology offers a means for understanding how populations have interacted with economic literacies advanced and diffused for professional as well as public consumption. It opens opportunities for investigating how populations have learned to live under what some now describe as a system of neofeudalism or oligarchy. The recent passage of the Trans-Pacific Partnership agreement, the multinational corporate interests privileged and protected in the deal, and the extreme ignorance-producing secrecy surrounding its passage—all of these have as much to do with an underexamined history of miseducation as with political economy. Agnotology also identifies cases that can be used to challenge those who dismiss economic inequality as moral panic alone or as uninteresting because it doesn't directly address cultural politics. If historians of education have "sharp eyes," their attention would necessarily turn toward developing a rich historical understanding of ignorance in relation to the greatest economic challenge in recent history.[9]

For those more interested in methods than in case studies, the agnotological frontier holds just as much promise. More categories and frameworks in the social construction of ignorance (e.g., beyond the native-passive-active heuristic) are needed to problematize and clarify what counts as a historical case of ignorance-making. Greater agnotological distinctions have the potential to yield more specific insights.

Consider, for instance, what this reflection proposes as the category of meta-agnotology. Meta-agnotology would involve examining how the study of agnotology comes into the service of perpetuating ignorance. The Cato Institute, a partisan think tank, has recently published *"Silent Spring" at 50: The False Crises of Rachel Carson*, a work that has the hallmarks of meta-agnotology. The book's self-described aim is to expose and analyze a "number of Carson's major arguments[, which] rested on what can only be described as deliberate ignorance." In its quest to expose ignorance production in Carson's *Silent Spring*, the Cato study has left in its wake a trail of controversy for the following: the financial connection between *"Silent Spring" at 50* and a libertarian organization funded by such corporate giants as Philip Morris, R.J. Reynolds, Chevron, Exxon, Shell, the American Petroleum Institute, Eli Lilly, Merck, and Pfizer; the timing of the book's publication to coincide with fiftieth-anniversary celebrations of Carson's groundbreaking book; the aim of creating doubt about Carson's research, which has inspired generations of mainstream scientists and ecologists; the role of business, law, and economics professors rather than scientists in editing a volume about science; and the longstanding tradition of distorting the historical, scientific, and policy record surrounding Carson to advance what appear to be contemporary regulatory interests.[10]

Whatever its reception among mainstream scientists, *"Silent Spring" at 50* will likely have an impact on how Carson is presented in educational settings. Publicly elected school board members and administrative officials seeking to minimize or obscure Carson's contributions can point to a publication that raises "serious questions" about her legacy. For these individuals, *"Silent Spring" at 50* will cast "doubts" about the merits of including mention of her work in the curriculum. As such, Cato has produced a "legitimate" controversy about achievements in the field of ecology, research on the environmental impact of products made by the chemical industry, and the creation and function of the US Environmental Protection Agency. Read this way, the Cato Institute is on the cutting edge of using agnotological analysis for the purpose of ignorance-making.[11]

While the need to *advance* our understanding of the substance and methods of ignorance production is pressing, so too is the need to *diffuse* awareness of systematic distortions of knowledge and the suppression of trustworthy information. The occasional academic conference shouldn't satisfy anyone who can envision the widespread appeal and provocative quality of the study of ignorance. What we need are university courses on agnotology, programs dedicated to the social construction of ignorance, and perhaps new interdisciplinary fields such as "fraud studies" with agnotology at their core. Such courses could be a boon for education historians, who have had uneven success in promoting their relevance in programs that prepare teachers, counselors, leaders, and other school and postsecondary professionals. Agnotology could position historians of education in an especially promising place to take the lead in the preparation of people seeking critical, historical, and policy skills. All signs point to a future likely to depend on those with such skills. Think tanks funded by special interests are dominating how public policy is shaped in such areas as finance, insurance, real estate, energy, defense, health care, the environment, and, of course, education. The historical and contemporary skills needed to detect fraud and active ignorance production will hold great value in the years ahead.[12]

Public schools have a role to play in diffusing the agnotology imperative as well. They've had trouble finding their footing ever since the Global Educational Reform Movement (or GERM, as some call it) of the late twentieth- and early twenty-first centuries. Corporate reformers, in particular, have successfully lobbied for increased testing and standardization. In the United States, No Child Left Behind put standardization at the forefront of schooling. When the American public school system has historically been at its best, it has stood for the common good. Lately, the common good has been replaced with common standards. Teachers have resisted the trend, school leaders have challenged it, but in the end an overwhelming number of schools have succumbed to the idea of holding the young with a scientistic hand. Agnotology could breathe new life into the flagging public mission of the common school. It could turn stale, fusty talk of critical thinking into a dynamic discussion of the forces that prevent access to vital knowledge and skills.[13]

Some, of course, will oppose such a vision. Recent politicians in Texas make no bones about their position on the matter. "I'm sorry," said David Bradley of the Texas State Board of Education, "this critical thinking stuff is gobbledygook." In 2012 the state's dominant political party stated emphatically: "We oppose the teaching of Higher Order Thinking Skills (HOTS) (values clarification), critical thinking, and similar programs." These initiatives, they argued, were distractions from the real point of education and "have the purpose of challenging the student's fixed beliefs and undermining parental authority." The message is clear: stick to basic instructional standards and don't question them.[14]

Nevertheless, the fight for the common good—through the fight against ignorance—hasn't disappeared. It has simply moved to different venues, a shift that speaks to the great demand for it and the promise agnotology has for schools today. Elements of the common good have reemerged in the faux news shows hosted by entertainers like Jon Stewart, Stephen Colbert, and John Oliver. The success of these programs is due in large measure to their ability to expose the steady stream of agnotological distortions generated by think tanks and media machines. These shows understand the educative and entertainment value of exposing the interests of the powerful and wealthy when they clearly conflict with the public good.[15]

In this demand for learning about agnotological distortions, schools have a mandate to reclaim mission and purpose. If colleges and universities began offering agnotology studies, schools systems could follow suit by providing the same at lower levels. A new mission rooted in agnotology could energize a generation of students to learn about how ignorance is produced, diffused, maintained, and regulated in domestic and foreign contexts. Some might claim that such a mission is a luxury, that there's no room for such thinking in today's curriculum, and that in the end it's just "gobbledygook." This book argues that the future of civil, sustainable democracies around the world depends on it.[16]

NOTES

1. Shannon Sullivan and Nancy Tuana, eds., *Race and Epistemologies of Ignorance* (Albany: SUNY Press, 2007); Robert N. Proctor and Londa Schiebinger, eds., *Agnotology: The Making and Unmaking of Ignorance* (Stanford, CA: Stanford University Press, 2008); Naomi Oreskes and Erik M. M. Conway, *Merchants of Doubt* (New York: Bloomsbury, 2011); Casey High, Ann H. Kelly, and Jonathan Mair, eds., *Anthropology of Ignorance: An Ethnographic Approach* (New York: Palgrave, 2012); Philip Mirowski, *Never Let a Crisis Go to Waste: How Neoliberalism Survived the Financial Meltdown* (New York: Verso, 2013); Matthias Gross and Linsey McGoey, eds., *Routledge International Handbook of Ignorance Studies* (New York: Routledge, 2015).

2. See William Reese and John Rury, *Rethinking the History of American Education* (New York: Palgrave, 2012).

3. For starters on the variety of (sometimes conflicting) views on the relationship between understanding and literacy, see E. D. Hirsch, *Cultural Literacy: What Every American Needs to Know* (New York: Vintage, 1988); Frank B. Golley, *A Primer for Environmental Literacy* (New Haven, CT: Yale University Press, 1998); Henry Milner, *Civic Literacy: How Informed Citizens Make Democracy Work* (Hanover, NH: University Press of New England, 2002); David Olson and Nancy Torrance, *The Cambridge Handbook of Literacy* (Cambridge: Cambridge University Press, 2009); and Jeffrey Nokes,

Building Students' Historical Literacies: Learning to Read and Reason with Historical Texts and Evidence (New York: Routledge, 2012).

4. In 1964, in *Jacobellis v. Ohio*, the US Supreme Court justice Potter Stewart used the phrase "I know it when I see it" to describe his standard for obscenity and pornography. For a classic on language and concepts, see George Lakoff and Mark Johnson, *Metaphors We Live By*, rev. ed. (Chicago: University of Chicago Press, 2003).

5. Historians, philosophers, and sociologists of knowledge have long analyzed the inherent political qualities of socially constructed knowledge. See, e.g., Thomas Kuhn, *The Structure of Scientific Revolutions* (Chicago: University of Chicago Press, 1962); Peter Berger and Thomas Luckmann, *The Social Construction of Reality: A Treatise in the Sociology of Knowledge* (Garden City, NY: Doubleday, 1966); and Mary Jo Nye, *Michael Polanyi and His Generation: Origins of the Social Construction of Science* (Chicago: University of Chicago Press, 2011). Manuela Fernández Pinto has recently argued that agnotologists need to clarify their "normative commitments" and articulate what counts as scientific knowledge. Some positions, Pinto suggests, rely on peer review, expert consensus, or scientific methods, but she problematizes these positions by highlighting, among other things, the influence of private research funding, the misuse of acceptable methods, and imperceptible biases. See Pinto, "Tensions in Agnotology: Normativity in the Studies of Commercially Driven Ignorance," *Social Studies of Science* 45 (2015): 294–315.

6. For a sampling of cognitive research on the related areas of "selective attention" and "belief bias," see Jon Driver, "A Selective Review of Selective Attention Research from the Past Century," *British Journal of Psychology* 92 (2001): 53–78; Dries Trippas, Simon J. Handley, and Michael F. Verde, "The SDT Model of Belief Bias: Complexity, Time, and Cognitive Ability Mediate the Effects of Believability," *Journal of Experimental Psychology: Learning, Memory, and Cognition* 39 (2013): 1393–1402; and Stephan Lewandowsky, Klaus Oberauer, and Gilles E. Gignac, "NASA Faked the Moon Landing—Therefore, (Climate) Science Is a Hoax: An Anatomy of the Motivated Rejection of Science," *Psychological Science* 24 (2013): 622–33.

7. Edward Clarke, *Sex in Education, or A Fair Chance for Girls* (Boston: James R. Osgood, 1875); Sue Zschoche, "Dr. Clarke Revisited: Science, True Womanhood, and Female Collegiate Education," *History of Education Quarterly* 29 (1989): 545–69; Alex Molnar, *School Commercialism: From Democratic Ideal to Market Commodity* (New York: Routledge, 2005); Susan Levine, *School Lunch Politics: The Surprising History of America's Favorite Welfare Program* (Princeton, NJ: Princeton University Press, 2008); Marion Nestle, *Food Politics: How the Food Industry Influences Nutrition and Health* (Berkeley: University of California Press, 2013); Edward Nik-Khah, "Neoliberalism and the Origins of Pharmaceutical Ignorance" (paper, annual meeting of the Society for Social Studies of Science, Cleveland, OH, 9 August 2014); Nik-Khah, "Neoliberal Pharmaceutical Science and the Chicago School of Economics," *Social Studies of Science* 44 (2014): 489–517.

8. Sam Harris, *End of Faith: Religion, Terror, and the Future of Reason* (New York: Norton, 2005); David Quammen, *Ebola: The Natural and Human History of a Deadly Virus* (New York: Norton, 2014); Daniel Bar-Tal, *Intractable Conflicts: Socio-Psychological Foundations and Dynamics* (New York: Cambridge University Press, 2013).

9. Thomas Piketty, *Capital in the Twenty-First Century* (Cambridge, MA: Harvard University Press, 2014); Chris Giles, "Data Problems with Capital in the 21st Century," *Financial Times*, 23 May 2014; Alan Reynolds, "Why Piketty's Wealth Data are Worthless," *Wall Street Journal*, 9 July 2014; Piketty, "Addendum: Response to FT,"

accessed 1 November 2014, http://piketty.pse.ens.fr/files/capital21c/en/Piketty2014TechnicalAppendixResponsetoFT.pdf; "Thomas Piketty Discusses, 'Capital in the 21st Century' with Ryan Grim and Alexis Goldstein," *Huff Post Live*, 16 April 2014, http://live.huffingtonpost.com/r/segment/5345cc4902a7600458000626; Emmanuel Saez and Gabriel Zucman, "Wealth Inequality in the United States since 1913" (working paper 20625, National Bureau of Economic Research, Cambridge, MA, 2014), 1–65, http://www.nber.org/papers/w20625; Joel Kotkin, *The New Class Conflict* (Candor, NY: Telos, 2014); Fred Bauer, "The New Feudalism: It's Not Too Late to Stop It from Undermining Our Liberal Republic," *National Review Online*, 9 July 2014, http://www.nationalreview.com/article/382266/new-feudalism-fred-bauer; Lori Wallach and Ben Beachy, "Obama's Covert Trade Deal," *New York Times*, 2 June 2013; Margot E. Kamisnski, "Don't Keep the Trans-Pacific Partnership Talks Secret," *New York Times*, 14 April 2015; Edward-Isaac Dovere, "Extreme Secrecy Eroding Support for Obama's Trade Pact," *Politico*, 4 May 2015; Michael McAuliff and Ryan Grim, "Senate Debate Reveals Absurd Level Of Trade Deal Secrecy," *Huffington Post*, 21 May 2015; Alan Morrison, "Is the Trans-Pacific Partnership Unconstitutional? Provisions that Allow Foreign Investors to Bypass the Federal Courts Could Undermind U.S. Legal Protections," *Atlantic*, 23 June 2015; Political economist Michael J. Thompson, in *The Politics of Inequality: A Political History of the Idea of Economic Inequality in America* (New York: Columbia University Press, 2007), is more stinging in his critique of the deafness in contemporary research to inequality: "Late twentieth and early twenty-first century leftist theory has been fragmented by issues such as 'identity' and 'culture' that are now given primacy. The corrosive effects of irrational, pseudo-intellectual fads such as poststructuralism, postmodernism, identity politics, and 'cultural studies' stand as barriers to conceiving ways of combating economic inequality" (xiii–xiv); Martin Gilens and Benjamin I. Page, "Testing Theories of American Politics: Elites, Interest Groups, and Average Citizens," *Perspectives on Politics* 12 (2014): 564–581; Ronald P. Formisano, *Plutocracy in America: How Increasing Inequality Destroys the Middle Class and Exploits the Poor* (Baltimore: Johns Hopkins University Press, 2015).

10. Roger Meiners, Pierre Desrochers, and Andrew Morriss, eds., *"Silent" Spring at 50: The False Crises of Rachel Carson* (Washington, DC: Cato Institute, 2012), 6; "Cato Institute," *Right Wing Watch*, accessed 22 November 2014, http://www.rightwingwatch.org/content/cato-institute; Thomas Medvetz, *Think Tanks in America* (Chicago: University of Chicago Press, 2012); Oreskes and Conway, *Merchants of Doubt*, 216–39.

11. Meiners, Desrochers, and Morriss, eds., *"Silent" Spring at 50*; Leo Hickman, "Rightwing Group Seeks to Strip Climate Change from US Classrooms," *Guardian*, 28 May 2010, http://www.theguardian.com/environment/blog/2010/may/28/rightwing-group-climate-change.

12. Laura Westoff, "The Historian's Role in Teacher Education," *Perspectives on History*, September 2006, http://www.historians.org/publications-and-directories/perspectives-on-history/september-2006/the-historians-role-in-teacher-education; "Where Do Historians of Education Live? Disciplines and Interdisciplines in the Academy," H-Net, accessed 23 November 2014, https://networks.h-net.org/node/14281/pages/14320/where-do-historians-education-live-disciplines-and-interdisciplines; Medvetz, *Think Tanks in America*.

13. Valerie Strauss, "How GERM Is Infecting Schools around the World," *Washington Post*, 29 June 2012, http://www.washingtonpost.com/blogs/answer-sheet/post/how-germ-is-infecting-schools-around-the-world/2012/06/29/gJQAVELZAW_blog.html;

Pasi Sahlberg, *Finnish Lessons: What Can the World Learn about Educational Change in Finland?* (New York: Teachers College Press, 2011); Diane Ravitch, *The Death and Life of the Great American School System: How Testing and Choice Are Undermining Education* (New York: Basic Books, 2011); Ravitch, *Reign of Error: The Hoax of the Privatization Movement and the Danger to America's Public Schools* (New York: Knopf, 2013).

14. Lisa Falkenberg, "Reading Plan Hard to Comprehend," *Houston Chronicle*, 22 April 2008, http://www.chron.com/news/falkenberg/article/Reading-plan-hard-to-comprehend-1644708.php; 2012 Republican Party of Texas, "Report of Platform Committee and Rules Committee," 12, accessed 23 November 2014), http://www.tfn.org/site/DocServer/2012-Platform-Final.pdf?docID=3201.

15. Bruce Hardy, Jeffrey A. Gottfried, Kenneth M. Winneg, and Kathleen Hall Jamieson, "Stephen Colbert's Civics Lesson: How Colbert Super PAC Taught Viewers about Campaign Finance," *Mass Communication and Society* 17 (2014): 329–53.

16. Falkenberg, "Reading Plan Hard to Comprehend."

Acknowledgments

The first time I encountered the word was in 2008. That's when *agnotology* made its formal debut in the title of Robert Proctor and Londa Schiebinger's edited volume. Ignorance as a topic of systematic study, of course, has been around for much longer than that. I've since learned of agnotology conferences held at the turn of the twenty-first century, an article with an equally funny term—*agniology*—from the seventies, and the original coining of the field and its peculiar language by nineteenth-century Scottish philosopher James Frederick Ferrier in his *Institutes of Metaphysics* (1854). But it was the slim collection of essays by Proctor and Schiebinger that first got me thinking about the problem of ignorance. I recall appreciating the volume's energy and enthusiasm for something so obvious. It made the familiar unfamiliar by recasting the significance of ignorance, making it as important as the air we breathe and lamenting that it was just as overlooked and taken for granted. The little volume, and the now substantial body of agnotologically oriented literature, inspired me to bring the research program to education. My goal was to recruit scholars to start building bridges between their areas of expertise and the historical dynamic between education and ignorance. For several years, the idea for this book sat on my desk while other projects soaked up my attention. Then came a sabbatical in fall 2013, and the wait came to an end.

Like all research projects, this one has accumulated many debts. I'm thankful first and foremost to the circle of historians of science—Proctor, Schiebinger, Oreskes, and others—who have stuck their scholarly necks out to write on matters concerning us all. They've published on the social construction of ignorance as it relates to such topics as public health and climate change. In doing so, they've received more than an arrow or two from deep-pocketed industrial interests. To my mind, their work serves as a model for those of us in education who face similarly outsized political and economic sources of influence. It offers a guide for analyzing the way ignorance production has affected policy and practice, past and present. The substantive and methodological possibilities afforded by the agnotology literature have the potential to inspire scholars who study education from historical, philosophical, political, and other disciplinary perspectives for many years to come. That's my hope at least.

During the sabbatical I lost a friend, colleague, and mentor to cancer, a person to whom I feel perpetually indebted: Thomas George Dyer. He was a first-rate scholar, educator, and administrator who held and continues to hold the respect of those fortunate enough to have worked with him. I marvel at the amount he taught me during a short stint in graduate school at the University of Georgia. We kept in touch from the time I graduated until his final months. I'm grateful to have had ongoing conversations with such a sharp mind about matters that animate historians, policymakers, and educational researchers. This book is dedicated to him.

I'm also indebted to each of the contributors for their fine chapters. When I sent out a targeted call for submissions to this volume in fall 2013, these seventeen authors stood out for good reason: they brought expertise in key areas and a glowing desire to do something new. They leapt at the chance to play with agnotology, education, and history in ways that would stretch us all. Their dedication and commitment to this study, along with their patience at each stage of the editing process, have made the collaboration a pleasure. Before I started on this journey, friends and colleagues warned me about the pitfalls and sometimes thanklessness of book editing. Their warnings often ended with some variation of "never again." From start to finish, my experience has been the opposite. I count this to be one the most rewarding and stimulating projects of my career, and I wish to thank the contributors for making it so.

Others who made working on this book a pleasure include Ashley Burnside, Thomas A. Underwood, Jonatha Vare, Greg Britton, and Joanne Allen. Burnside, my graduate assistant, played a critical role in supporting the logistics of manuscript editing and processing. She not only prove to be dependable and reliable on assignment but she also demonstrated a knack for improving all she touches thanks to her excellent training and scholarly instincts. Underwood, a long-time colleague, gave excellent advice during the final leg of the journey. His insights on language, writing, and scholarship have always come at precisely the right time. Vare, my former chair at Winthrop University, deserves acknowledgment for every book project I've completed to date. Her exceptional skills as an administrator, as well as her appreciation for basic research, distinguishes her as a genuine scholar-leader. Despite external pressures we all face from the technocratic and scientistic, she maintains a humanist's outlook. Without her support, this book wouldn't have been possible. Britton, my editor at Johns Hopkins University Press, has a remarkable ability to take in big ideas with a single gulp. His enthusiasm from the start, as well as his ability to separate the worthwhile from the unworthy, has pushed this study to greater heights. And Allen gave the entire manuscript the kind of scrubbing only an expert line editor could give. Just before leaving for a trip to Scandinavia, this book received her close, careful attention and I'm grateful to her for all the improvements she made.

Winthrop's Dacus Library and History Department both deserve recognition for contributing to the advancement of this undertaking. Dean Mark Herring of Dacus and Chair Gregory Crider of History merit special thanks for their assistance with licensing the book's cover art. Their support helped bring years of scholarly work to a satisfying conclusion with just the right image to match the contributors' penetrating analyses of miseducation.

Along the way, I've incurred debts of a nonscholarly kind too. I wish to thank Derek Frank for his music, particularly his album *Let the Games Begin*, and his relentless creative energy. Visiting Frank in Los Angeles during the editing process was one of the best decisions I've made in recent years. The Notso Hostel in Charleston, South Carolina, has long been an important destination for writing and editing. Manager Vikki Matsis and the staff there have always made it a positive, productive, and friendly place to play with ideas. And finally, I wish to acknowledge Kaitlyn Herthel, my partner and inspiration. She has patiently lived with this and other projects scattered around the house for years. Her compassion and intellectual curiosity helped bring this book to life.

Contributors

A. J. Angulo is the Elizabeth Singleton Endowed Professor of Education and Professor of History (Courtesy) at Winthrop University. His publications include the award-winning *William Barton Rogers and the Idea of MIT* (Johns Hopkins University Press, 2009) and *Empire and Education* (Palgrave, 2012). Angulo's articles have appeared in *History of Education Quarterly*, *Historical Studies in Education*, *History of Science* (UK), *Perspectives on the History of Higher Education*, *Vitae Scholastica*, *World History Encyclopedia*, *Education and Law Journal*, and *Innovative Higher Education*.

Daniel Bar-Tal is the Branco Weiss Professor of Research in Child Development and Education at the School of Education, Tel Aviv University, Israel. His research interests are political and social psychology, including the sociopsychological foundations of intractable conflicts and peace building, as well as the development of political understanding among children and peace education. He has published twenty books and more than two hundreds articles and chapters in major social and political psychological journals, books, and encyclopedias. He served as president of the International Society of Political Psychology and received various awards for his work, including the Lasswell Award and the Nevitt Sanford Award of the International Society of Political Psychology.

Kevin C. Elliott is associate professor of philosophy at Michigan State University, with appointments in Lyman Briggs College, the Department of Fisheries and Wildlife, and the Department of Philosophy. His book *Is a Little Pollution Good for You? Incorporating Societal Values in Environmental Research*, was published by Oxford University Press in 2011. He has also published numerous articles on the philosophy of science, environmental ethics, and research ethics in journals such as *Philosophy of Science*, *History and Philosophy of Science*, *Perspectives on Science*, *Cell*, *Human and Experimental Toxicology*,

Environmental Ethics, *Ethics, Policy, and Environment*, *Science and Engineering Ethics*, and *Accountability in Research*.

E. Thomas Ewing is associate dean for graduate studies, research, and Diversity in the College of Liberal Arts and Human Sciences and a professor in the Department of History at Virginia Polytechnic Institute and State University. He teaches courses in Russian, European, Middle Eastern, and world history, gender/women's history, and historical methods. His books include, as author, *Separate Schools: Gender, Policy, and Practice in the Postwar Soviet Union* (2010) and *The Teachers of Stalinism: Policy, Practice, and Power in Soviet Schools in the 1930s* (2002); as editor, *Revolution and Pedagogy: Transnational Perspectives on the Social Foundations of Education* (2005); and as coeditor, with David Hicks, *Education and the Great Depression: Lessons from a Global History* (2006). His articles on Soviet educational history have appeared in *Russian Review*, *Journal of Women's History*, *Gender and History*, and *History of Education Quarterly*.

Karen Graves is a professor and chair of the Department of Education at Denison University. Her research addresses twentieth-century schooling in the United States with a focus on gender and sexuality, and legal policies concerning education. Her most recent book, *And They Were Wonderful Teachers: Florida's Purge of Gay and Lesbian Teachers*, was awarded a 2010 Critics Choice Book Award from the American Educational Studies Association. Other publications include *Girls' Schooling during the Progressive Era: From Female Scholar to Domesticated Citizen* (Garland, 1998) and the co-edited volume *Inexcusable Omissions: Clarence Karier and the Critical Tradition in History of Education Scholarship* (Peter Lang, 2001), with Timothy Glander and Christine Shea. She is a past president of the History of Education Society and a former vice president of the American Educational Research Association, Division F: History and Historiography.

Dongping Han is a professor of history and political science at Warren Wilson College and a guest professor at Hebei University in China. His publications include "The Scientific and Technological Development in Chinese Agriculture during the Cultural Revolution Decade," in *Mr. Science and Chairman Mao's Cultural Revolution*, ed. Darryl E. Brock and Chunjuan Nancy Wei (Lexington Books, 2012); "Farmers, Mao, and Discontent in China, from the Great Leap Forward to the Present," *Monthly Review*, December 2009, 21–38; “Impact of the Cultural Revolution on Rural Education and Economic Development,” *Modern China* 27 (2001); and *The Unknown Cultural Revolution—Education Reform and Rural Development* (Garland, 2000).

Lisa Jarvinen is an associate professor of history at La Salle University in Philadelphia. She holds a PhD in history from Syracuse University. She researches and publishes on the history of cultural relations in the Spanish-speaking world and on the history of education. Her books include *The Rise of Spanish-Language Filmmaking: Out from Hollywood's Shadow, 1929–1939* (Rutgers University Press, 2012) and the coedited volume (with Richard Garlitz) *Teaching America to the World and the World to America: Education and Foreign Relations since 1870* (Palgrave, 2012). She is currently researching the intersections between education, territorial expansion, and citizenship in the United States.

Adam Laats is a historian in the Graduate School of Education at Binghamton University (State University of New York). He is the author, most recently, of *The Other School Reformers: Conservative Activism in American Education* (Harvard University Press, 2015). He is currently working on a book with philosopher Harvey Siegel about the history and philosophy of evolution education for the University of Chicago Press.

Daniel Perlstein is an associate professor of education at University of California, Berkeley. He is a historian committed to scholarship that promotes the creation of more equitable and humane schools. His research focuses on the relationship of democratic aspirations to social inequalities in American schools and life. His book *Justice, Justice: School Politics and the Eclipse of Liberalism* (Peter Lang, 2004) explores the place of race and class conflicts in the politics of urban education. He has also written on such topics as gender and school violence, progressive pedagogy, and the history of African American education. His work can be found in such academic journals as *History of Education Quarterly*, *Paedagogica Historica*, and *Teachers College Record*, as well as in such professional journals as *Educational Leadership* and such activist publications as *Transforming Teacher Unions* and *Putting the Movement Back Into Civil Rights Teaching.*

Jennifer Burek Pierce is an associate professor in the School of Library and Information Science at the University of Iowa. Fellowships from the American Antiquarian Society and Winterthur Museum, Libraries and Garden have funded her study of children's books and related materials. Her books include *What Adolescents Ought to Know: Sexual Health Texts in Early 20th Century America* (University of Massachusetts Press, 2011) and *Sex, Brains, and Video Games: A Librarian's Guide to Teens in the 21st Century* (ALA Editions, 2009). She developed the Youth Matters column for *American Libraries* and has also written for the *Chronicle of Higher Education*. She has served on the advisory boards of the University of Iowa Center for the Book and the Iowa Initiatives for Sustainable Communities.

Matt Pierce is a lecturer in the Department of Telecommunications at Indiana University–Bloomington. He earned his JD from the Indiana University School of Law–Bloomington in 1987 and his BA in Telecommunications/Political Science from Indiana University in 1984. He teaches courses in telecommunications law and policy and telecommunications industry and management. His experience includes work as an intern for the US House of Representatives Subcommittee on Telecommunications, Consumer Protection and Finance and as a legal clerk at the National Association of Broadcasters. He was chief of staff for Congressman Baron Hill from 1999 to 2002. He was elected to the Indiana House of Representatives in 2002. Pierce holds an Extra Class amateur radio license and is an adviser for the IU amateur radio club (K9IU).

Lisa Pine is a reader in history at London South Bank University. She is a graduate of the Department of International History at the London School of Economics and Political Science. She obtained her doctorate from the University of London in 1996. Her main research interests are the social history of Nazi Germany and the Holocaust. Her major publications are *Education in Nazi Germany* (Berg, 2010), *Hitler's "National Community": Society and Culture in Nazi Germany* (Hodder Arnold, 2007); and *Nazi Family Policy, 1933–1945* (Berg, 1997).

Adam R. Shapiro is a lecturer in intellectual and cultural history in the Department of History, Classics, and Archaeology at the University of London. His principal publications include *Trying Biology: The Scopes Trial, Textbooks, and Antievolution in American Schools* (University of Chicago Press, 2013); "Between Training and Popularization: Regulating Science Textbooks in Secondary Education," *Isis* 103 (2012): 99–110; "State Regulation of the Textbook Industry," in *Education and Print Culture in Modern America*, ed. Adam Nelson and John L. Rudolph (University of Wisconsin Press, 2010), 173–90; "The Scopes Trial: Beyond Science and Religion," in *Science and Religion: New Historical Perspectives*, ed. Thomas Dixon, Stephen Pumfrey, and Geoffrey Cantor (Cambridge University Press, 2010), 382–424; "William Paley's Lost 'Intelligent Design,'" *History and Philosophy of the Life Sciences* 31 (2009): 55–78; and "Civic Biology and the Origins of the School Antievolution Movement," *Journal of the History of Biology* 41 (September 2008).

Stephen Samuel Smith is a professor of political science at Winthrop University. He holds a PhD from Stanford and is coeditor (with Roslyn Arlin Mickelson and Amy Hawn Nelson) of *Yesterday, Today, and Tomorrow: School Desegregation and Resegregation in Charlotte* (Harvard Education Press, 2015) and author of *Boom for Whom? Education, Desegregation, and Development in Charlotte* (SUNY Press, 2004). Among his other publications are "Still Swimming against the Resegregation Tide? A Suburban Southern School District in the Aftermath of *Parents Involved*," *North Carolina Law Review* 88 (2010), and the coauthored articles (with Jessica Kulynych) "It May Be *Social*, But Why Is It *Capital*? The Social Construction of Social Capital and the Politics of Language," *Politics & Society* 30 (2002), and (with Karen M. Kedrowski and Joseph M. Ellis) "Electoral Structures, Venue Selection, and the (New?) Politics of School Desegregation," *Perspectives on Politics* 2 (2004). He also writes about antiwar movements and resistance by US soldiers to military authority during the wars in Iraq and Afghanistan.

Eileen H. Tamura is a professor of history of education and chair of the Department of Educational Foundations in the College of Education at the University of Hawai'i. Her publications include *In Defense of Justice: Joseph Kurihara and the Japanese American Struggle for Equality* (University of Illinois Press, 2013); as editor, *The History of Discrimination in U.S. Education: Marginality, Agency, and Power* (Palgrave, 2009); as editor, *Asian and Pacific Islander American Education: Social, Cultural, and Historical Contexts* (Pacific Asia Press, 2002); and *Americanization, Acculturation, and Ethnic Identity: The Nisei Generation in Hawaii* (University of Illinois Press, 1994). Tamura is past president of the History of Education Society and the recipient of the Franklin Buchanan Award (Association for Asian Studies) and the James Harvey Robinson Prize (American Historical Association).

Kim Tolley is a historian of education and professor at Notre Dame de Namur University. Her publications have appeared in such journals as *History of Education Quarterly*, *Teachers College Record*, *Social Science History*, *North Carolina Historical Review*, and *Journal of Curriculum Studies*. She has received the History of Education Society's prize for best article published in a refereed journal (2008). Her book *The Science Education of American Girls: A Historical Perspective* (RoutledgeFalmer, 2003) received the Outstanding Academic Title Award from the Association of College and Research Libraries (2003), Honorable Mention for the Outstanding Book Award from the History of Education Society (2004), and Honorable Mention for Best Book by a New Scholar (Division F, American Educational Research Association, 2005). She was coeditor (with Nancy Beadie) of *Chartered Schools: Two Hundred Years of Independent Academies in the United States, 1727–1925* (RoutledgeFalmer, 2002) and editor of *Transformations in Schooling: Comparative and Historical Perspectives* (Palgrave, 2007). Tolley has served on the editorial board of the *History of Education Quarterly* and on the board of the History of Education Society. She recently completed *Heading South: Education, Family, and Slavery in the Second Great Awakening* (University of North Carolina Press, 2015).

Soli Vered is a doctoral student at the School of Education, Tel Aviv University, Israel. She holds a master's degree in organizational sociology from the University of Haifa. Her PhD research focuses on the policy of the Israeli Ministry of Education in recent years in regard to peace education and the challenges that peace educators face in the field. Her other research interest is in critical thinking in education, culture of conflict, and peace-building.

Donald Warren is a professor emeritus in the history of education and policy programs at Indiana University–Bloomington and university dean of education emeritus at IUB, serving from 1990 to 2000. His previous academic and administrative appointments were at the University of Maryland, College Park; the University of Hawaii at Manoa; the University of Illinois, Chicago; and Chicago State University. He holds degrees from the University of Texas at Austin, Harvard University, and the University of Chicago, where he completed the PhD in education and history. His books, monographs, chapters, and articles have focused on the history of federal roles in US education, contributions of historical research to policy analysis, and the intersecting histories of teacher education, teachers, and teaching. He has been president of the American Educational Studies Association, the History of Education Society, and the Council of Learned Societies in Education and vice president of the American Educational Research Association for the Division of History and Historiography.

Index

Aborigines, 231
abortion, 91
academic achievement, 129, 140, 147
academic freedom, 53
Acanfora v. Board of Education of Montgomery County, 65
Adam, Peter, 254
Adams, Charles Francis, 222, 225
administrators: government, 141; school, 53, 58, 153, 155, 225, 258, 275, 321
Advancement of Sound Science Coalition, The (TASSC), 104–5, 110, 112
advocacy, 40, 64–66, 75–76, 84, 90, 189, 224
Afghanistan, 220, 235
African Americans, 4, 124–45, 147
Africans, 29, 140
Age of Fracture, 187
agnotology, 5, 7–10, 13, 54, 85–86, 89, 91, 96–97, 132, 263, 272, 296, 320–21, 323, 332, 340–41, 343–48
Aguinaldo, Emilio, 232, 234
Alabama, 14, 27, 187
Alaska, 207
aliyah, 302, 303. *See also* immigration; Zionism
All God's Dangers: The Life of Nate Shaw, 186–87
American Anti-Imperialist League, 222
American Association of Christian Schools (AACS), 162
American Association of University Professors, 53
American Cancer Society, 99
American Civil Liberties Union of South Carolina, 53
American College of Epidemiology, 105
American Indian: Algonquian, 199–200; Comanche, 202, 208; Crow, 208; Hopi, 167; Iroquois, 184–85, 196, 199, 202–3, 208; Mohawk, 184, 196; Onondaga, 196; Plains nations, 202; Powhatan, 199; Pueblo, 167; Seneca, 196; Sioux, 202, 208; Six Nations, 195–97; tribes, 172, 174, 193, 200, 202, 204–5, 207–9; US Bureau of Indian Affairs, 205
Americanization, 156, 225
American Library Association, 52
American Lung Association, 99
American Mercury, 43
American Meteorological Society, 108
American Revolution, 164, 185
American Social Hygiene Association, 34, 49
American Umpire, 236
ancient period: Greece, 284; Rome, 218, 284
Andreas, Joel, 328
And Tango Makes Three, 52
Anglo-Saxons, 224
Answers in Genesis, 78, 81–82, 85, 91
anthropology, 196, 199–201, 208, 339
anthropomorphism, 37
anti-American thought, 221
antidiscriminatory legislation, 60
antievolution, 73–78, 80, 85–87, 89–93
antifamily, 52
antigay: backlash, 60; discrimination, 60
anti-imperialist, 217–19, 221–25, 228–31, 233–34, 238
anti-life, 62
antiregulatory, 104
anti-Semitism, 245–46, 259, 263, 298
anti-Slav, 255
antislavery, 14, 26, 179
April Theses, 272
Arabian, 269, 286
archaeology, 196, 199, 201
Arikara, 208

arithmetic, 128–29, 258–59
Arkansas, 14–15, 75, 86–87
army, 150, 270, 299
Arnove, Robert, 13
arrest. *See* imprisonment
Arrow, Dennis, 64–65
art: cubist, 254; Dadaist, 254; Expressionist, 253; futurist, 254
asbestos: exposure, 102; industry, 5, 97, 102, 103, 110
Asia, 7, 140–48, 150–58, 173, 192, 232, 331, 340, 342
Asian American: identity evolution, 140; studies, 146
Asian American Student Alliance, 145
assimilation, 141, 205–7, 233–34, 295
Association of Christian Schools International (ACSI), 162
Atlantic, 184, 190
Austria, 2, 250
authority, 40, 44–45, 66, 89, 128, 140–41, 194, 228, 279, 311, 332, 346
autocratic power, 222
auto-policing, 252
auto-surveillance, 252
Ayçoberry, Pierre, 250
Aylesworth, M. H., 43, 51

Bachmann, Rep. Michele, 161–62, 179
Bailyn, Bernard, 5, 188–95, 198–201
Baker, David V., 25
Bangladesh, 146
Barbarous Years, The: The Conflict of Civilizations, 1600–1675, 192, 199–200
Barnes, Julian, 191
Barrett, James Judge, 64
Barton, David, 161
Bartz, Paul, 167
bastardization, 260
Bauer, Gary, 167
Beck, Glenn, 161, 165
Beder, Sharon, 105
Bedford, Daniel, 97
behavior, 55, 61, 90, 140, 143, 149, 156, 185
Behe, Michael, 79
Being George Washington, 165
Beisner, Robert, 219
Bemis, Samuel Flagg, 231
Ben-Amos, Avner, 304
Bering Strait, 173
Beth-El, Ilana, 304
Beveridge, Albert, 223–24, 228–32, 237
biases, 83, 191
Bible, 15, 73–74, 82–83, 86, 90, 165, 167, 170, 172–73, 175
Bible-Science Newsletter, 167
biblical figures: Adam, 73, 83, 91; Canaan, 29; Christ, 26, 164, 166; Eve, 73, 83, 91; Ham, 29; the Holy Spirit, 164; the Lord, 167, 170; Noah, 173; Peleg, 173
biblical interpretation: allusions, 172; faith, 166; hermeneutics, 176; narratives, 74; revelation, 165; text, 83
biblical locations: Babel, 172; Eden, 91; Upper Galilee, 303
biblical text: Deuteronomy, 175; Galatians, 164; Genesis, 74–75, 78, 81, 86, 91, 172; Noachian flood, 82; revelation of scripture, 90; Romans, 164; 2 Thessalonians, 175; the virgin birth, 90
bilingualism, 152, 154
biology, 75–76, 78–80, 93, 126, 128, 199, 257–61
Bismarck, Otto von, 272
blacklisted, 257
Blackmon, Douglas A., 192–93
black power, 144–45
Blessings-of-Civilization Trust, 217
bloodline theory, 325
Blue Angel, 255
Blunck, Hans Fredrich, 255–56
Board of Education of Oklahoma City v. National Gay Task Force, 65
Boers, the, 230
Boer War, The, 221, 272
Bolshevism, 259
bourgeois, 132, 272
Boxer Rebellion, 217
Brennan, Justice, 64–65, 68
Briggs, John, 61–64, 66
Brink, Jack, 201–2
Brinkley, Alan, 168
Britain: colonial experiments, 173; experience in India, 225; government, 2, 196, 199, 205, 305; precedent, 225; tyranny, 164
Brohmer, Paul, 260
Brown v. Board of Education, 56
Brulle, Robert, 109

Bryan, William Jennings, 90, 222, 232–33
Bryant, Anita, 59–64, 67
bureaucratization, 322
Bush, President George H. W., 107, 112, 151
Bush administration, 107
Butts, R. Freemans, 203–4

California, 45, 60–61, 63–64, 66–67, 76, 86, 100, 135, 145, 151, 164
California Society of the Sons of the Revolution, 228
Caligula, 286
Cambodia, 146
Camp David Summit (2000), 297
Canada, 207
cancer, 5, 98–103, 105, 111, 219
Capital in the Twenty-First Century, 187, 344
capitalism, 124, 127, 135, 173–74, 179, 322, 331
Carlin, George, 44
Carnegie, Andrew, 165, 222, 225
Carter, President Jimmy, 165
Caucasians, 140
censorship, 34, 36–44, 52–53, 244–45, 252–64
Central America, 236
Central Committee of the Soviet Communist Party, 8, 274–77, 281–82
"Cheer for American Imperialism, A," 234
chemical companies: BASF, 106; Monsanto, 106, 110
Chemical Manufacturers Association, 105
chemistry, 6, 79, 90, 98, 103, 105–6, 110, 114, 126
ChemRisk, 103, 110
Chicago, Illinois, 41, 48–49, 142
Chicago Daily News, 142
Chicano studies, 146
China, 8, 141, 146, 175, 201, 217, 231, 268, 319–44
Chinese Communist Party (CCP): Central Committee of, 322–23; Resolution of Historical Issues (1981), 326
Chinese Exclusion Act of 1882, 141
Chinese Cultural Revolution as History, The, 330
chlorine, 105–6, 110
Chlorine Chemistry Council, 105
Christianity, 7, 15, 26, 60, 67, 161–79, 217, 263
Christianity and the Constitution: The Faith of Our Founding Fathers, 165
Christian publishers: Accelerated Christian Education (ACE), 169–70; Alpha Omega, 169; A Beka Book, 169–76; Bob Jones University Press, 169–70; Christian Light Education, 169; Mott Media, 169; Rod and Staff, 169
Christian Schools International (CSI), 162–63
chromium, 97, 103, 110
civil rights, 54, 56, 59–61, 66, 142–45, 148, 153, 155
class, 4, 22, 28, 123–37, 141, 144–46, 148, 185, 249, 272, 286, 322, 324, 333
class struggle, 3, 124, 130, 132–33, 136, 274, 278, 319–20
Cleopatra, 286
Cleveland, President Grover, 229
climate change, 5, 7, 77, 96–97, 106–9, 111–12, 115–16, 219, 339
coercion, 224, 325
Cohen, Sol, 189
Cohen v. California, 45
Cold War, 54–56, 67, 112, 344
collectivism, 168, 179
colonialism, 8, 15, 19, 164–65, 170–71, 173–74, 184–86, 188, 190, 192, 196–207, 217–38
Columbus, 207
Communications Act of 1934, 39
communism, 8, 55, 113, 168, 170, 173–74, 268–89, 319–34
compulsory education, 62, 75, 311
confidential business information (CBI), 114
conflicts: American Revolution, 164, 185; Arab Revolt, 304; Battle of Kulikovo Fields, 282; Boer War, 221, 272; Chinese Revolution, 321, 324, 326; civil war, 14–15, 29, 187, 192, 268; cultural, 185; Deir Yassin, 305, 308; Imperial War, 272; Indian Wars, 232; Israeli-Arab, 295–312; Kfar Qasim, 306; Metacom's War, 164; Philippine-American War, 217, 230; Qibya, 306; Russian Revolution of 1917, 273; second intifada, 297; Sinai War of 1956, 305; Spanish Civil War, 186–87; Thirty Years War, 256; Vietnam War, 145–46, 178; War of 1898, 8, 217–18, 220, 237; World War II, 55, 141, 145, 148–49, 153, 187, 251, 298

Congo, 186
conservatism, 36, 43, 60, 109, 161–79
conservative think tanks: American Enterprise Institute, 109; Cato Institute, 109, 345; Heritage Foundation, 109; Hoover Institute, 109; Manhattan Institute, 109
Conway, Erik, 78, 85, 100–101, 104, 106–7, 109, 111–13
cosmology, 78, 197
Council for Tobacco Research (CTR), 98–100
craniology, 260
creationism, 73–79, 82–89, 91
Cremin, Lawrence A., 184–86, 189–93, 201, 203
Crow, Jim, 91, 144, 193
Cuba, 8, 13, 221–22, 225–26, 230
Cubberley, Ellwood P., 190, 198, 201
Cullinane, Michael Patrick, 223
cultural capital, 148, 320, 333
Cultural Revolution, 8, 319–34
culture: Arabian, 269; annihilation, 144, 247; asset, 191, 204, 208; authority, 89; blinders, 208; change, 61, 88, 191, 196–99, 254; chauvinism, 207; continuity, 185, 191, 194, 196, 199, 207–8; differences, 189; exchange, 254; expression, 252; formation, 187, 192, 201, 205, 207; legitimacy, 207; minority, 178; occlusion, 161, 176–79, 259, 263; purification, 253; respect, 88; responsive teaching, 146
curriculum, 53–54, 63–64, 66, 86, 123, 129–31, 134–35, 168–69, 196, 202–4, 206, 257–60, 263–64, 269, 275–76, 287, 295, 302, 326, 329, 345, 347
Curti, Merle, 199
Cyprus, 300

Damocle's sword, 54
Darwin, Charles, 75, 79, 90–91
Darwinianism, 88–91
Darwin on Trial, 84
Darwin's Black Box, 79–80
dazibao, 321, 333
Declaration of Independence, 53, 222
Degenerate Art Exhibition, 253–54
Dekanisora, 195
Dembski, William, 84, 88
Democrats, 222
demography, 199
destruction: by adventurism, 218, 229; of books, 323; of culture, 330; by fire, 17–18, 21, 28; of property, 300, 306–7; weapons of mass, 2
development: agency, 195; of child, 310; cultural, 192, 275; curricular, 75; national, 114, 253, 272, 276, 322–23, 331; social, 194, 295, 297; US, 206
Dewey, John, 123, 125–27, 129, 133, 134, 168
dictatorship, 251, 253, 264, 273, 327
Dinur, Ben-Zion, 295
dioxins, 97, 104–7, 110
disciplines. *See specific disciplines*
discrimination: education, 140; employment, 54, 60–62, 68, 155; housing, 60; racial, 141, 144, 153
disease prevention, 34, 48
disinformation, 7, 140, 244–45, 247, 249, 252, 263–64
disloyalty, 20, 233
displacement, 307. *See also* exodus; expulsion
dissent, 64–65, 68, 149–51, 163, 223, 321
distortions, 186, 264, 320, 344–45, 347
diversity, 7, 41, 54–55, 67, 74, 77, 146–48, 155, 174, 196, 204, 233
Dix, Otto, 253–54
DNA, 79
doctrinal beliefs, 167, 260
Dollars, Doctors, and Disease, 34
donations, 88, 109
doubt, 5, 18, 26, 73–74, 76–78, 81–82, 84–87, 89, 92, 96, 99, 100, 101–3, 106–7, 109, 111, 201, 219, 221, 235, 287–88, 330, 344–45
Dover, PA, 75, 78
draft resisters, 150–51
dropout rates, 147
Du Bois, W. E. B., 3–4, 193
DuPont, 102, 110
Dutch, 185, 206
duty, 18, 143, 150–51, 224, 229, 232, 234

earth, 74, 78, 81, 82, 86, 86, 90, 97, 107–8, 172–73
economy, 3, 130–31, 175, 185, 199, 248, 250, 331–32, 339, 344
education: access, 2, 8, 14, 36, 39, 41–42, 45–46, 52–53, 57, 61, 114, 134, 329, 331–32, 346; administrators, 53, 58, 141, 153, 155,

170, 225, 258, 275, 319, 321; attainment, 208, 333; canon, 187, 194, 311; common, 189, 192, 346; dangerous, 15, 36, 89, 171, 178, 340; differentiated, 172, 195; discordant, 185, 197, 203; elementary, 2, 8, 257, 274–77, 285; expansion, 75, 190, 271, 329, 341; formal, 279, 301; graduate, 2, 194, 326, 328; history, 161, 168, 174, 176–77; intentional, 202; as a lived experience, 195, 327–28, 330; middle, 319, 326, 328–29; opportunity, 7, 329, 331; primary, 53, 260, 281, 319; public, 37, 66, 97, 110, 113, 116, 204, 218, 256; secondary, 2, 257, 274–75, 277–78, 281, 329; slave, 195; technical, 147, 198, 323; undergraduate, 237, 326; white-imposed, 188, 206. *See also* higher education
educational outreach, 63, 198
Edwards v. Aguillard, 87
Egypt, 201, 296, 305, 309
Eidsmoe, John, 161, 165
Einstein, Albert, 261
emissions, 104, 107–8
empire, 207–8, 218–21, 223, 227–30, 232–37
Engels, Friedrich, 283
English language: learners, 146; limited proficiency, 152, 154
ENVIRON, 103
environment, 80, 96–116, 125, 127–28, 342, 345–46
Environmental Protection Agency (EPA), 345
epidiascope, 284
epistemology, 77–80, 82–84, 90–91
Epperson v. Arkansas, 75, 86
equality, 4, 29, 54, 143, 175, 203–4, 206
ethnicity, 133, 140–44, 146–56, 200, 203
eugenics, 35, 91, 247, 260
Eurocentrism, 189, 191, 194
evangelicalism, 161–79, 184, 191, 197, 199
evolution, 7, 73–93, 342–43
examinations, 2, 186, 272–73, 277, 285–86, 326, 329, 333, 346
exile, 149, 253, 255, 269, 297–98, 304, 308
existential threat, 298, 310
exodus, 255, 296, 307, 309
expansionism, 190, 207–8, 217–38
exploitation, 60, 101, 136, 185, 220–21, 259, 277, 285
expression, 38, 45, 53, 55, 67, 227, 252–54, 269, 280, 322
expulsion, 306–7
extremism, 343
ExxonMobil, 109, 345

fallacy, 79, 81, 144
Farnam, Henry W., 14
FCC v. Fox Television Studios, 45
fedayyin, 305
Federalist Papers, The, 53
Filipino organizations, 152–53
Fenton, William N., 195–99, 208
Ferguson, Niall, 235
Field, Henry, 35, 41
Field Museum of Natural History, 41
Filipinos, 152–55, 218, 224, 231–34
film, 187, 245–49, 260, 264, 270
"Filthy Words," 44
First Amendment, 53, 65–66, 68, 87
Florida, 14, 55–61, 186, 205
folk culture, 132, 185, 204, 258
foreign policy, 55, 184, 196, 224, 236, 250, 283, 285, 309, 321–22
foundations: Bradley, 109; Carnegie, 165; Donors Capital, 109; Donors Trust, 109; Ford, 194; Heritage, 109; Howard, 109; Koch, 109; Pacifica, 44–45; Pope, 109; Scaife, 109; Searle, 109
442nd Regimental Combat Team, 150
four "olds," the, 323–33
Fourteenth Amendment, 60, 64, 68
Four Year Plan, 249
France, 19, 154, 196–97, 206, 268, 305
Franconia, 246
Frankfurter, Justice, 66
Franklin, Benjamin, 198, 202–3
fraud, 106, 108, 112, 346
Freemasons, 261
Fremont, Walter, 171
From Darwin to Hitler, 91
fundamentalism, 163, 166
funding, 53, 96, 99–100, 104, 108–9, 113–14, 189, 332. *See also* money

Gallup poll, 92
Gao, Mobu, 328–29
Gaylord, James, 52, 65
Gaylord v. Tacoma School District No. 10, 65

gays and lesbians (LGBTQ), 52–68, 91, 175, 250, 253
Gaza Strip, 309
gender, 54–55, 186, 205–6, 343
General Motors, 102
genetics, 79, 84, 260, 339
genocide, 207, 261
geography, 54–55, 63, 73, 257–59, 277
geology, 78, 82
geopolitics, 298
George C. Marshall Institute, 107–9, 111–12
Georgia, 8, 13–29
Germany, 2, 8, 187, 244–64, 340–42
Gilmer, George R., 13, 27–29
Ginsberg v. New York, 45
globalization, 205, 341
Goebbels, Josef, 244, 248
Gould, Stephen Jay, 85, 88
government: federal, 7, 35–44, 53–55, 64, 75, 123, 130, 136, 150, 153, 192; officials, 53–54, 99, 101, 109, 131; regulators, 7, 36, 38, 106, 110, 112, 114–15, 340, 345; restrictions, 44; state, 13, 41, 53–68, 76, 86, 130–31, 144–45, 147, 152–54, 192–93, 197, 233, 346
Grand Rapids, MI, 47, 227
Grant, Ulysses S., 223
Greece, 2, 284
gun ownership, 168
gypsies, 259, 263

Hagana, 305, 308
Haley, Gov. Nikki, 53
Han, Dongping, 328–29
Hawai'i, 151–56, 207, 222, 226–27
health: activists, 40–41; associations, 41; boards, 48–49; care, 331, 346; commissioners, 34, 41, 46–48
heart disease, 100
Helm, Mary, 64, 67
herbicide, 103
heredity, 74, 99, 260–61, 263
heritage, 109, 132–33, 162, 165, 169, 171, 176–77, 204–5, 299
Herodotus, 268
heroism, 201, 298, 301, 304
heterosexuality, 54, 59, 175
higher education: admissions, 147; Beijing University, 321–23; Bob Jones University (BJU), 166–67, 169–71, 174–76; Brown University, 29; California State University, Long Beach, 145; College of Charleston, 53; Hampton Institute, 225; Harvard University, 66, 200, 343; history of, 332; Oklahoma City University, 64; Rockefeller University, 111; Stanford University, 41, 107; Tsinghua University, 323; Tuskegee Institute, 4, 225; University of California, Berkeley, 103; University of California, San Francisco, 100; University of California, Santa Cruz, 145; University of Florida, 56; University of Heidelberg, 258; University of Pittsburgh, 103; University of South Carolina–Upstate, 53; William and Mary, 202
Himmler, Heinrich, 248, 251
Hindostan, 225
historiography, 36, 188, 220, 270, 307
history: ancient, 3, 173, 202, 204–5, 268, 274, 284, 298; authentic, 168; bad, 176; Communist Party, 272; defense, 258; economics, 3, 7, 13, 22, 55, 104, 123–27, 129, 134–36, 142, 145–46, 148, 172, 174, 187, 193, 206–7, 221, 236, 246, 250, 277, 298, 321, 330–31, 344–45; education, 7, 156, 184–209, 332; European, 277; external, 7, 140–41; internal, 7, 99, 103, 140, 196, 221; rape of, 165; of sexuality, 54; of slavery, 133, 142, 192; Stalinist, 269, 271, 275; world, 272, 319
Hitler, Adolf, 91, 244, 246–51, 254, 257–58, 262, 264
Hmong, 146
Hoffman, Elizabeth Cobbs, 236–37
homeschool, 163
homogeneity, 132, 204, 245, 257
homosexuality, 52–68, 91
Homosexuality and Citizenship in Florida, 55, 58–59
Hoover, Herbert, 38
horticulture, 208
Horton, Ronald, 166–67
Hotel Florida, 186
human nature, 200, 343
Hutchison, Susan Nye, 16–18, 20–25, 29
Hutus, 300
hydrology, 82
hypocrisy, 198, 224, 250

identity, 54, 62, 65, 75, 89, 126, 140–56, 163, 195, 295–98, 300–301, 310–11, 320, 342, 344
ideology, 59, 164, 168, 245–46, 248, 252–53, 255–59, 261, 263–64, 271, 296, 319–20, 322
ignorance: active, 6, 77, 346; coerced, 224, 275, 287, 324, 325; construction of ignorance, 5–8, 35, 54, 59, 76, 78, 81, 96, 144, 150, 156, 193, 219, 244, 250, 269, 271, 296, 302, 320, 323, 324, 332–33, 339, 342–44, 346; density of, 224; first-order, 73; geography of, 54; naïve, 5–6, 73, 85, 161, 179, 271, 342; native, 5–6, 161, 271, 284, 344; selective ignorance, 271; strategic, 37
illiterate, 29, 268, 276, 323
illness, 248, 261
immigration, 146, 152, 154, 302
immorality, 63, 65, 91
imperialism, 168, 186, 198–99, 217–38, 272, 274, 319, 341–42
Imperial Republic, The, 229
impressionability, 61, 259
imprisonment, 13, 19–24, 26, 131, 141, 145, 148–51, 244, 252, 255, 269–71, 326, 328
incentives, 23, 113–15, 174, 322
inclusion, 60, 189, 195, 217, 263, 342
Indiana, 224
Indians, 133, 147, 167, 172–74, 184–209, 224–25, 232, 343
Indian Wars, 232
indoctrination, 269, 271, 273, 282–84, 286–87, 309
inequality, 130, 187, 319–20, 322, 344
infallibility, 245, 251
infants, 22, 42
in-group, 297–300, 309
innocence, 20, 24, 29, 36, 38, 46, 73, 150
Inouye, Daniel K., 144–45
Institute of Early American History and Culture, 188
instruction, 1, 8, 13–15, 26–28, 35, 52–54, 62, 73, 76, 86–87, 90, 97, 123–24, 129, 133–34, 136, 144, 146, 164, 168, 170, 176, 185, 187–90, 192, 196–97, 202–5, 208–9, 219, 224–26, 237, 248, 257–63, 268–89, 295–96, 301, 310–12, 323, 332–34, 346
insurgents, 231–32
intelligence, 39, 75–76, 78–79, 81, 83–84, 86, 88–90, 105, 131, 140
interest groups, 97–98, 103, 105, 110, 113, 115. *See specific interest groups*
interrogations, 21, 56, 252
interventions, 81, 90, 130, 218, 220
intimidation, 109, 252
invasions, 199, 205
investigations, 6, 15, 56–59, 99, 104, 111, 115, 193–95, 199, 201–2, 209, 223, 339, 343–44
Iraq, 220, 235
Ireland, 2, 132–33, 192, 300
irreducible complexity, 79–80
islands, 154–55, 193, 207, 217, 221–28, 231–34, 237
Israel, 8, 175, 295–312, 340, 342
Italy, 2, 206, 303

Jackson, President Andrew, 223
jail. *See* imprisonment
Japan, 145, 151, 272
Japanese American, 140–55
Japanese American Citizens League, 142
Japanese Americans: Oppression and Success, 142
Japanese myths, 269
Jastrow, Robert, 107, 111
Jefferson, Thomas, 127, 223, 230
Johnson, Philip E., 84, 88
Johnson, President Lyndon, 2, 153
Jones, Elizabeth, 22–23
Jordan, 297, 309
journalism, 4, 100–101, 106, 108, 111, 142, 168, 187, 227, 234, 302
judgments, 38–39, 76, 83, 87, 91, 164, 235, 248, 342

Kentucky, 15, 77
Kershaw, Ian, 249, 251
Kitzmiller v. Dover Area School District, 75
Klee, Paul, 253–54
knowledge: absence of, 5, 78, 288; competing systems, 161; the moralization of, 77, 90; nonobtained, 85; officially sanctioned, 280, 326, 328; real, 319; the tree of, 83, 91
Korea, 146–48, 152
Kramer, Paul, 221, 223
Kraus, Richard Curt, 331–32
Kuhn, Thomas, 332
Kuomintang, 319

LaHaye, Tim, 164–65
Laos, 146
Latin, 202
Latin America, 7, 145, 147, 221
Lawrence v. Texas, 60
lead, 97, 102–3, 110
Lebec, California, 76
Left Behind series, 164
Lehi (Freedom Fighters of Israel), 305
Lenard, Philipp, 261
Lenin, Vladimir, 268, 270, 272, 274–75, 284
Light and the Glory, The, 164
literacy, 7–8, 13–16, 26–29, 41, 46, 97, 110, 115–16, 144, 146, 192, 268, 276, 328, 340–41
literature, 7, 14, 28, 36, 186, 188, 200, 209, 235, 245, 252–63, 268, 278, 283, 285, 296, 320, 327
lobbying, 76, 97, 99, 101–2, 105, 108
London, England, 184
Louisiana, 14, 223
Lowenthal, David, 162, 176–77

Maier, Charles, 236
Malaysian, 146
manipulation, 3, 103, 106, 109, 113, 205, 245, 251, 264, 269–70
Mann, Charles, 200, 208
Maoists, 8, 168, 319–34
"March of the Flag, The" 224, 228–29
marginalization, 88, 96, 100, 103, 108, 134–35, 195, 253, 304, 309, 332
marriage, 54, 167, 194
Marshall, Justice, 64–65, 68
martial law, 19, 151
Marxism, 3, 268, 274–75, 283–84, 286–87, 322, 324
Maryland, 15, 193
Massachusetts, 67, 114
Massachusetts Toxic Use Reduction Act (TURA), 114
massacre. *See* conflicts
McAtee, John, 229–30
McKinley, President William, 222–24, 227, 231, 233
McLean v. Arkansas Board of Education, 75, 87, 89, 92
media, 1, 35–38, 40–41, 44, 57, 76, 88, 97, 101, 103, 132, 149, 245, 311, 327, 341, 343–44, 347
memory, 140, 149, 155–56, 165, 184, 191, 282, 296–97, 306–7, 310–11
"mercy killings," 248
Metacom's War, 164
methodology, 6, 84–85, 90, 101, 185, 187–88, 191, 193, 334, 341
Mexico, 207–8
Michaels, David, 102–3, 110–11
Miegel, Agnes, 255–56
migration, 146, 173, 202, 206, 208
military, 6, 18, 37, 55, 141, 150, 185, 205, 207–8, 219, 221–22, 228–29, 231, 234, 237–38, 262, 298, 304–5, 339
Milwaukee Journal Station, 42, 48
Ministry of Education, 8, 257–58, 263, 295–96, 301–7, 309–10
misconduct, 65, 103, 113
misinformation, 8, 97, 149, 156, 249
misrepresentation, 109, 171, 185
missing links, 79, 81
Modern Language Association, 53
Monaghan, E. Jennifer, 13, 15
money, 88, 99, 108, 113, 129, 247–48
Mongols, 282
"Mouse House" research center, 99
mufti, 303
Mugwumps, 222
multiculturalism, 165
music, 34, 44, 129, 132–33, 253

naïvete, 5–6, 73, 85, 161, 179, 271, 342
narratives, 2, 74, 77, 82, 89, 91, 136, 163, 166, 170, 173, 185–86, 188, 197, 199–201, 205, 264, 270, 282, 285, 296–312, 325–28, 330, 332–33
National Academy of Sciences (NAS), 114
National Advisory Council on Radio in Education (NACRE), 34, 40, 42, 44
National Association for the Advancement of Colored People (NAACP), 56
National Association of Evangelicals, 167
National Cancer Institute (NCI), 100
National Center for Education Statistics, 163
National Coalition Against Censorship, 53
National Council for Social Studies, 333
National Council of Teachers of English, 53
National Gay Task Force, 64, 66–67
National Institute of Occupational Safety and Health (NIOSH), 102–3

nationalism, 207, 245, 248, 258, 306, 340–41
National Military Organization, 305
National Socialism, 246, 249, 254–55, 257–64
National Socialist Teachers' Association, 259
Native Americans, 145, 188, 192, 197, 204, 206, 224. *See also* Indians
Native Hawaiians, 152, 154
Nazism, 8, 98, 244–64, 298, 303, 342
Needs and Opportunities for Study, 195, 199
New Deal, 123–25, 135–36, 168
New England, 127, 197, 205–6, 222
New Mexico, 40, 224
New York, 25, 34, 38, 40–45, 52, 98, 193, 327
New York Post, 43
New York Times, 90, 142, 227, 233–34
nicotine, 99
Nierenberg, William, 107, 111–12
Nisei, 150–51, 153–56
Nixon, President Richard, 152
North America, 171–73, 185, 192, 199–201, 208, 217, 232
North American Review, 217
North Dakota, 208
Numbers, Ronald L., 86, 88

Oahu Filipino Community Council, 152
objectivity, 101, 112, 166, 246, 324
occlusion, 161, 176–77, 179, 259, 263
Occupational Safety and Health Administration, 103
Office of Science and Technology Policy (OSTP), 104
oil industry, 103, 108
Oklahoma, 60, 64–67, 229
Omaha, NE, 234
Oneida, 196
One Nation Under God, 165
Oreskes, Naomi, 78, 85, 96, 100–101, 104, 106–7, 109, 111–13
Organization of Educational Historians, 189
Origin of Species, 75
Orwell, George, 271
Oslo Accords, 296
Ottoman Palestine, 302
Out Loud: The Best of Rainbow Radio, 53

Pakistan, 146
paleontology, 81
Palestine, 8, 296–312
Palestinian Liberation Organization, 296
Parran, Thomas, 34–37, 39–46
paternalism, 125, 207
patriotism, 150, 167, 197, 232, 295, 299
pedagogical strategies. *See* instruction
Pensacola Christian Academy, 169–71
Peopling of British North America, The, 192
Pericles, 268
Peshkin, Alan, 169
Petersburg, KY, 77
Petersen, William, 142–43
petrochemicals, 103
Pfeiffer, Albert, 39, 47
Philadelphia, PA, 153, 164, 229
Philip Morris, 99–100, 104, 110–12, 345
Philippines, 8, 152–55, 217–18, 220–26, 230–35
philosophy, 3, 66, 76, 87–89, 164, 168, 170–71, 191, 339
Physical Science for Christian Schools, 169
physics, 113, 258, 261–62, 277–79
Pickering balance test, 66
Piketty, Thomas, 187, 193, 344
pluralism, 53, 170–71
Portugal, 2, 192
poverty, 2, 123–37, 140, 143, 145, 193, 320, 340, 342
Powell, Justice, 66
prejudice, 140, 145, 168, 246
prison. *See* imprisonment
private interests, 97–98, 110, 113–14, 116
Proctor, Robert N., 5, 52, 54–55, 73, 96, 98–99, 161, 176, 178, 271
propaganda, 8, 52, 199, 244–49, 251–52, 262, 264, 303, 308, 343
Prussian Academy of Poetry, 255
pseudoscience, 88–89
psychology, 25, 57, 67, 90, 104, 131, 140, 167, 193–94, 296, 299–300, 307, 310
public education. *See* education
public relations, 97–98, 100–102
public relations firms: Burson-Marsteller, 110; Foundation for Public Affairs, 110; Hill and Knowlton, 101, 110; Mongoven, Biscoe, & Duchin (MBD), 105, 110
Puerto Rico, 8, 221–22, 225–26, 230
Puritans, 44, 175, 205
Pushkin, A. S., 269, 277–78
push polling, 92

race, 20–21, 29, 153, 186, 225, 229–30, 233, 246–47, 256–57, 260–61, 263
Racial Composition of the Jews, The, 261
radio, 1, 5, 7, 34–45, 52–53, 132, 249, 340
Reagan, President Ronald, 2, 66, 104, 107, 112
Red Book, the, 324, 327, 334
Red Guards, 321–22, 327
Rediscovering God in America, 165
Red Scarf Girl, 333
Reformation, 13, 167
regulation, 2, 5–8, 13, 34, 36–40, 42, 44–45, 80, 104–6, 110–12, 114–15, 155, 233, 297, 309, 320, 339–41, 345, 347
Regulatory Toxicology and Pharmacology, 111
Reich Reader, 258
religion, 3, 6, 27, 35, 53, 60–61, 64, 73–93, 161–79, 191, 198, 205, 208, 261, 303, 340, 342–43. *See also* evangelicalism
republicanism, 55, 108, 218, 222–25, 228, 230–31, 233, 237
revisionism, 172, 174–75, 187–89, 191, 204–6
Rice, John L., 41, 43
Rio Grande River, 208
Rippa, S. Alexander, 204
R. J. Reynolds, 99, 111, 345
Rock Springs, WY, 141
Rodgers, Daniel T., 187
Roman republic, 201, 229
Rome, Italy, 218, 284
Roosevelt, Theodore, 222, 224, 233
Root, Elihu, 231–33
Rosengarten, Theodore, 186–87, 192
Rowland v. Mad River Local School District, 65, 68
Rury, John, 205–7
Rushdoony, Rousas, 166–67
Rust, Bernhard, 257–58
Rwanda, 300

Saint Domingue, 19
Salem, MA, 164
Samoa, 152, 154
Sandinistas, 168
San Francisco, 48, 63, 100, 141–42
San Salvador, El Salvador, 234
Sansei, 150
Santer, Ben, 108, 112
Schindler's List, 177
Schlafly, Phyllis, 167
Schweninger, Hermann, 248
Scientific American, 106
segregation, 4, 56, 144
Seitz, Fredrick, 107, 111–12
semantics, 219, 230, 233–34, 237
settlements, 124–25, 128–29, 134, 141, 303–5, 310
sex, 7, 8, 34–49, 167–68, 342
sexuality, 7, 8, 52–68, 91, 175, 250, 253, 342
Shell Oil, 110, 345
Shoemaker, Troy, 170–71
Short History of the Communist Party, 282
Singer, Fred, 104, 112
Sinology, 320, 331
Sinti, 253
"Sixteen Points," 324–25
skepticism, 80–81, 88, 112–13, 200
slavery, 4, 13–29, 91, 132–33, 142, 144, 161, 179, 192–93
Slavery by Another Name, 192
Slavs, 263
Smith, Captain John, 170, 173, 175
smoking, 5, 21, 98–101, 111, 178, 219
Snow, William F., 34, 41, 43, 45
Snowden, Edward, 2
Social Education, 333–34
socialism, 131, 135, 170, 245–46, 249, 245, 255, 257–64, 273–74, 278, 282–83, 285, 287, 321–22, 324, 331–32
socialization, 201, 206, 262–64, 311
Society for Environmental Journalism, 108
Society for the Propagation of the Gospel in Foreign Parts, 184
sociology, 109, 140, 163, 191, 194, 328
Söderbaum, Christina, 246
sodomy, 60, 64, 66
Some of Us: Chinese Women Growing Up in the Mao Era, 327
South Carolina, 14, 19, 28, 53, 193
Southern History of Education, 189
sovereignty, 198, 206, 221–22, 227–28, 233, 309–10
Soviet Union (USSR), 221, 251, 268–89, 305, 331, 340–42
Spain, 186, 192, 197, 200, 205–6, 217, 221, 227–28
Sri Lanka, 146
Stalinism, 8, 268–76, 279–80, 287–89

Stark, Johannes, 261–62
"Star of Empire, The," 230
stereotypes, 140, 146, 149, 151, 247, 261, 263, 302
sterilization, 91, 247, 260
Stevens, Justice, 65
stigma, 35, 150, 248
Stonehenge, 201
Stonewall Riots of 1969, 64
Storr, Richard, 190
strategic defense initiative (SDI), 107, 112
Strauss, Emil, 255–56
strikes, 130–31, 151
Stuart, James, 16, 24
subjugation, 223, 227, 237
subordination, 83, 135, 145, 225, 257, 273
Suez Canal, 305
Sununu, John, 107
supernaturalism, 81, 84, 90. *See also* religion
Supreme Court, 45, 54, 56, 60, 64–68, 86–87, 141, 229, 233, 341
Suri, Jeremy, 236
surveillance, 251–52
Switzerland, 255
Sword of the Lord, 167
Syngenta, 103–4

taboo, 42
Taiwan, 146
teaching. *See* instruction
Temptation of Despair, The, 187
Tennessee, 14–15, 75, 86
terrorism, 64, 229, 237, 251, 264, 305, 308
testing. *See* examinations
Texas, 14, 60, 208, 346
textbooks, 5, 7, 29, 83, 150, 161–63, 165, 167–79, 203–6, 226, 245, 256–64, 270, 272, 74–76, 280–82, 284–85, 288, 301–8, 326
Thanksgiving, 167, 228
theistic evolution, 90
theology, 79, 82, 166, 172, 192
think tanks, 101, 107, 109–10, 116, 346–47. *See specific think tanks*
third-person effect, 52
Third World, 145, 331
Thirty Years War, 256
tobacco, 5, 77, 96–102, 104–5, 107, 110–12, 115, 178, 219, 339
Tobacco Industry Research Committee (TIRC), 98–100
torture, 252
totalitarianism, 271
training. *See* education
transparency, 109, 113–14
treason, 232–33
treaties, 206, 217, 221, 227–28, 230, 250, 296–97
Trotsky, Leon, 270
Truman administration, 55
Tsarist Russia, 272
Tuana, Nancy, 271
Turner, Nat, 15, 26
tutelage, 223, 225
Twain, Mark, 217–18, 222, 225, 237

Ukraine, 283
Uncle Sam, 226–27
unemployment, 123, 131, 250
Union of Concerned Scientists, 111
United States Congress, 37–39, 55, 104, 106–8, 111, 135, 150, 154, 161–62, 179, 223–24, 231–32, 31, 237
United States Constitution, 53
United States History for Christian Schools, 169, 171
United States History in Christian Perspective, 169, 171
universities. *See* higher education
Urban, Wayne, 205–6
urbanization, 27–28, 125, 132, 143, 145, 148, 204, 256, 327–28, 329
US Civil War, 14–15, 29, 187, 192
Uyematsu, Amy, 145

Vesey, Denmark, 19, 21
Vietnam, 145–47, 177–78
Virginia, 14–15, 19, 25, 27, 46, 188, 202–3
visual aids, 268, 285–86
Vogel, Alfred, 260–61

Wagoner, Jennings, 205–6
Wall Street Journal, 108, 112
War of 1898, 8, 217–18, 220, 237
Washington, Booker T., 3–4
Washington, DC, 46, 134
Wayland, Francis, 29
"wedge" strategy, 60, 88

Weimar Republic, 247, 252–53, 255, 257
Weinberg Group, Exponent, Inc., 110
West Bank, the, 306, 309
Wineburg, Sam, 161, 168, 175–78
Winthrop, John, 165
Wistrich, Robert, 254
women, 22–25, 35, 101, 128, 130, 144, 151, 191–92, 202–3, 230, 249, 254, 263, 327, 331, 343
Wood, Leonard, 233
World War I, 123, 259
World War II, 55, 141, 145, 148–49, 153, 187, 251, 298

Xiaoping, Deng, 326, 329, 331–32

Yellow Peril, 141, 155
Yishuv period, 305
Yonsei, 150
young-earth interpretation, 78, 86

Za kommunisticheskoe prosveshchenie (For communist education), 274, 276, 279, 281
Zedong, Mao, 8, 168, 319–27, 331–32, 334
Zeno's paradox, 333
Ziegler, Adolf, 253–54
Zionism, 296–99, 301–11